Studies in Applied Philosophy, Epistemology and Rational Ethics

Volume 41

Studies in Applied Philosophy, Epistemology and Rational Ethics (SAPERE) publishes new developments and advances in all the fields of philosophy, epistemology, and ethics, bringing them together with a cluster of scientific disciplines and technological outcomes: from computer science to life sciences, from economics, law, and education to engineering, logic, and mathematics, from medicine to physics, human sciences, and politics. It aims at covering all the challenging philosophical and ethical themes of contemporary society, making them appropriately applicable to contemporary theoretical, methodological, and practical problems, impasses, controversies, and conflicts. The series includes monographs, lecture notes, selected contributions from specialized conferences and workshops as well as selected Ph.D. theses.

Advisory Board

More information about this series at http://www.springer.com/series/10087

David Danks · Emiliano Ippoliti
Editors

Building Theories

Heuristics and Hypotheses in Sciences

 Springer

Editors
David Danks
Department of Philosophy
Carnegie Mellon University
Pittsburgh, PA
USA

Emiliano Ippoliti
Dipartimento di Filosofia
Sapienza University of Rome
Rome
Italy

ISSN 2192-6255 ISSN 2192-6263 (electronic)
Studies in Applied Philosophy, Epistemology and Rational Ethics
ISBN 978-3-319-72786-8 ISBN 978-3-319-72787-5 (eBook)
https://doi.org/10.1007/978-3-319-72787-5

Library of Congress Control Number: 2017962552

Printed on acid-free paper

This Springer imprint is published by Springer Nature
The registered company is Springer International Publishing AG
The registered company address is: Gewerbestrasse 11, 6330 Cham, Switzerland

Preface

The development of rich, subtle, powerful theories of the world is a signature aspect of the sciences. We do not, in the sciences, simply measure and catalog the world, but rather aim to develop more systematic pictures of parts of the world around us. As one might expect, there are corresponding traditions in the philosophy, history, sociology, and psychology of science whose focus has been, at least nominally, the construction and emergence of scientific theories. At the same time, these traditions have, in many cases, been somewhat disconnected from one another, and from much actual scientific practice. For example, a significant part of the philosophy of science tradition starts with Hans Reichenbach and Karl Popper's sharp division between the "context of discovery"—the conditions and processes in which scientists come up with new theories—and the "context of justification"—the methods that scientists do and should use to evaluate those theories, and then focuses almost exclusively on the latter context. Research in this area of philosophy of science has led to tremendous advances in our understanding of the diverse ways in which scientists integrate information from many sources to evaluate the plausibility of scientific theories, or to determine previously unrecognized limitations of those theories. However, it has provided little-to-no insight into the origins or generation processes of the theories that are the target of scientists' justificatory practices.

More generally, there has been relatively little research on the construction and development of novel theories, and the continual rethinking of old theories. Instead, work on the nature and development of science has focused primarily on experimental and data collection practices, as well as methods for theory confirmation and justification. These are clearly important aspects of the development of scientific knowledge and understanding, but they are equally clearly not the whole of science. Moreover, this prior work has largely and uncritically accepted the orthodox view that the scientific process can be divided into four, logically and temporally distinct stages: (1) generation of new hypotheses; (2) collection of relevant data; (3) justification of possible theories; and (4) selection from among equally confirmed theories. While these are all important aspects of scientific practice, the assumption that they can be cleanly separated—conceptually, logically, or temporally—is a substantive one that is not necessarily borne out by close examination of actual scientific practice. Of course, we inevitably must delimit the scope of our investigations, but focusing on just one of these four aspects arguably threatens to provide a misleading understanding of the sciences.

The papers in this volume start to fill these lacunae in our understanding of the development of novel scientific theories and the adjustment of previous theories; in short, they give us a better picture of how and why scientific theories get built. They spring from a conference at Sapienza University of Rome in June 2016, and represent a number of different perspectives on the nature and rationality of theory construction. They address issues such as the role of problem-solving and heuristic reasoning in theory building; how inferences and models shape the pursuit of scientific knowledge; the relation between problem-solving and scientific discovery; the relative values of the syntactic, semantic, and pragmatic view of theories in understanding theory construction; and the relation between ampliative inferences, heuristic reasoning, and models as a means for building new theories and knowledge. These individual investigations all involve close examination of case studies in multiple scientific fields, including logic, mathematics, physics, biology, and psychology, and each provides insight into a particular aspect of theory construction. Moreover, although each chapter stands on its own, there are consistent themes that emerge across the chapters.

First, we find numerous rejections of the Reichenbachian-Popperian distinction between the contexts and discovery and justification (e.g., the chapters by Gillies, Nickles, Ippoliti, Cellucci, and Sterpetti). These chapters each argue, in their own ways, against a sharp distinction between the construction and evaluation of a particular scientific theory. Instead, they examine the myriad ways in which scientists can, and should, use evaluative or justificatory information and processes to develop novel scientific theories, or adjust the ones that we already have. Second, and relatedly, multiple chapters provide models of scientific discovery and theory building that fall within a middle ground between deductive logic and unjustified guesswork (Morrison, Nickles, Ippoliti, Danks, Darden, Cellucci, Sterpetti, Magnani, and Gillies). Novel theories almost never arise through logical reasoning, and there are typically many logically possible ways to revise our theories in light of conflicting data. These absences of logical guarantees have sometimes been taken to mean that no rational defenses can be given for theory building practices, and so they are tantamount to lucky guesses (perhaps glossed as "scientific insight" or "inspiration"). In contrast, a consistent theme throughout this volume is that theory building is guided by defensible principles and practices that do not guarantee success, but are also not completely arbitrary.

Third, across many chapters, there is widespread use of case studies to both guide and evaluate accounts of theory building (Gillies, Morrison, Nickles, Danks, Darden, Ippoliti, Ulazia, and Longo & Perret).[1] Many investigations of aspects of scientific theories have proceeded from the armchair: the investigator considers

---

[1] Some authors thus practice what they preach, as they blur the line between theory construction and theory evaluation, though in this case, for philosophical theories rather than scientific ones.

what seems plausible about the science, or evaluates matters entirely based on the final, published scientific record. In contrast, these authors examine the details of particular scientific case studies, both large and small. Importantly, they do not thereby fall into the trap of merely reporting on the science; rather, the particular details provide evidence for rich models of theory construction. Fourth, and more specifically, several chapters emphasize the importance of psychological processes in understanding the ways in which scientists develop and adjust theories (Nickles, Ippoliti, Ugaglia, Ulazia, and Longo & Perret). These psychological models of theory building are partly inspired by the case studies, but are equally informed by relevant cognitive science; they are not simple applications of naïve folk psychology. Moreover, these chapters highlight the focus in much of this volume on both normative and descriptive aspects. Psychological processes are not employed solely to provide descriptive models of how scientific theories are actually built, but also to give a principled basis for evaluation of the rationality (or not) of theory construction processes.

Theory building is a critical aspect of the scientific enterprise, but there have been only sporadic attempts to develop coherent pictures of the relevant processes and practices. The chapters in this volume aim to provide answers to long-standing questions about the possibility of a unified conceptual framework for building theories and formulating hypotheses, as well as detailed examinations of the key features of theory construction. The diverse perspectives, disciplines, and backgrounds represented by the chapters collectively shed significant light on previously under-explored aspects of these processes. We thank the authors for their valuable contributions, and Springer for valuable editorial assistance. The two editors contributed equally to this volume; our names are simply listed in alphabetical order.

Pittsburgh, USA   David Danks
Rome, Italy   Emiliano Ippoliti

Contents

Part III   New Models of Theory Building

Part I
Understanding Theory Building

Building Theories: The Heuristic Way

Emiliano Ippoliti

Abstract Theory-building is the engine of the scientific enterprise and it entails (1) the generation of new hypotheses, (2) their justification, and (3) their selection, as well as collecting data. The orthodox views maintain that there is a clear logical and temporal order, and distinction, between these three stages. As a matter of fact, not only is this tenet defective, but also there is no way to solve these three issues in the way advocated by traditional philosophy of science. In effect, what philosophy of science tells us is that (a) there is not an infallible logic, in the sense of a simple set of logical rules, to justify and confirm a hypothesis, and (b) the process of generation of hypotheses is not unfathomable, but can be rationally investigated, learned and transmitted. So, as an alterative, I discuss the heuristic approach to theory-building, especially the one based on problems, and I argue that it offers a better way of accounting for theory-building than the traditional ways.

Keywords Theory-building · Problem-solving · Heuristics · Discovery Justification

1 Introduction

Theory-building is the engine of the scientific enterprise, whose aim is advancing knowledge about the cosmos we live in. In order to achieve this goal, theories, that is sets of hypotheses that account for problems starting from sets of data, are continually produced, refined, and eventually abandoned.

Building theories entails the generation of new hypotheses, their justification, and selection. These three features, even if traditionally they are kept separated, as a matter of fact are conceptually closely interrelated (e.g. see also Ippoliti and Cellucci 2016, Chap. 4). While the first aspect, the generation of a new hypothesis,

E. Ippoliti (✉)
Sapienza University of Rome, Rome, Italy
e-mail: emiliano.ippoliti@uniroma1.it

has been analyzed at most descriptively, the other two, justification and, to some extent, selection, have been approached normatively.

As a subject, the generation of a hypothesis in theory-building is still underdeveloped both theoretically and practically. In the latter case, even today, "most graduate programs [...] require multiple courses in research methodology to ensure that students become equipped with the tools to test theories empirically", but "the same cannot be said for theory construction" (Jaccard and Jacobi 2010, 3). In the former case, the crucial step to build a theory is the generation (discovery) of new hypotheses and many scientists and philosophers argue that that there is no rational way for it.

This fact suggests that there is still a lot of work to do in order to teach and reflect upon this aspect of theory-building in a more systematic way, even if in the last decades an increasing amount of work has been devoted to it.[1]

The use of gerund of the verb 'to build' in the title of this paper might give the impression of a biased approach, since it might suggest that in order to pursue the search for new knowledge, we need to refine our understanding of how we construct theories. In effect, such a view argues for the adoption of a dynamical and a rational viewpoint on the subject. So the use of the gerund aims at emphasizing that:

(a) theory-building is a process, and in this sense it embeds a dynamic viewpoint. The static approaches to the issues—the more traditional ones—look only at the final outcome of the process.
(b) this process is rational, in the sense that the construction of a new theory can be guided by reason (and reasoning), and is not a 'black box'.

This view contrasts with the static views, which explicitly draw on the idea that what really matters, and can be investigated in a rational and logical way, is only the final outcome—the structure of a theory and its consequences. Thus, the process of building a theory can be 'safely' ignored. The main reason for this tenet is that for those views what really matters about theories is:

(1) the selection and the justification of a theory (which can be pursued only after it has been produced), and,
(2) the fact that the procedures, if any, employed in the construction of a theory are completely irrelevant for its justification—that is its evaluation and testing.

Such an asymmetry characterizes all the traditional views (see Sect. 3).

Therefore building theories is a twofold study: on one side it aims at examining and deepening the study of the very construction of theories, on the other side it aims at understanding procedure and conditions that enable us to assemble theories that are *really* capable of building new knowledge, and to put forward and extend the comprehension of the cosmos we live in. Thus the term 'building' refers both to *what* enables us to construct a theory and *what* a theory enables us to build *really*.

---

[1]See in particular: Cellucci (2013), Ippoliti (2006, 2008, 2016), Ippoliti and Cellucci (2016), Magnani (2001), Magnani et al. (2010), Clement (2008), Jaccard and Jacobi (2010), Grosholz (2007), Grosholz and Breger (2000), Lakatos (1976, 1978), Laudan (1977, 1981, 1996), Nersessian (2008), Nickles (1980a, b, 1981, 2014), Nickles and Meheus (2009), Gillies (1995), Darden (2006), Ulazia (2015).

2 Theory-Building: The Basics

Theory-building can be broken down into three main issues, all of which are essential for the construction and extension of knowledge:

1. the generation of a hypothesis (or set of hypotheses);
2. the justification of a hypothesis (or set of hypotheses);
3. the selection of a hypothesis (or set of hypotheses).

Of course, all three of these stages also require collecting data, which could be considered a fourth dimension in theory-building. The orthodox views (see Hempel 1966) maintain that there is also a logical and temporal order, and distinction, between these three stages.

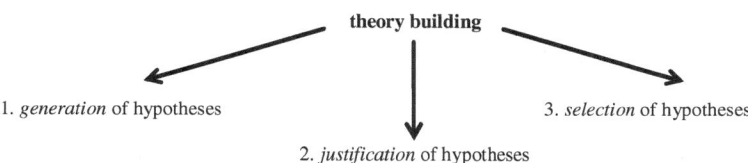

The first issue (1) investigates the procedures by which to generate a new hypothesis, if any. The second (2) investigates the procedure by which a hypothesis, or a theory, can be justified. The third (3) examines the criteria or procedure by which to select, or choose among, multiple competing hypotheses.

Thus, (2) and (3), which are intertwined, examine procedures about options (i.e. hypotheses) that are *already* available, whilst (1) examines procedures for producing *new* options (hypotheses).

The key point here is that none of the three issues can be answered only by logic (in the narrow sense) or probability theory. In addition, (1), (2) and (3) are not separate processes, as argued by the orthodox position: as a matter of fact they are temporally and logically interdependent. They are distinguished mostly for the sake of conceptual simplicity and for educational purposes. The main counter-example to this contention is the case of the analogy, which is a *rational* procedure that is used (even simultaneously, i.e. the very *same* analogy) both in order to generate new hypotheses and to confirm them and, in some cases, to choose between rival hypotheses (see e.g. Shelley 2003). Therefore we have a less linear and stylized relation between these three stages than the one outlined by the static view: mutual dependences, and no strict temporal or logical order (see also Sect. 2.4).

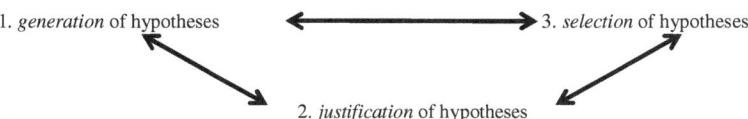

2.1 Confirming Hypotheses

The fact that a confirmation of a theory is not achievable in pure logical or probabilistic terms can be sketched in the following way. First, the problem of confirmation is usually summarized as follows:

If a given hypothesis or theory, say h, implies a fact f, and it turns out that f is
true by comparing it with experience, then the credibility, the degree of confirmation, in h increases. In more formal terms:

$$\frac{\begin{array}{l} h \rightarrow f \\ f \end{array}}{h \text{ is more credible}}$$

Now, since confirmation is an implication from a logical viewpoint, we have
several problems. First, of course we must avoid the mistake of *affirming the
consequent* ($h \rightarrow f, f$ is true, then h is true), that is the fact that we cannot draw the
conclusion that h is true on the basis of the truth of several of its consequences,
f. The only way to establish the truth of h in this way would be to check the truth of
all consequences f of h. But since all the possible consequences of h are infinite in
number, this task cannot be achieved within the physical limits of the universe we
live in. This implies that, as human beings, we can never establish the truth of h in a
definitive way, beyond any doubt, by comparing the consequences f of h with the
experience.

Second, a logical implication generates a series of paradoxes when it is
employed to provide an inductive support—known as the paradoxes of confirmation. These paradoxes undermine the very idea that it is possible to provide a direct
confirmation of a single hypothesis (see e.g. the problem of irrelevant conjunctions
and irrelevant disjunctions). As is well known, the paradoxes of confirmation stem
from the fact that a proposition such as (a) 'all raven are black'—which has the
logical form $\forall x \, (R(x) \rightarrow B(x))$—, is logically equivalent to a proposition (b) 'all
things that are not black are not a raven'—which has the form $\forall x \, (\neg B(x) \rightarrow \neg R(x))$. Since (a) and (b) are mutually dependent, an observation (experience) like
(c) 'this apple (not a raven) is red (not black),' which confirms (b), should also
support (a). The paradox obviously follows from the fact that it seems impossible,
or very strange, to get knowledge about crows by making use of apples.

Those paradoxes can be answered only in part by means of logic and probability
theory. In effect classical logic and probability theory do not provide a full solution to
the problem of confirmation, or inductive support. In particular the several approaches put forward to solve the puzzle (the probabilistic ones, the non-probabilistic and
the informational) do not offer a fully satisfactory response to the counter-arguments
like the one provided by Popper (see Popper and Miller 1983).

This holds also for the probabilistic (and Bayesian) measures of confirmation
(see e.g. Schippers 2014), that is a new version of the idea that probability offers a
way to answer the problem of inductive support. This idea can be broken down into

two connected tenets. The first states that inductive reasoning is reducible to probability, and the second states that inductive reasoning is a means of justification —a way to determine, also by mechanical procedure, the logical probability, or degree of confirmation of a hypothesis.

The former case is defective since probability and induction are distinct. To show this,[2] it is enough to show that there are hypotheses obtained by induction and that are not probable, but are plausible; or that are probable, but are not plausible. Hypotheses obtained by induction from a single case are not probable in the sense of the classical concepts of probability (i.e., ratio between winning and possible cases or subjective interpretations): the probability of a conclusion of an induction from a single case becomes zero when the number of possible cases is infinite. Similarly, hypotheses obtained by induction from multiple cases have turned out to be false. Such is the case of the famous hypothesis that all swans are white, which should have non-zero probability when the number of possible cases is not infinite (as is the case for the hypothesis that all swans are white).

The latter case, induction as a means of justification, or better yet, the Carnapian concept of degree of confirmation of a hypothesis (or conclusion) on the basis of some given evidence (or premises), is defective too. This is not only because we are generally unable to determine the probability of our hypotheses about the world, let alone to determine it by a mechanical procedure, but especially because: applying the calculus of probability to our hypotheses about the world would involve making substantial assumptions about the working of the world. Then, if our hypotheses were justified by showing that they are probable, they would not be really justified by induction, but rather by our assumptions about the working of the world. Such assumptions could not be justified in terms of the calculus of probability, because it would be impossible to account for them in purely mathematical terms. (Cellucci 2013, p. 336)

Bottom line: we do not have a way of establishing in pure logical and probabilistic terms when a hypothesis is confirmed. There is no way to solve this puzzle in the way advocated by the traditional philosophy of science and the answer to this issue of theory-building can only be pragmatic and provisional. As a matter of fact, what philosophy of science tells us is that there is no infallible logic, in the sense of a simple set of logical rules, to justify and confirm a set of hypotheses.

2.2 Selecting Hypotheses

Like the problem of confirmation, the issue of the selection of hypotheses cannot be definitely solved by employing logic and probability. This is a well-known issue: given a set of hypotheses (h_1, h_2, ..., h_n) that account for a problem p starting from

---

[2]See Cellucci (2013) for a discussion of these points.

the same set of data e, we have to rely on very special criteria in order to answer the question about which hypotheses to choose.

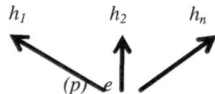

More precisely, the only solution that can be given is a pragmatic one. Thus, also the answer to this issue of theory-building can be only tentative and provisional and the problem of selection of a hypothesis between rivals, again, cannot be solved in the way advocated by the traditional philosophy of science: there is no infallible logic in the sense of a simple set of logical rules, to choose a hypothesis among rivals. As a matter of fact, to approach this issue we need to employ extra-logical criteria, such as scope, plausibility, fruitfulness, simplicity or parsimony. Another option, often favored by mathematicians, is the use of aesthetic criteria, like 'beauty'.

For instance, the scope of a theory can be a guide in order to decide which of the rivals has to be preferred over the others. In effect it seems reasonable to prefer a fairly general theory, namely one that accounts for a wider range of phenomena: the broader in scope a theory is, the more it should be preferred over rival theories.

Fruitfulness, that is the ability of solving problems and generating new ones, is another important factor when we have to decide which hypothesis to select.

A crucial and often-employed criterion is the simplicity of a theory (a typical form is Occam's razor). It states that a simpler theory, that is, one with the fewer assumptions and hypotheses, should be preferred to a less simple one, provided that the two offer equally acceptable explanations of the same pieces of evidence. The use of simplicity is justified on the basis of two well-known properties of the theories: falsifiability and under-determination. The former states that a theory, to be scientific, has to be falsifiable, that is, its consequences have to compare with experience and in the case of disagreement the theory has to be rejected. The latter states that the same set of data can be explained by an infinite number of theories. Combining the two properties, we obtain that in principle there is always an extremely large number of possible and more and more complex rival hypotheses capable of accounting for the same set of data. A corollary of it is the fact that is always possible to defend a theory by means of the introduction of ad hoc hypotheses in order to prevent its falsification. Therefore, a simpler theory is preferable since it is easier to control, and to falsify, since a smaller number of hypotheses is easier to manage. More precisely, it is easier to go through a smaller number of hypotheses and determine which one is the wrong one, i.e., the one that generate the derivation of a false consequence—it is a less demanding task.

Of course very often these criteria are not used alone: a combination, or a balance of them, is used in order to make a more appropriate selection of the better 'prospect'. A notion that tries to merge all these criteria is so-called plausibility. In effect it is another crucial property in selecting a hypothesis. The definition of this notion is not stable and fixed. In any case, here I employ the one put forward by Cellucci (see in particular Cellucci 2013, Chap. 4), which explicitly draws on the notion of Aristotle's *endoxa* (see also Ippoliti 2006). A plausible hypothesis in this sense is such that arguments for the hypothesis must be stronger than those against it on the basis of available experience and data. In balancing arguments *pro* and *contra* a given hypothesis the procedure takes into consideration all the criteria mentioned above—simplicity, fruitfulness, etc. We can also specify a 'test of plausibility' (see Cellucci 2013, p. 56). Nevertheless, Plausibility cannot be reduced to either logic or probability, since there are many hypotheses that are plausible even if their probability is low, or zero (in the technical sense of ratio of the number of observations of the event to the total numbers of the observations).

Since the selection of hypotheses is not reducible to logic or probability theory, a toolbox of heuristic procedures has been identified as a rational way of accomplishing this task. Here the notion of rationality is interpreted in ecological terms, not in logical ones. In particular the F&F approach (see e.g. Gigerenzer and Todd 1999) provides us with a rational way for selecting hypotheses. In greater detail, this tradition describes a 'selective heuristics' (see Ippoliti and Cellucci 2016): heuristic procedures that can be applied to options (hypotheses) that are *already known* and *accessible*, and they can be especially effective to draw conclusions and to make choices under conditions of a shortage of temporal, informational, or computational resources. The appeal of the F&F approach is that these heuristic tools offer a better way of selecting an option than the optimizing tools (i.e. classical logic or probability theory): a heuristics does not produce second-best results, and optimization is not always better.

2.3 Generating New Hypotheses

The generation of new hypotheses is the core of theory-building. Tellingly, it also by far the most underdeveloped issue. The main reason for this is what could be called 'a strange story of rationality and discovery'. An influential and long-standing tradition—of which the idea of Romantic Genius is a stock example—argues that the procedures for generating and discovering new hypotheses are not only outside the realm of logic or probability, but also that of rationality. The act that leads to new ideas and hypotheses is unfathomable, and entirely subjective. This tenet is shared both by famous philosophers and scientists (see e.g. Einstein 2002, 2010), and essentially provides an explanation of this issue by employing notions like intuition, creativity, or insight. Put in another way, they believe that there is nothing like a logical process (also in a very weak sense of logic) to build new hypotheses or ideas. It is a kind of 'black box'. Most of time this attempt ends up with an *obscurum per obscurius;* nonetheless it shapes the traditional approaches to theory building, that is the syntactic, the semantic and, in part, the pragmatic one.

A non-secondary issue here is the notion of novelty: what we mean by 'new' and what are, if any, the means for obtaining such a *new* knowledge. This opens an interesting debate, since the orthodoxy (i.e. analytical philosophy) argues that deduction and deductive rules are means for obtaining new knowledge [see also

Ippoliti (2014) on this point and Cellucci (2013)]. Roughly, this view maintains that knowledge in several scientific domains (e.g. mathematics) can be obtained, and advanced, by starting from a set of propositions already known and taken for granted (to be true, it should be noted) by means of the application of deductive rules Such a viewpoint, unfortunately, is untenable, both in a strong (or strict) and a weak sense.

Let us consider the former case: a deduction is such that its conclusion cannot contain more information than is contained in the premises. In a deductive rule the conclusion is literally included in the premises. For instance in the deductive rule *modus ponens* (A, A $\rightarrow$ B $\therefore$ B), the conclusion (B) is literally part of the premises (the consequent in the conditional A $\rightarrow$ B). Thus, no new information is given in the conclusion. Bottom line: no 'new' knowledge can be obtained by the application of a deductive rule. From a logical viewpoint, deduction is not a way of amplating knowledge.

The 'novelty power' of deductive reasoning could be defended in another way, that is a weak sense. A stock example is the *argument of labor*, put forward for instance by Frege (1960). In a nutshell, this argument is based on a botanical analogy (plant:seeds::conclusion:premises) and goes like this (see also Cellucci 2008 Ippoliti 2014): the fact that there is no logical novelty *strictu* sensu in the conclusion of deductions does not imply that these rules do no generate new knowledge, since the conclusions can also require a lot of labor and inferential work in order to be drawn. The conclusion of a deduction is contained in the premises *taken together*, but their conclusion cannot be seen in advance and has to be extracted. This extraction requires an inferential effort, at the end of which we get something that was not known or foreseen *beforehand*.

This account of the novelty power of deductions is also unsatisfactory. A straight answer to the argument of labor usually consists of two main points:

(1) it is *not* true that to draw a conclusion requires labor; it only requires a computer. As a matter of fact, there is an algorithmic method, that is, a pure mechanical procedure, which enables us to enumerate all the deductive consequences of given premises—the set of all the conclusions from those premises. So given *enough* temporal and spatial resources, *if* a conclusion can be drawn, *then* the algorithmic method will find a proof for it.

(2) The argument of labor totally disregards the distinction between logical or objective novelty on one side, and psychological novelty on the other. Of course a deductive conclusion can be surprising or 'new' from a psychological viewpoint for a reasoner, but it does not imply that the conclusion is an objective ampliation of our knowledge.

It is worth noting that the fact that deductive reasoning cannot extend our knowledge does not imply that it is useless. On the contrary, it is essential in several ways for our knowledge—for instance in compacting our knowledge (see in particular Cellucci 2008, 2013; Ippoliti 2014; Ippoliti and Cellucci 2016). It simply implies that it cannot provide an answer to the problem of generation of hypotheses, for it cannot introduce new information other than that contained in the premises and thus it cannot enable us to solve interesting problems.

2.4   The Distinction of the Three Phases

The traditional approaches to theory-building build upon Reichenbach's distinction between the context of discovery and context of justification (see Reichenbach 1938).

This distinction has been questioned by the 'friends of discovery' (Nickles 1980a) and the heuristic tradition, who have shown that it is essentially untenable (see in particular Nickles 1980a; Shelley 2003; Cellucci 2013). They argue that not only are discovery, justification and selection of hypotheses not separate processes, but that they employ both ampliative and non-ampliative reasoning. In essence, they show that justification is not simply a matter of classical logic, as the traditional approaches would hope, but that it requires heuristic reasoning too.

In order to illustrate this point, I will recall a few examples. The first one is given by Polya (Polya 1954, Vol. II), who provides us with several examples of heuristic reasoning that can be used as a way of both generating and confirming a hypothesis. Here I simply outline one of them, a pattern of plausible inference based on analogies and similarities. Suppose we have formulated the hypothesis A to solve a problem P, and let us suppose that A implies a set of consequences B_1, B_2, ..., B_n, all of which have been successfully confronted with the existing knowledge and experience, that is, such consequences are perfectly compatible with what we already know. In addition let us suppose that A implies also a further consequence B_{n+1}, which is similar to previous ones, that is, there are similarities between B_{n+1} and B_1, B_2, ..., B_n. Then this pattern of plausible inference concludes that A is even more credible after the verification of this new result B_{n+1}.

$A \rightarrow B_{n+1}$
B_{n+1} is very similar to the consequences of A previously verified B_1, B_2, ..., B_n
B_{n+1} is true
$\overline{\phantom{A \text{ is more credible(plausible)}}}$
A is more credible(plausible)

Moreover, the heuristic tradition has provided us with the discussion of other historical examples of heuristics employed both in the formulation and confirmation of a hypothesis. A nice example comes from mathematics: the solution advanced by Euler to Mengoli's problem (see Ippoliti 2006, 2008 for a discussion of this case). Here the analogy, the stock example of heuristic reasoning, is employed both in the formulation of the hypothesis that solves the problem, and in its justification.

Another interesting example is the one discussed by Shelley (see Shelley 2003) and taken from archaeology: the solution advanced by Talalay (see Talalay 1987) to the problem of determining the feature of 18 unusual fragments found at five different sites in the Peloponnese, and dating back to the Neolithic period. Here the *very same* analogy is used both to generate and justify the several hypotheses, or better, the hypothesis is at the same time generated and justified by the same heuristic inferences (analogies).

3   Building Theories: The Heuristic Way

The heuristic view maintains that a theory is an open collection of problems about the world and an open set of hypotheses to solve those problems. Just like the pragmatic view (see in particular Cartwright 1983, 1999; Cartwright et al. 1995; Suarez and Cartwright 2008), it argues that the notion of truth does not play a role in the scientific enterprise and progress and that theory-building is carried out bottom-up, that is, starting from a problem and possibly other data, and then searching for a hypothesis that enables us to deduce a solution for the problem. Moreover, several versions of the heuristic view explicitly challenge the main tenet of the traditional approaches and argue that theory-building is a rational process.

In effect, the traditional approaches to theory building—the syntactic, semantic and pragmatic view—shed little light, if any, on this process.

The heuristic view draws upon the idea that theory-building is problem-solving and that it is a rational enterprise which admits a method. This idea dates back to the works of Plato (see Cellucci 2013) and Aristotle (see Quarantotto 2017; Cellucci 2013), who account for building theories in terms of problem-solving. Thus not only the idea that the construction of theories is essentially problem-solving is not new, but, as a matter of fact, it has several traditions. Even if these lines of thought share the idea that theories are the outcome of problem-solving, they disagree with the way the solutions to problems are discovered.

The main approaches based on problems are:

(1) The Popper-Laudan tradition.
(2) The Poincaré-Simon tradition.
(3) The Lakatos-Cellucci tradition.

The tradition supported by Popper and Laudan (see Popper 1999; Laudan 1981) moves first of all from the contention that truth is not the goal of science. We cannot know that science is true or is approximating truth, but we know that we can approach and answer local issues, that is, problems. We cannot obtain a criterion for truth, but we can answer the question of whether a given theory solves a problem or not. So, this line of thought argues that knowledge advances through–and in the end is–problem-solving, but at the same time it denies that we can build or reach something like a rational account, much less a 'logic', of the generation of hypotheses and of scientific discovery. Discovery is a subjective and psychological process, and cannot be rationally or logically investigated, reconstructed and transmitted. Bottom line: this tradition ends up simply accepting Frege's contention (see Frege 1960) that philosophy and logic cannot say anything about the way new results are discovered.

These tenets are questioned by the other traditions, which consider them unjustified, since a guide for solving problems can be built.

As a matter of fact, the Poincaré-Simon tradition maintains that knowledge is built and advances by solving problems, and that we can find processes to account for it, but does not produce a cogent methodology for it.

Poincaré (see Poincaré 1908) draws on the idea of expressing and formalizing the unconscious processes that underlie the search for a solution to a problem. He argues that in order to solve problems our unconscious first combines and then selects ideas: when we face a problem and our conscious attempt to solve it does not succeed and stalls, the unconscious processes step in and work by creating all the possible combinations starting from certain ideas.

Since a psychological account of the generation of hypotheses is the received view on theory-building, and since "it may not be an exaggeration to say that modern psychological theorizing about *it* [...] is built solidly on the foundation of Poincaré's theory" (Weisberg 2006, p. 396), I will discuss it more closely. Hadamard (Hadamard 1945) provides a deeper characterization of Poincaré's description. In particular, he offers a reconstruction of this process as consisting of four stages:

(i) **preparation**, which denotes the initial and conscious work on the problem, characterized by an immersion in the problem and several attempts to solve it. If this phase does not produce a solution, the process of problem-solving stalls, and then the conscious activity on the problem is stopped (it goes 'off-line').

(ii) **incubation**, the conscious activity goes 'off-line' but the unconscious keeps working on it. The production of new and potentially useful combinations by the unconscious leads us to the next phase.

(iii) **illumination**, the moment when the hypothesis breaks the threshold of awareness.

(iv) **verification**, the very last step, which aims at determining the adequacy and plausibility of the hypothesis.

Poincaré maintains that such a pure unconscious combinatorial activity, by working in a parallel way (while the conscious processes are serial in nature), will produce a lot of ideas—a combinatorial explosion. A large part of these combined ideas are of course useless, while another part will offer candidates for the solution to the problem at hand, if any. This raises the question of the selection of these candidates. Poincaré provides a kind of metaphysical solution to the issue by maintaining that the main criterion to make this selection, especially in mathematics, is 'beauty', where 'beauty' means 'harmony', 'symmetry', or 'simplicity'. It is worth recalling that the idea that the generation of hypotheses is unfathomable and totally subjective is now questioned also in cognitive sciences (see in particular Weisberg 2006 on this point), and it is argued that a kind of rational reconstruction of it can be pursued and offered. Anyway, Poincaré's account shapes the one put forward by Simon (see e.g. Simon 1977). Simon explicitly states that his goal is to explain and even formalize the stages (ii) and (iii) of the generation of hypotheses, namely incubation and illumination. This is realized by Simon's conceptualization of the 'problem-space' (as given by initial state, end-state, operators and constraints). Moreover, Simon ends up advancing a 'mechanization' of the scientific generation of hypotheses, and scientific discovery in general.

More specifically, the seminal idea here is the analogy between the human brain and a computer that shapes the whole approach put forward by Simon: the human brain can be regarded as "an information-processing system, whose memories hold interrelated symbol structures and whose sensory and motor connections receive encoded symbols from the outside via sensory organs and send encoded symbols to motor organs" (Simon et al. 1987, 8). Under this analogical hypothesis, thinking is equated to "copying and reorganizing symbols in memory, receiving and outputting symbols, and comparing symbol structures for identity or difference" (Ibid.). As a consequence, problem-solving is carried out by creating a:

symbolic representation of the problem (called the problem space) that is capable of expressing initial, intermediate, and final problem situations as well as the whole range of concepts employed in the solution process, and using the operators that are contained in the definition of the problem space to modify the symbol structures that describe the problem situation (thereby conducting a mental search for a solution through the problem space) (Ibid).

That means that the generation of hypotheses can be guided, and it is here that the heuristics step in: "the search for a solution to a problem is not carried on by random trial and error, but it is guided in the direction of a goal situation (or symbolic expressions describing a goal) by rules of thumb, called heuristics". Simon provides two sets of heuristics, namely the *weak* and the *strong* one, which "make use of information extracted from the problem definitions and the states already explored in the problem space to identify promising paths for search" (Ibid.). But the most interesting feature of Simon's approach is his idea of a mechanization of discovery, as put in use with the several versions of the BACON software: "our method of inquiry will be to build a computer program (actually a succession of programs) that is capable of making nontrivial scientific discoveries and whose method of operation is based on our knowledge of human methods of solving problems-in particular, the method of heuristic selective search" (Ibid, p. 5). One example provided by Simon is the discovery of Kepler's law.

Unfortunately, this view is defective and the very idea of a pure mechanical process of discovery seems untenable for several reasons (see in particular Nickles 1980a; Kantorovich 1993; Gillies 1996; Weisberg 2006). Here I summarize some of these reasons.

The very first one is about the choice of the relevant variables and data. In this sense the BACON software has all the hard work done for it: the data and the relevant variables are chosen and provided by the programmers, which enable the software to trigger its heuristics. A crucial step in the generation of a hypothesis is to decide what to concentrate on, which made the data relevant and opened up the opportunity for a calculation: "the computer program is fed relevant data, whereas the original researcher had to determine in the first place exactly which data were relevant to the problem he or she was facing." (Weisberg 2006, 89).

In effect, the BACON software is told "(i) which two variables to relate and (ii) the general form of the law it should look for. The really difficult part of Kepler's discovery was finding out (i) and (ii), and this BACON.1 does not do at all. Once (i) and (ii) are given, the problem reduces to estimating two parameters from the data" (Gillies 1996, 24). Therefore, "BACON.1 does no more than provide an alternative solution to this elementary problem. It is thus not surprising that it is not able to discover any new and significant generalizations" (Ibid.).

Second, the only thing that a computer can do, and actually does better than humans, is to (1) find regularities in datasets and, in doing this, to (2) extend our computational skills in the exploration of these datasets. In effect BACON does discover regularities hidden in the data, when fed the *right* data, and it shows how we can extend our data-processing power.

Third, BACON is a reconstruction, *ex-post*, of a historical discovery. It re-discovers something known, but it does not discover something new and unknown. In doing this it benefits from the knowledge of the final result, that is, the advantage of knowing what the problem is, and the advantage of knowing that the data can be approached by certain heuristics, and that the problem is solvable.

Weisberg (see Weisberg 2006) argues that a partial defense of this view can be advanced by noting that there are similarities between Kepler's situation and that of the BACON software. In effect, in the specific case of the discovery of Kepler's third law we can see that:

(1) Kepler, like BACON, did not have "to raise the critical questions concerning the motions of the planets and what they might mean; those questions had already been put forth by others and were well known to educated people" (Weisberg 2006, 152).

(2) the data were provided to, and not by, Kepler (they were collected by Brahe), "so here again we see that the table was already set when he began his work" (Ibid.).

Bottom line: "one could conclude that Kepler's status in the enterprise was not all that different from that of the BACON program when the researchers gave it the data and turned it on" (Ibid.).

Of course this objection is reasonable to the extent that we look at single fragments of the search for a solution to a problem. But if we look at the whole, dynamic process of problem-solving, the limitation of a computer still stands: it cannot do all the conceptualization required to solve a problem. All that a computer can do are the very last few, computational, steps of the process that eventually leads to the generation of a hypothesis. For example, as noted by Nickels, a computer "does not have to ask itself conceptually deep questions or consider deep reformulations of the problem, because the primary constraints (and therefore the problem itself) are programmed in from the beginning. Only certain empirical constraints remain to be fed in" (Nickles 1980b. I, 38).

Of course also the third tradition, the Lakatos-Cellucci one (see also Magnani 2001; Nersessian 2008), starts from the idea that problem-solving is the key to understanding theory-building. In addition, this tradition argues that the process that leads to a solution to a problem can be examined, and transmitted and learned in a rational and logical way. In its more recent developments (see e.g. Cellucci 2013; Ippoliti and Cellucci 2016) this view argues for a (more) unitary and systematic way for both solving and finding problems (problem-solving and problem-finding). In particular, Cellucci argues for an analytic version of the heuristic approach (see also Cellucci's paper in this volume).

The crucial step in theory-building in accordance with the analytic version is the discovery of solutions to problems, whose core is the generation of hypotheses by means of ampliative rules (induction, analogy, metaphor, etc. see in particular Cellucci 2013, Chaps. 20–21). A hypothesis, once introduced, must satisfy the requirement of 'plausibility', that is, "the arguments for the hypothesis must be stronger than the arguments against it, on the basis of experience" (Cellucci 2013, 56). This of course implies that solutions to problems are not absolutely certain, but only plausible.

The method to carry out solutions to problems is a revised version of the analytic method as put forward by Plato, and it states that:

> to solve a problem, one looks for some hypothesis that is a sufficient condition for solving the problem, that is, such that a solution to the problem can be deduced from it. The hypothesis is obtained from the problem, and possibly other data already available, by some non-deductive rule and must be plausible. But the hypothesis is in its turn a problem that must be solved, and is solved in the same way. That is, one looks for another hypothesis that is a sufficient condition for solving the problem posed by the previous hypothesis, it is obtained from the latter, and possibly other data already available, by some non-deductive rule, and must be plausible. And so on, ad infinitum. (Ibid, 55)

Thus, this version of the analytic method explicitly draws on an 'infinitist' solution to the 'regress argument' and argues that knowledge is provisional and can be revised as new data come out. Of course once a hypothesis is introduced, its plausibility has to be established with the help of deductive reasoning and heuristic reasoning. More precisely a hypothesis undergoes the following test (see Cellucci 2013, 56):

(1) Deduce conclusions from the hypothesis.
(2) Compare the conclusions with each other, in order to check that the hypothesis does not produce a contradiction.
(3) Compare the conclusions with other hypotheses already known to be plausible, and with results of observations or experiments, in order to see that the arguments for the hypothesis are stronger than those against it on the basis of experience.

This test shows how deductive reasoning and heuristic reasoning both contribute to the advancement of knowledge.

4 The Heuristic View: A Critical Evaluation

The heuristic view offers a robust account of theory-building, with several epistemic advantages over the other approaches (see in particular Laudan 1981 on this
point). Below I outline some of them.

The first one is the fact that a theory can be accepted even if it is not cumulative.
A new theory does not have to preserve all the results of previous theories;, it is
sufficient that the new theory's 'effectiveness' in problem-solving exceeds that of
the previous theories. Of course here, by effectiveness, we do not mean naively the
'number of solved problems'. This way of characterizing the issue is defective,
since it tries to put in use a kind of 'metrics' and in order to do this, it needs not only
to commit to a realist position, but also the notion of Truth and of approximation of
truth. That is, we have to accept the idea that theories would be an approximation to
reality. Nevertheless truth (in science) is just what the heuristic approach is trying to
show as an inadequate concept. As a matter of fact a new theory conceptually
reshapes a specific field (that is, the relations between problems and hypotheses) in
such a way that what counted as a problem in the old theory could no longer be
such in the new theory. If it is so, such a metric comparison between the two
theories in terms of number of solved problems would be totally misleading.

Secondly, the principles of rationality underlying the development of science are
local and flexible. Advancing knowledge is not simply a question of rejection or
acceptance of a theory, but requires the assessment of its effectiveness and of the
way a theory extends our knowledge and the understanding of the world.

Third, it explains why it is rational for scientists to also accept theories that
display anomalies, and why scientists are sometimes reluctant to accept theories
that apparently seem well confirmed.

Moreover, this approach does not assign to science transcendent or unattainable
aims.

Furthermore, it accounts for the coexistence of rival theories, and shows how
and why theoretical pluralism contributes to scientific progress.

Moreover, the analytic account offers a robust account of problem-solving, with
several epistemic advantages over the other approaches based on problems.

First, the analytic account does not assume a *closure*, since it states that a theory
is an *open* set of hypotheses, not given once and for all. Theory-building consists in
starting from a problem, introducing hypotheses and deducing a solution to the
problem from them. This fact makes the analytic view compatible with Gödel's first
incompleteness theorem, as it does not commit to a single system of hypotheses to
solve all the problems of a given domain.

Second, the analytic view:

(a) provides a rational account of theory-building, in terms of hypotheses obtained
by non-deductive rules and validated by their plausibility.
(b) It does explain theory change, as hypotheses are subject to modification or
replacement when they become implausible as new data emerge. The modified

or new hypotheses are obtained through an analysis of the reasons why the former hypotheses have become implausible.

In addition, the analytic view offers more epistemic benefits. Below I outline some of them.

First, it does not require the problematic notion of 'intuition', which does not have a role in the formulation or the justification of a hypothesis. As a matter of fact, unlike other problem-oriented approaches or the axiomatic method, it relies on an inferential and rational view. No black boxes have to be assumed. In effect, on one hand the hypothesis for the solution of a problem is inferred from the problem, and other data, by the application of non-deductive rules—so by means of inferences. On the other hand, the justification, and the plausibility, of the hypothesis is established by comparing the reasons for and the reasons against the hypothesis, therefore not appealing to intuition, but in a rational fashion.

Second, it does not require the controversial logical and temporal separation between discovery and justification. The generation of a hypothesis embeds its justification.

Third, the fact that a hypothesis can only be plausible, and never true or certain, is not detrimental for the analytical view. As a matter of fact this perfectly fits the core idea of modern science and its conceptual turn, that is the idea that the aim of science is not to know the essence of natural substances, but some 'affections' of them.

Moreover, hypotheses are flexible and effective objects from an ontological and methodological viewpoint, unlike axioms. While the former are local, provisional and meant to solve specific problems, the latter are meant to be such to prove all the propositions that are 'true' in a given domain.

Last but not least, in the analytical view not only do different problems in general require different hypotheses, but also the same problem can be solved using different assumptions. This follows from the fact that every problem is multi-faceted and therefore can be viewed from different perspectives, each of which can lead to different hypotheses, and then to different solutions.

Acknowledgements I would like to thank David Danks, Carlo Celluci, and the two anonymous referees for their valuable comments on an early version of the paper.

References

Cartwright, N. (1983). *How the laws of physics lie*. New York: Oxford University Press.
Cartwright, N. (1996). Models and the limits of theories: Quantum Hamiltonians and the BCS model of superconductivity. In M. Morgan & M. Morrison (Eds.), *Models as mediators: Perspectives on natural and social science* (pp. 241–281). Cambridge: Cambridge University Press.
Cartwright, N. (1999). *The dappled world: A study of the boundaries of science*. Cambridge: Cambridge University Press.

Cartwright, N., Shomar, T., & Suarez, M. (1995). The toolbox of science. In W. Herfel, W. Krajewski, I. Niiniluoto, & R. Wojcicki (Eds.), *Theories and models in scientific processes* (pp. 137–149). Amsterdam: Rodopi.

Cellucci, C. (2008). *Le ragioni della logica*. Bari-Roma: Laterza.

Cellucci, C. (2013). *Rethinking logic*. Dordrecht: Springer.

Clement, J. J. (2008). *Creative model construction in scientists and students: The role of imagery, analogy, and mental simulation*. New York: Springer.

Darden, L. (Ed.). (2006). *Reasoning in biological discoveries: Essays on mechanisms, interfield relations, and anomaly resolution*. New York: Cambridge University Press.

Einstein, A. (2002). Induction and deduction in physics. In Albert Einstein (Ed.), *Collected papers* (Vol. 7, pp. 108–109). Princeton: Princeton University Press.

Einstein, A. (2010). *Ideas and opinions*. New York: Crown.

Frege, G. (1960). *The foundations of arithmetic. A logic-mathematical enquiry into the concept of number*. New York: Harper.

Gigerenzer, G., & Todd, P. M. (1999). *Simple heuristics that make us smart*. New York: Oxford University Press.

Gillies, D. (1995). *Revolutions in mathematics*. Oxford: Clarendon Press.

Gillies, D. (1996). *Artificial intelligence and scientific method*. Oxford: Oxford University Press.

Grosholz, E. (2007). *Representation and productive ambiguity in mathematics and the sciences*. New York: Oxford University Press.

Grosholz, E., & Breger, H. (2000). *The growth of mathematical knowledge*. Dordercht: Springer.

Hadamard, (1945). *An essay on the psychology of invention in the mathematical field*. Princeton: Princeton University Press.

Hempel, Carl Gustav. (1966). *Philosophy of natural science*. Englewood Cliffs: Prentice-Hall.

Ippoliti, E. (2006). *Il vero e il plausibile*. Morrisville (USA): Lulu.

Ippoliti, E. (2008). *Inferenze ampliative*. Morrisville (USA): Lulu.

Ippoliti, E. (Ed.). (2014). *Heuristic reasoning*. Berlin: Springer.

Ippoliti, E. (Ed.). (2016). *Models and inferences in science*. Berlin: Springer.

Ippoliti, E., & Cellucci, C. (2016). *Logica*. Milano: EGEA.

Jaccard, J., & Jacoby, J. (2010). *Theory construction and model-building*. New York: Guilford Press.

Kantorovich, A. (1993). *Scientific discovery: Logic and tinkering*. New York: State University of New York Press.

Lakatos, I. (1976). *Proofs and refutations: The logic of mathematical discovery*. Cambridge: Cambridge University Press.

Lakatos, I. (1978). *The methodology of scientific research programmes*. Cambridge: Cambidge Univesity Press.

Laudan, L. (1977). *Progress and its problems*. Berkeley and LA: University of California Press.

Laudan, L. (1981). A problem-solving approach to scientific progress. In I. Hacking (Ed.), *Scientific revolutions* (pp. 144–155). Oxford: Oxford University Press.

Laudan, L. (1996). *Beyond positivism and relativism: Theory, method, and evidence*. Oxford: Westview Press.

Magnani, L. (2001). *Abduction, reason, and science: Processes of discovery and explanation*. New York: Kluwer Academic.

Magnani, L., & Magnani, W. (Eds.). (2010). *Model-based reasoning in science and technology: Abduction, logic, and computational discovery*. Heidelberg: Springer.

Nersessian, N. (2008). *Creating scientific concepts*. Cambridge (MA): MIT Press.

Nickles, T. (Ed.). (1980a). *Scientific discovery: Logic and rationality*. Boston: Springer.

Nickles, T. (Ed.). (1980b). *Scientific discovery: Case studies*. Boston: Springer.

Nickles, T. (1981). What is a problem that we may solve it? Scientific method as a problem-solving and question-answering technique. *Synthese, 47*(1), 85–118.

Nickles, T. (2014). Heuristic appraisal at the frontier of research. In E. Ippoliti (Ed.), *Heuristic reasoning* (pp. 57–88). Berlin: Springer.

Nickles, T., & Meheus, J. (Eds.). (2009). *Methods of discovery and creativity*. New York: Springer.

Poincaré, H. (1908). L'invention mathématique. *Enseignement mathématique, 10,* 357–371.

Polya, G. (1954). *Mathematics and plausible reasoning* (Vol. I e II). Princeton : Princeton University Press.

Popper, K. (1999). *All life is problem solving*. London: Routledge.

Popper, K., & Miller, D. (1983). A proof of the impossibility of inductive probability. *Nature, 302,* 687–688.

Quarantotto, D. (2017). Aristotle's problemata style and aural textuality. In R. Polansky & W. Wians (Eds.), *Reading aristotle* (pp. 75–122). Leiden: Brill.

Reichenbach, H. (1938). *Experience and prediction: An analysis of the foundations and the structure of knowledge*. Chicago: The University of Chicago Press.

Schippers, M. (2014). Probabilistic measures of coherence: From adequacy constraints towards pluralism. *Synthese, 191*(16), 3821–3845.

Shelley, C. (2003). *Multiple analogies in science and philosophy*. Amsterdam: John Benjamins B.V.

Simon, H. (1977). *Models of discovery*. Dordrecht: Reidel.

Simon, H., Langley, P., Bradshaw, G., & Zytkow, J. (1987). *Scientific discovery: Computational explorations of the creative processes*. Boston: MIT Press.

Suarez, M., & Cartwright, N. (2008). Theories: Tools versus models. *Studies in History and Philosophy of Modern Physics, 39,* 62–81.

Talalay, L. E. (1987). Rethinking the function of clay figurine legs from neolithic Greece: An argument by analogy. *American Journal of Archaeology, 91*(2), 161–169.

Ulazia, A. (2015). Multiple roles for analogies in the genesis of fluid mechanics: How analogies can cooperate with other heuristic strategies. *Foundations of Science*, https://doi.org/10.1007/s10699-015-9423-1.

Weisberg, R. (2006). *Creativity: Understanding innovation in problem solving, science, invention, and the arts*. Hoboken: Wiley.

Building Theories: Strategies Not Blueprints

Margaret Morrison

Abstract Views of theory structure in philosophy of science (semantic and syntactic) have little to say about how theories are actually constructed; instead, the task of the philosopher is typically understood as *reconstruction* in order to highlight the theory's essential features. If one takes seriously these views about theory structure then it might seem that we should also characterize the practice of building theories in accordance with the guidelines they set out. Examples from some of our most successful theories reveal nothing like the practices that conform to present accounts of theory structure. Instead there are different approaches that partly depend on the phenomena we want to account for and the kind of theory we desire. At least two strategies can be identified in high energy physics, (1) top down using symmetry principles and (2) bottom up strategy beginning with different types of models and gradually embedding these in a broad theoretical framework. Finally, in cases where methods and techniques cross disciplines, as in the case of population biology and statistical physics, we see that theory construction was largely based analogical considerations like the use of mathematical methods for treating systems of molecules in order to incorporate populations of genes into the theory of natural selection. Using these various examples I argue that building theories doesn't involve blueprints for what a theory should look like, rather the architecture is developed in a piecemeal way using different strategies that fit the context and phenomena in question.

Keywords Theories · Models · Electrodynamics · Electroweak theory
Population genetics · Symmetries · Mathematical structures

1 Introduction

When we speak about "building theories" the first question that comes to mind is: What *is* a theory anyway? Typically, we have a rough and ready account that allows us to identify the successful theories science has produced—Newtonian mechanics,

---

M. Morrison (✉)
Philosophy Department, University of Toronto, Toronto, Canada
e-mail: mmorris@chass.utoronto.ca

© Springer International Publishing AG 2018
D. Danks and E. Ippoliti (eds.), *Building Theories*, Studies in Applied Philosophy,
Epistemology and Rational Ethics 41, https://doi.org/10.1007/978-3-319-72787-5_2

spec.al and general relativity, quantum mechanics and the standard model, to name a few. But this kind of roll call is not sufficient to satisfy the philosophically curious, those who want to know what characteristics a theory has, how we identify it, what distinguishes it from models and hypotheses, and the processes involved in its construction.

The two most prominent accounts of theory structure in the philosophy of science have been the syntactic and semantic views. Both involve a reconstruction of the working notion of a theory used by scientists. However, when evaluating the pros and cons of these views the natural question to ask is whether such reconstructions in fact help us to answer questions about the nature of theories—their construction, function, and how we use them. In the first section of the paper I want to briefly examine these accounts. What I hope my examination will reveal is that unless we're interested in the logical structure of theories, these philosophical accounts have little to offer in helping us understand the nature of theories. In other words, they fail to illuminate the foundational features of the more concrete versions of theories we find in scientific text books. But, as we shall also see, that fact needn't prevent us from providing a philosophical analysis (rather than a reconstruction) of these theories and the processes that led to their construction.

One of the underlying assumptions of both the syntactic and semantic views is that theories have a definite structure which ought to entail, at least to some extent, the existence of specific rules for their construction. I argue in the remaining sections of the paper that this is, in fact, not the case, nor is a reconstruction along those lines useful. Rather than having blueprints for formulating theories, what we have are a number of rather loosely defined strategies, strategies that work in different contexts depending on the phenomena we want to understand and the what we expect from the theory itself. I identify three such strategies: bottom up from phenomena to theory via models; top down using mathematical frameworks to generate dynamical theory/explanations, and sideways using analogies from other theories. For each of the strategies there are well documented examples of how they are used to arrive at successful theories. What is interesting about these strategies is how they are targeted to specific types of problem solving, another reason to think that a monolithic approach to theory structure is largely unhelpful.

Before going on to discuss these strategies let me begin by looking at the semantic and syntactic views of theory structure, what I take to be the problems with each and why, in general, these rigidly defined approaches to identifying theories seem to collapse under their own weight.

2   Theories—Some Philosophical Background

One of the consequences of the emphasis on models in philosophy of science has been a shift away from focusing on the nature of theories. Indeed, the semantic view, is, in most of its guises, not about theories at all but about models because the former are defined solely in terms of the latter (van Fraassen 1980, 1989, 2008; Giere 1988, 2004; Suppes 1961, 1962, 1967, 2002; Suppe 1989). Although the semanticists stress that their view is primarily a logico-philosophical account of theory structure, they also emphasize its ability to capture scientific cases. Because of this dual role, it is important to evaluate the model-theoretic features of the account to determine its success in clarifying the nature of theory structure as well as its merits in dealing with the "scientific" or practical dimensions of modeling.

The semantic view was formulated as an alternative to the difficulties encountered by its predecessor, the syntactic, or received view. The former defines theories in terms of models while the later defines models in terms of theories, thereby making models otiose. The history and variations associated with these views is a long and multifaceted story involving much more technical detail than I can (or a reader would want me to) rehearse here.[1] What these views do have in common, however, is the goal of defining what a theory is, and hence how it should be constructed. On the syntactic view the theory is an uninterpreted axiomatized calculus or system—a set of axioms expressed in a formal language (usually first-order logic)—and a model is simply a set of statements that interprets the terms in the language. While this is one way of interpreting the language, the more common approach has been to use correspondence rules that connected the axioms (typically the laws of the theory) with testable, observable consequences. The latter were formulated in what was referred to as the "observation language" and hence had a direct semantic interpretation. Many of the problems associated with the syntactic view involved issues of interpretation—specifying an exact meaning for correspondence rules, difficulties with axiomatization, and the use of first-order logic as the way to formalize a theory. But there were also problems specifically related to models. If the sets of statements used to interpret the axioms could be considered models, then how should one distinguish the intended from the unintended models?

The difficulties associated with axiomatization and the identification of a theory with a linguistic formulation gave rise to the semantic view whose advocates (Suppes 1961, 1967, 2002; Suppe 1989; Giere 1988) appeal, in a more or less direct way, to the notion of model defined by Tarski (1953). Although van Fraassen (1980) opts for the state-space approach developed by Weyl and Beth, the underlying similarity in these accounts is that the model supposedly provides an interpretation of the theory's formal structure but is not itself a linguistic entity. Instead of formalizing the theory in first-order logic, one simply defines the intended class of models directly.

Suppes' version of the semantic view includes a set-theoretic axiomatization that involves defining a set-theoretical predicate, (i.e., a predicate like "is a classical particle system" that is definable in terms of set theory), with a model for the theory being simply an entity that satisfies the predicate. He claims that the set-theoretical model can be related to what we normally take to be a physical or scientific model by simply interpreting the primitives as referring to the objects associated with a physical model. Although he acknowledges that the notion of a physical model is important in physics and engineering, the set-theoretical usage is the "fundamental"

---

[1]For an extended discussion see da Costa and French (2003).

one. It is required for an exact statement of any branch of empirical science since it illuminates not only "the exact statement of the theory" but "the exact analysis of data" as well (2002: 24). Although he admits that the highly physical or empirically minded scientists may disagree with this, Suppes also claims that there seems to be no point in "arguing about which use of the word model is primary or more appropriate in the physical sense" (22).[2]

Van Fraassen specifically distances himself from Suppes' account, claiming that he is more concerned with the relation between physical theories and the world than with the structure of physical theory (1980: 67). To "present a theory is to specify a family of structures, its models" (64); and, "any structure which satisfies the axioms of a theory [...] is called a model of that theory" (43). The models here are state-spaces with trajectories and constraints defined in the spaces. Each state-space can be given by specifying a set of variables with the constraints (laws of succession and coexistence) specifying the values of the variables and the trajectories their successive values. The state-spaces themselves are mathematical objects, but they become associated with empirical phenomena by associating a point in the state-space with a state of an empirical system.[3]

One of the problems with Suppes' account it that it lacks a clearly articulated distinction between the primitives used to define particle mechanics and the realization of those axioms in terms of the ordered quintuple.[4] Moreover, if theories are defined as families of models there is, strictly speaking, nothing for the model to be true of, except all the other models. In other words, the models do not provide an interpretation of some distinct theory but stand on their own as a way of treating the phenomena in question. While there may be nothing wrong with this in principle, it does create a rather peculiar scenario: there is no way of identifying what is "fundamental" or specific about a particular theoretical framework since, by definition, all the paraphernalia of the models are automatically included as part of the theory. But surely something like perturbation theory, as a mathematical technique,

---

[2]What this suggests, then, is that as philosophers our first concern should be with the exact specifications of theoretical structure rather than how the models used by scientists are meant to deliver information about physical systems.

[3]But once this occurs the state-space models take on a linguistic dimension; they become models of the theory in its linguistic formulation. Similarly, in Suppes' account, when it comes to specifying the set theoretical predicate that defines the class of models for a theory, we do need to appeal to the specific language in which the theory is formulated. And, in that context, which is arguably the one in which models become paramount, they cease to become nonlinguistic entities. But as long as no specific language is given priority at the outset, we can talk about models as nonlinguistic structures.

[4]We can axiomatize classical particle mechanics in terms of the five primitive notions of a set P of particles, an interval T of real numbers corresponding to elapsed times, a position function s defined in the Cartesian product of the set of particles and the time interval, a mass function m and a force function F defined on the Cartesian product of the set of particles, the time interval and the positive integers (the latter enter as a way of naming the forces). A realization or model for these axioms would be an ordered quintuple consisting of the primitives $P = <P, T, s, m, f>$. We can interpret this to be a physical model for the solar system by simply interpreting the set of particles as the set of planetary bodies, or the set of the centers of mass of the planetary bodies.

should not be identified as part of quantum mechanics any more than the differential calculus ought to be included as part of Newton's theory.

Moreover, if a theory is just a family of models, what does it mean to say that the model/structure is a realization of the theory? The model is not a realization of the theory because there is no theory, strictly speaking, for it to be a realization of. In other words, the semantic view has effectively dispensed with theories altogether by redefining them in terms of models. There is no longer anything to specify as "Newtonian mechanics" except the models used to treat classical systems. While there may be nothing wrong with this if one's goal is some kind of logical/ model-theoretic reconstruction—in fact it has undoubtedly addressed troublesome issues associated with the syntactic account—but, if the project is to understand various aspects of how models and theories are related, and how they function in scientific contexts, then reducing the latter to the former seems unhelpful. And, as a logical reconstruction, it is not at all clear how it has enhanced our understanding of theory structure—one of its stated goals.

Van Fraassen's focus on state-spaces does speak to features of the "scientific practice" of modeling. As I noted earlier, the state-space approach typically involves representing a system or a model of a system in terms of its possible states and its evolution. In quantum mechanics a state-space is a complex Hilbert space in which the possible instantaneous states of the system may be described by unit vectors. In that sense, then, we can see that the state-space approach is in fact a fundamental feature of the way that theories are represented. The difficulty, however, is that construed this way the theory is nothing more than the different models in the state-space. While we might want to refer to the Hilbert space formulation of QM for a mathematically rigorous description, such a formalism, which has its roots partly in functional analysis, can be clearly distinguished from a theory that gives us a more "physical" picture of a quantum system by describing its dynamical features.

On Giere's (2004) version of the semantic view we have nothing that answers to the notion of "theory" or "law." Instead we have principles that define abstract objects that, he claims, do not directly refer to anything in the world. When we refer to something as an "empirical" theory, this is translated as having models (abstract objects) that are structured in accordance with general principles that have been applied to empirical systems via hypotheses. The latter make claims about the similarity between the model and the world. The principles act as general templates that, together with particular assumptions about the system of interest, can be used for the construction of models. Although the notion of a general principle seems to accord well with our intuitive picture of what constitutes a theory, the fact that the principles refer only to models and do not describe anything in the world means that what we often refer to as the "laws of physics" are principles that are true only of models and not of physical objects themselves. This leaves us in the rather difficult position of trying to figure out what, if any, role there could be for what we typically refer to as quantum theory, Newtonian mechanics, evolutionary theory, and what those theories say about the world. We need to also ask what is gained by the kind of systematic elimination of theory that characterizes the various formulations of the semantic view.

Giere claims that one of the problems that besets the notion of "theory" is that it is used in ambiguous and contradictory ways. While this is certainly true, there are other, equally serious problems that accompany the semantic reconstruction. One of these, mentioned at the outset, involves specifying the content of the theory if it is identified strictly with its models. Another, related issue concerns the interpretation of that content. In particular, the models of many of our theories typically contain a good deal of excess structure or assumptions that we would not normally want to identify as part of a theory. Although van Fraassen claims that it is the task of theories to provide literal descriptions of the world (1989: 193), he also recognizes that models contain structure for which there is no real-world correlate (225–228). However, the issue is not simply one of determining the referential features of the models, even if we limit ourselves to the "empirical" data. Instead I am referring to cases where models contain a great deal of structure that is used in a number of different theoretical contexts, as in the case of approximation techniques or principles such as least time. Because models are typically used in the application of higher level laws (that we associate with theory), the methods employed in that application ought to be distinguished from the content of the theory (i.e., what it purports to say about physical systems).

Consider the following example (discussed in greater detail in Morrison 1999). Suppose we want to model the physical pendulum, an object that is certainly characterized as empirical. How should we proceed when describing its features? If we want to focus on the period, we need to account for the different ways in which it can be affected by air, one of which is the damping correction. This results from air resistance acting on the pendulum ball and the wire, causing the amplitude to decrease with time while increasing the period of oscillation. The damping force is a combination of linear and quadratic damping. The equation of motion has an exact solution in the former case but not in the latter case, since the sign of the force must be adjusted each half-period to correspond to a retarding force. The problem is solved using a perturbation expansion applied to an associated analytic problem where the sign of the force is not changed. In this case, the first half-period is positively damped and the second is negatively damped with the resulting motion being periodic. Although only the first half-period corresponds to the damped pendulum problem, the solution can be reapplied for subsequent half-periods. But only the first few terms in the expansion converge and give good approximations— the series diverges asymptotically, yielding no solution.

All of this information is contained in the model, yet we certainly do not want to identify the totality as part of the theory of Newtonian mechanics. Moreover, because our treatment of the damping forces requires a highly idealized description, it is difficult to differentiate the empirical aspects of the representation from the more mathematically abstract ones that are employed as calculational devices. My claim here is not just that the so-called empirical aspects of the model are idealized since all models and indeed theories involve idealization. Rather, the way in which the empirical features are interconnected with the non-empirical makes it difficult to isolate what Newtonian mechanics characterizes as basic forces if one characterizes the theory as simply a family of models.

Why do we want to identify these forces? The essence of Newtonian mechanics is that the motion of an object is analyzed in terms of the forces exerted on it and described in terms of the laws of motion. These core features are represented in the models of the theory, as in the case of the linear harmonic oscillator which is derived from the second law. Not only are these laws common to all the models of Newtonian mechanics, but they constrain the kind of behavior described by those models and provide (along with other information) the basis for the model's construction. Moreover, these Newtonian models embody different kinds of assumptions about how a physical system is constituted than, say, the same problem treated by Lagrange's equations. In that sense, we identify these different core features as belonging to different "theories" of mechanics.[5] For example, when asked for the basic structure of classical electrodynamics, one would immediately cite Maxwell's equations. They form a theoretical core from which a number of models can be specified that assist in the application of these laws to specific problem situations. Similarly, an undisputed part of the theoretical core of relativistic quantum mechanics is the Dirac equation, and in the case of non-relativistic quantum mechanics we have the Schrodinger equation, the Pauli principle, wave particle duality, the uncertainty principle and perhaps a few more. In none of these cases is it necessary to draw rigid boundaries about what should be included. The fact is we can identify basic assumptions that form the core of the theory; that's all we need.

Admittedly there may be cases where it is not obvious that such a theoretical core exists. Population genetics is a good example. But even here one can point to the theory of gene frequencies as the defining feature on which many of the models are constructed. My point is simply that by defining a theory solely in terms of its many models, one loses sight of the theoretical coherence provided by core laws, laws that may not determine features of the models but certainly constrain the kind of behaviors that the models describe. Indeed, it is the identification of a theoretical core rather than all of the features contained in the models that enables us to claim that a set of models belongs to Newtonian mechanics. Moreover, nothing about this way of identifying theories requires that they be formalized or axiomatized.

In light of these various difficulties it seems that any analysis of theory construction needs to move beyond the current philosophical frameworks if we are going to capture interesting features of the way theories are built and used in scientific contexts. If we identify a theory with the kinds of core laws/equations

---

[5]The notion that these are different theories is typically characterized in terms of the difference between forces and energies. The Newtonian approach involves the application of forces to bodies in order to see how they move. In Lagrange's mechanics, one does not deal with forces and instead looks at the kinetic and potential energies of a system where the trajectory of a body is derived by finding the path that minimizes the action. This is defined as the sum of the Lagrangian over time, which is equal to the kinetic energy minus the potential energy. For example, consider a small bead rolling on a hoop. If one were to calculate the motion of the bead using Newtonian mechanics, one would have a complicated set of equations that would take into account the forces that the hoop exerts on the bead at each moment. Using Lagrangian mechanics, one looks at all the possible motions that the bead could take on the hoop and finds the one that minimizes the action instead of directly calculating the influence of the hoop on the bead at a given moment.

I referred to above, then the interesting question becomes how those are arrived at —how is the theory in question built? In an attempt to provide an answer, I want to look at some specific strategies, strategies for which there are no specified rules or algorithms but that depend partly on the kind of phenomena being investigated and the available methods for treating those phenomena. As we shall see, the strategies bear little, if any, relation to the philosophical accounts of theory discussed above, except for the fact that models sometimes play a role in the initial stages of construction. The overarching question of course, given these different strategies, is whether we even need more formal philosophical reconstructions discussed above in our attempts to understand the nature and structure of theories?

3   Bottom Up—Phenomena to Theory via Modelling

Some examples of this type of strategy include the development of the Bardeen, Cooper and Schreiffer (BCS) theory of superconductivity (see Morrison (2007) although it wasn't analysed as such in that paper) as well as Maxwell's formulation of electrodynamics. I call this bottom-up because the strategy is often implemented when one has a specific type of phenomena that needs to be embedded into a theory, as in the case of electromagnetic phenomena understood in terms of a field, but for which no satisfactory theoretical framework exists. In other words, we have phenomena that suggest the need for a new conceptual understanding with no obvious way to construct such a theory. Typically, the process begins with some fundamental observations or data that need to be given a theoretical foundation but are incapable of being absorbed into or are inconsistent with theories presently available. As I noted above, electrodynamics presents us with a clear case of this strategy.

So, what are the specific features that make the development of electrodynamics an example of the bottom up approach? To answer that question, we need to first see what type of phenomena had to be accounted for. In the mid-nineteenth century, the received view in electrodynamics was that electrical and magnetic phenomena could be explained in terms of the interaction of charges in an electric fluid according to an inverse square law. However, the experimental work of Faraday had revealed new relationships between electric and magnetic phenomena and light which were not easily understood using the prevailing theory. In 1831 he showed that one could produce an electrical current from a changing magnetic field, a phenomenon known as **electromagnetic induction**. When an electrical current was passed through a coil, another very short current was generated in a nearby coil. This marked a new discovery about the relationship between electricity and magnetism. Faraday's law, as it became known, stated that the induced electromotive force was equal to the rate of change of the magnetic flux passing through a surface whose boundary is a wire loop.

Faraday also discovered that a magnetic field influenced polarized light—a phenomenon known as the **magneto-optical effect or Faraday effect**. The plane of vibration of a beam of linearly polarized light incident on a piece of glass rotated when a magnetic field was applied in the direction of propagation of the beam. This was certainly the first clear indication that magnetic force and light were related to each other and hence that light is related to electricity and magnetism. Faraday's explanation of these phenomena was in terms of a field which involved lines of force filling space. He conceived the whole of the space in which electrical force acts to be traversed by lines of force which indicate at every point the direction of the resultant force at that point. He also assumed that the lines could represent the intensity of the force at every point, so that when the force is great the lines might be close together, and far apart when the force is small.

In his early papers on the subject Maxwell describes how Faraday's account was at odds with the prevailing mathematically based action at a distance views—where Faraday saw lines of force traversing all space the prevailing view hypothesised centres of force attracting at a distance; Faraday saw a medium as opposed to empty space; and he assumed the seat of electromagnetic phenomena was the result of actions in the medium as opposed to a power of action at a distance impressed on the electric fluids. But Faraday's views were just that—views based on experimental investigations without any theoretical foundation. Yet his experimental work had uncovered some important and interesting electromagnetic relationships that needed to be taken account of; something current theory seemed unable to do in a systematic way. In order to have any impact this notion of a 'field' had to be developed into a systematic theory. That task was left to Maxwell who went on to construct a theory of electromagnetism grounded in Faraday's notion of lines of force. So, the interesting question for our purposes is how was this theory "built" from the bottom up?

Maxwell's describes his 1861–62 paper, "On Physical Lines of Force," as an attempt to "examine" electromagnetic phenomena from a mechanical point of view and to determine what tensions in, or motions of, a medium were capable of producing the observed mechanical phenomena (Maxwell 1965, 1: 467). In other words, how could one provide a mechanical account of these experimental findings grounded in a medium rather than action at a distance? At the time Kelvin had developed a model of magnetism that involved the rotation of molecular vortices in a fluid aether, an idea that led Maxwell to hypothesize that in a magnetic field the medium (or aether) was in rotation around the lines of force; the rotation being performed by molecular vortices whose axes were parallel to the lines. In order to explain charge and to derive the law of attraction between charged bodies, Maxwell constructed an elastic solid aether model in which the aetherial substance formed spherical cells endowed with elasticity. The cells were separated by electric particles whose action on the cells would result in a kind of distortion. Hence, the effect of an electromotive force was to distort the cells by a change in the positions of the electric particles. That gave rise to an elastic force that set off a chain reaction. Maxwell saw the distortion of the cells as a displacement of electricity within each molecule, with the total effect over the entire medium producing a "general displacement of electricity in a given direction" (Maxwell 1965, 1: 491). Understood literally, the notion of displacement meant that the elements of the dielectric had changed positions. Because changes in displacement involved a motion of electricity, Maxwell argued that they should be "treated as" currents (1965, 1: 491).[6]

Because the phenomenological law governing displacement expressed the relation between polarization and force, Maxwell was able to use it to calculate the aether's elasticity (the coefficient of rigidity), the crucial step that led him to identify the electromagnetic and luminiferous aethers. Once the electromagnetic medium was endowed with elasticity, Maxwell relied on the optical aether in support of his assumption: "The undulatory theory of light requires us to admit this kind of elasticity in the luminiferous medium in order to account for transverse vibrations. We need not then be surprised if the magneto-electric medium possesses the same property" (1965, 1: 489). After a series of mathematical steps, which included correcting the equations of electric currents for the effect produced by elasticity, calculating the value for e, the quantity of free electricity in a unit volume, and E, the dielectric constant, Maxwell went on to determine the velocity with which transverse waves were propagated through the electromagnetic aether. The rate of propagation was based on the assumption described above—that the elasticity was due to forces acting between pairs of particles.

Using the formula $V = \sqrt{m/\rho}$, where m is the coefficient of rigidity, ρ is the aethereal mass density, and μ is the coefficient of magnetic induction, we have

$$E^2 = \pi m$$

$$\mu = \pi \rho$$

giving us $\pi m = V^2 \mu$, which yields $E = V\sqrt{\mu}$. Maxwell arrived at a value for V that, much to his astonishment, agreed with the value calculated for the velocity of light (V = 310,740,000,000 mm/s), which led him to remark that: The velocity of transverse undulations in our hypothetical medium, calculated from the electro-magnetic experiments of Kohlrausch and Weber, agrees so exactly with the velocity of light calculated from the optical experiment of M. Fiseau that we can *scarcely avoid the inference that light consists in the transverse undulations of the same medium which is the cause of electric and magnetic phenomena* (1965, 1: 500).

Maxwell's success involved linking the equation describing displacement ($R = -4\pi E^2 h$) with the aether's elasticity (modeled on Hooke's law), where displacement produces a restoring force in response to the distortion of the cells of the

---

[6]Displacement also served as a model for dielectric polarization; electromotive force was responsible for distorting the cells, and its action on the dielectric produced a state of polarization. When the force was removed, the cells would recover their form and the electricity would return to its former position (Maxwell 1965, 1: 492). The amount of displacement depended on the nature of the body and on the electromotive force.

medium. However, ($R = -4\pi E^2 h$) is also an electrical equation representing the flow of charge produced by electromotive force. Consequently, the dielectric constant E is both an elastic coefficient and an electric constant. Interpreting E in this way allowed Maxwell to determine its value and ultimately identify it with the velocity of transverse waves traveling through an elastic aether.

In modern differential form, Maxwell's four equations relate the Electric Field (**E**) and magnetic field (**B**) to the charge (ρ) and current (**J**) densities that specify the fields and give rise to electromagnetic radiation—light.

Gauss's law $\nabla \cdot D = \rho_f$

Gauss's law for magnetism $\nabla \cdot B = 0$

Faraday's law of induction $\nabla \times E = -\frac{\partial B}{\partial t}$

Ampère's circuital law with displacement $\nabla \times H = J_f + \frac{\partial D}{\partial t}$

D is the displacement field and **H** the magnetizing field. Perhaps the most important of these from the perspective of "theory building" is Ampère's law which states that magnetic fields can be generated by both electrical currents (the original "Ampère law") and changing electric fields (Maxwell's correction). This amendment to Ampère's law is crucial, since it specifies that both a changing magnetic field gives rise to an electric field and a changing electric field creates a magnetic field. Consequently, self-sustaining electromagnetic waves can propagate through space. In other words, it allows for the possibility of open circuits of the sort required by a field theory.[7]

What Maxwell had in fact shown was that given the specific assumptions employed in developing the mechanical details of his model, the elastic properties of the electromagnetic medium were just those required of the luminiferous aether by the wave theory of light. His bottom up approach beginning with electromagnetic phenomena described by Faraday and culminating in a model from which he derived the wave equation was undoubtedly a theoretical triumph, but there was no theoretical justification for many of the modelling assumptions nor for the existence of electromagnetic waves. Hence, the link from phenomena to the equations describing the field theoretic approach, while a successful exercise in theory building, lacked the experimental confirmation required for acceptance. At first this might seem rather odd; how could a bottom up theory grounded in experimental results lack the requisite confirmation? The answer lay in the mechanical description of how the phenomena were produced. While Maxwell's model certainly "saved" the phenomena there was no independent evidence for the existence of the field or the displacement current that facilitated the formulation of the wave equation. In other words, there was no independent evidence for the main assumptions upon which the theory was built.

---

[7]For a more extensive discussion of this point, see Morrison (2008).

Recognizing the problem, Maxwell developed a different version of the theory "A Dynamical Theory of the Electromagnetic Field" (1865) (DT) based on experimental facts and general dynamical principles about matter in motion as characterized by the abstract dynamics of Lagrange. But, unlike his initial formulation the later version was a thoroughly top down affair. The experimental facts, as conceived by Faraday, were embedded in a theory whose starting point was a mathematical framework that ignored the mechanical forces describing the field or aether. The aim of Lagrange's *Mécanique Analytique* (1788) was to rid mechanics of Newtonian forces and the requirement that we must construct a separate acting force for each particle. The equations of motion for a mechanical system were derived from the principle of virtual velocities and d'Alembert's principle. The method consisted of expressing the elementary dynamical relations in terms of the corresponding relations of pure algebraic quantities, which facilitated the deduction of the equations of motion. Velocities, momenta, and forces related to the coordinates in the equations of motion need not be interpreted literally in the fashion of their Newtonian counterparts. This allowed for the field to be represented as a connected mechanical system with currents, and integral currents, as well as generalized coordinates corresponding to the velocities and positions of the conductors. In other words, we can have a quantitative determination of the field without knowing the actual motion, location, and nature of the system itself.

Using this method Maxwell went on to derive the basic equations of electromagnetism without any special assumptions about molecular vortices, forces between electrical particles, and without specifying the details of the mechanical structure of the field. This allowed him to treat the aether (or field) as a mechanical system without any specification of the machinery that gave rise to the characteristics exhibited by the potential-energy function.[8]

It becomes clear, then, that the power of the Lagrangian approach lay in the fact that it ignored the nature of the system and the details of its motion; the emphasis was on energetic properties of a system, rather than its internal structure. Maxwell claimed that all physical concepts in "A Dynamical Theory," except energy, were understood to be merely illustrative, rather than substantial. Displacement constituted one of the basic equations and was defined simply as the motion of electricity, that is, in terms of a quantity of charge crossing a designated area. But, if electricity was being displaced, how did this occur? Due to the lack of a mechanical foundation, the idea that there was a displacement of electricity in the field (a charge), without an associated mechanical source or body, became difficult to motivate

---

[8]His attachment to the potentials as primary was also criticized, since virtually all theorists of the day believed that the potentials were simply mathematical conveniences having no physical reality whatsoever. To them, the force fields were the only physical reality in Maxwell's theory but the formulation in DT provided no account of this. Today, of course, we know in the quantum theory that it is the potentials that are primary, and the fields are derived from changes in the potentials.

theoretically. These issues did not pose significant problems for Maxwell himself, since he associated the force fields with the underlying potentials.[9]

Unlike the initial bottom up strategy, the theory cast in the Lagrangian formalism is in some ways typical of a type of "top down" mathematical strategy. The latter is often used when one is unsure how construct a theoretical model that incorporates specific features of the phenomena and instead general structural features such as the least action principle, or symmetry principles, become the starting point. However, because Maxwell's Lagrangian formulation of the theory was a reconstruction of its original model-based account it differs from the kind of top down strategy we see with the electroweak theory discussed below. Perhaps the biggest difference is that once deprived of its models Maxwell's electrodynamics lacked real explanatory power due to the absence of specific theoretical details about how electromagnetic waves could travel in free space or how the field was constituted. Although the field equations could account for both optical and electromagnetic processes in that they gave accurate values for the velocity of electromagnetic waves traveling through space, there was no theoretical foundation for understanding of how that took place. In that sense, devoid of mechanical models Maxwell's theory really was nothing other than Maxwell's equations, as Hertz so aptly remarked years later.

But, the top down approach needn't be limited to providing only a set of mathematical equations. By contrast the electroweak theory, while clearly a top down product of the mathematics of group theory, nevertheless provides a robust account of the theoretical foundations required for understanding how the weak and electromagnetic forces could be unified. In particular, what this example of theory building shows is how the mathematics itself can provide an explanatory foundation for the phenomena. Although the electroweak theory began with the goal of unifying the weak and electromagnetic forces it was soon realized that the phenomena in question were dramatically different and required a new conception of how the fields might be combined. Neither electromagnetic theory nor the existing account of weak interactions was capable of being modified to include the other. And, because of the incompatibility of the phenomena themselves, a bottom up approach also seemed unable to provide the foundation for a unified theory.

---

[9]The methods used in "A Dynamical Theory" were extended and more fully developed in the *Treatise on Electricity and Magnetism* (TEM), where the goal was to examine the consequences of the assumption that electric currents were simply moving systems whose motion was communicated to each of the parts by certain forces, the nature and laws of which "we do not even attempt to define, because we can eliminate [them] from the equations of motion by the method given by Lagrange for any connected system" (Maxwell 1873, Sect. 552). Displacement, magnetic induction and electric and magnetic forces were all defined in the *Treatise* as vector quantities (Maxwell 1873, Sect. 11, 12), together with the electrostatic state, which was termed the vector potential. All were fundamental quantities for expression of the energy of the field and were seen as replacing the lines of force.

4   Top Down—From Mathematics to Physics

The electroweak theory, a fundamental part of the standard model of particle physics, brings together electromagnetism with the weak force in a single relativistic quantum field theory.[10] However, as I noted above, from the perspective of phenomenology, the weak and electromagnetic forces are very different. Electromagnetism has an infinite range, whereas the weak force, which produces radioactive beta decay, spans distances shorter than approximately 10^{-15} cm. Moreover, the photon associated with the electromagnetic field is massless, while the bosons associated with the weak force are massive due to their short range. Given these differences there was little hope for constructing a theory that could combine the weak force and electromagnetism in a unified way. Despite these seeming incompatibilities, there are some common features: both kinds of interactions affect leptons and hadrons; both appear to be vector interactions brought about by the exchange of particles carrying unit spin and negative parity, and both have their own universal coupling constant that governs the strength of the interactions.

The challenge then was to build a theory that could unify the phenomena while accommodating their inconsistencies. Prima facie this seems like an impossible task; how can we construct a unified account of phenomena that appear fundamentally different? If we think back to the bottom up strategy employed in electrodynamics we quickly see that no such approach seems possible in the electroweak case. This is because the phenomena that need to be incorporated at the outset—the particles and their various properties—cannot be combined in a single model where the phenomenology or empirical data is the foundation on which the theory is built. We also saw in the top down version of electromagnetism that much of the theoretical detail was ignored in favour of a more abstract mathematical approach; so the question is whether a similar type of strategy might be useful here.

Schwinger's (1957) work marked the first significant connection between the weak and electromagnetic forces. His approach utilized some basic principles of symmetry and field theory, and went on to develop a framework for fundamental interactions derived from that fixed structure. The impetus came from quantum electrodynamics where it was possible to show that from the conservation of electric charge, one could, on the basis of Noether's theorem, assume the existence of a symmetry, and the requirement that it be local forces one to introduce a gauge field which turns out to be just the electromagnetic field. The symmetry structure of the gauge field dictates, almost uniquely, the form of the interaction; that is, the precise form of the forces on the charged particle and the way in which the electric charge current density serves as the source for the gauge field. The question was how to extend that methodology beyond quantum electrodynamics to include weak interactions.

---

[10]This section draws on work presented in Morrison (2000).

The top down character of this approach to theory building is especially interesting because we see how the mathematics generates, in some sense, a story about physical dynamics. At first sight this might seem a strange claim to make but, as will become clear below, the story takes shape as a result of the powerful role that symmetry principles play not only in constructing physical theories but in relating the pure mathematics of group theory to the physical features of conserved quantities Weinberg (1967). All of this takes place via what is known as a gauge theory. In very simple language, a gauge theory involves a group of transformations of the field variables (gauge transformations) that leaves the basic physics of the field unchanged. This property is called gauge invariance which means that the theory has a certain symmetry that governs its equations. In short, the structure of the group of gauge transformations in a particular gauge theory entails general restrictions on the way in which the field described by that theory can interact with other fields and elementary particles. In mathematics, a group is a set of elements that has associative multiplication [$a(bc) = (ab)c$ for any a, b, c], an identity element and inverses for all elements of the set. In pure mathematics groups are fundamental entities in abstract algebra.

In physics what this means is that a gauge theory is simply a type of field theory where the Lagrangian is invariant under a continuous group of local transformations known as gauge transformations. These transformations form a Lie group, which is the symmetry group or the gauge group with an associated Lie algebra of group generators. The generators are elements such that repeated application of the generators on themselves and each other is capable of producing all the elements in the group. Each group generator is associated with a corresponding vector field called the gauge field. Simply put: in a gauge theory there is a group of transformations of the field variables (gauge transformations) that leaves the basic physics of the quantum field unchanged. This condition, called gauge invariance, gives the theory a certain symmetry, which governs its equations. Hence, as I mentioned above, the structure of the group of gauge transformations in a particular gauge theory entails general restrictions on how the field described by that theory can interact with other fields and elementary particles. This is the sense in which gauge theories are sometimes said to "generate" particle dynamics—their associated symmetry constraints specify the form of interaction terms.

The symmetry associated with electric charge is a local symmetry where physical laws are invariant under a local transformation. But, because of the mass differences between the weak force bosons and photons a different kind of symmetry was required if electrodynamics and the weak interaction were to be unified and the weak and electromagnetic couplings related. Indeed, it was the mass differences that seemed to rule out a bottom up strategy that began with similarities among the phenomena themselves. The phenomenology suggested that the two theories were in no way compatible. However, by employing a top down approach it was possible to search for different symmetry principles that might be able to accommodate the differences in the particles. Due to the mass problem, it was thought that perhaps only partial symmetries—invariance of only part of the Lagrangian under a group of infinitesimal transformations—could relate the massive bosons to the massless photon. This was because, even with a top down approach, there didn't appear to be a single symmetry group that could govern both the massive and massless particles.

In 1961 Glashow developed a model based on the SU(2) x U(1) symmetry group, a pasting together of the two groups governing the weak force and electromagnetism. This symmetry group required the introduction of an additional particle, a neutral boson Z_s. By properly choosing the mass terms to be inserted into the Lagrangian, Glashow was able to show that the singlet neutral boson from U(1) and the neutral member of the SU(2) would mix in such a way as to produce a massive particle B (now identified as Z^0) and a massless particle that was identified with the photon. But, in order to retain Lagrangian invariance gauge theory requires the introduction of only massless particles. As a result the boson masses had to be added to the theory by hand, making the models phenomenologically accurate but destroying the gauge invariance of the Lagrangian, thereby ruling out the possibility of renormalization. Although gauge theory provided a powerful tool for generating an electroweak model, unlike electrodynamics, one could not reconcile the physical demands of the weak force for the existence of massive particles with the structural, top down, demands of gauge invariance. Both needed to be accommodated if there was to be a unified theory, yet they were mutually incompatible. In order for the electroweak theory to work, it had to be possible for the gauge particles to acquire a mass in a way that would preserve gauge invariance.

The answer to these questions was provided by the mechanism of spontaneous symmetry breaking. From work in solid state physics, it was known that when a local symmetry is spontaneously broken the vector particles acquire a mass through a phenomenon that came to be known as the Higgs mechanism (Higgs 1964a, b). This principle of spontaneous symmetry breaking implies that the actual symmetry of a system can be less than the symmetry of its underlying physical laws; in other words, the Hamiltonian and commutation relations of a quantum theory would possess an exact symmetry while physically the system (in this case the particle physics vacuum) would be nonsymmetrical. In order for the idea to have any merit one must assume that the vacuum is a degenerate state (i.e., not unique) such that for each unsymmetrical vacuum state there are others of the same minimal energy that are related to the first by various symmetry transformations that preserve the invariance of physical laws. The phenomena observed within the framework of this unsymmetrical vacuum state will exhibit the broken symmetry even in the way that the physical laws appear to operate. Although there is no evidence that the vacuum state for the electroweak theory is degenerate, it can be made so by the introduction of the Higgs mechanism, which is an additional field with a definite but arbitrary orientation that breaks the symmetry of the vacuum.

The Higgs field (or its associated particle the Higgs boson) is really a complex SU(2) doublet consisting of four real fields, which are needed to transform the massless gauge fields into massive ones. A massless gauge boson like the photon has two orthogonal spin components transverse to the direction of motion while massive gauge bosons have three including a longitudinal component in the direction of motion. In the electroweak theory the W^{+-} and the Z^0, which are the carriers of the weak force, absorb three of the four Higgs fields, thereby forming their longitudinal spin components and acquiring a mass. The remaining neutral Higgs field is not affected and should therefore be observable as a particle in its own right. The Higgs field breaks the symmetry of the vacuum by having a preferred direction in space, but the symmetry of the Lagrangian remains invariant. So, the electroweak gauge theory predicts the existence of four gauge quanta, a neutral photon-like object, sometimes referred to as the X^0 and associated with the U(1) symmetry, as well as a weak isospin triplet W^{+-} and W^0 associated with the SU(2) symmetry. As a result of the Higgs symmetry breaking mechanisms the particles W^{+-} acquire a mass and the X^0 and W^0 are mixed so that the neutral particles one sees in nature are really two different linear combinations of these two. One of these neutral particles, the Z^0, has a mass while the other, the photon, is massless. Since the masses of the W^{+-} and Z^0 are governed by the structure of the Higgs field they do not affect the basic gauge invariance of the theory. The so-called "weakness" of the weak interaction, which is mediated by the W^{+-} and the Z^0, is understood as a consequence of the masses of these particles.

We can see from the discussion above that the Higgs phenomenon plays two related roles in the theory. It explains the discrepancy between the photon and the intermediate vector boson masses—the photon remains massless because it corresponds to the unbroken symmetry subgroup U(1) associated with the conservation of charge, while the bosons have masses because they correspond to SU(2) symmetries that are broken. Second, the avoidance of an explicit mass term in the Lagrangian allows for gauge invariance and the possibility of renormalizability. With this mechanism in place the weak and electromagnetic interactions could be unified under a larger gauge symmetry group that resulted from the product of the SU(2) group that governed the weak interactions and the U(1) group of electrodynamics.[11]

From this very brief sketch, one can get at least a sense of the role played by the formal, structural constraints provided by gauge theory/symmetry in the development of the electroweak theory. We can see that the use of symmetries to categorize various kinds of particles and their interaction fields is much more than simply a phenomenological classification; in addition, it allows for a kind of particle dynamics to emerge. In other words, the symmetry group provides the foundation for the locally gauge-invariant quantum field theory in that the formal restrictions of the symmetry groups and gauge theory can be deployed in order to produce a formal model showing how these gauge fields could be unified. Although the Higgs mechanism is a significant part of the physical dynamics of the theory the crucial feature that facilitates the interaction of the weak and electromagnetic fields is the non-Abelian structure of the group rather than something derivable from

---

[11]In order to satisfy the symmetry demands associated with the SU(2) group and in order to have a unified theory (i.e., have the proper coupling strengths for a conserved electric current and two charged W fields), the existence of a new gauge field was required, a field that Weinberg associated with a neutral current interaction that was later discovered in 1973. For a discussion of the difficulties surrounding the neutral current experiments, see Galison (1987) and Pickering (1984).

phenomenology of the physics. Even the Higgs mechanism which was clearly seen as a physical particle/field was postulated as a result of symmetry considerations related to the boson mass problem.

This kind of top down approach is common in mathematical physics where there is often difficulty constructing theories when the initial assumptions are specific properties of particles or fields. Often these characteristics are not known and are simply hypothesized in accordance with mathematical features of the theory's structure and symmetries. This in turn results in difficulties determining the dividing line between what counts as the "physics" and what is simply mathematical structure. The consequences of this type of approach are very much present in contemporary high energy physics in contexts like string theory, the multiverse, etc. In the case of string theory, the entire structure is simply an abstract mathematical framework capable of "saving" some of the phenomena without making any empirical predictions. In the end experiment is the deciding feature in establishing the physical foundations of any theory, as in the case of the Higgs discovery in 2012 but that often comes many years after a theory has enjoyed a great deal of explanatory and predictive success. Although symmetry constraints are now considered the foundation for theory building in high energy physics, the role they played in establishing the standard model was crucial for cementing the top down approach as the method of choice in particle physics and cosmology.[12] And, unlike the top down approach in Maxwell's electrodynamics, in particle physics we have an explanatory framework that describes the relevant interactions.

It is important to note here, especially in light of the discussion in section one, that no amount of philosophical reconstruction in terms of models or first order logic will enable us to capture the way these mathematical theories are "built" or how they function in an explanatory and predictive capacity. Moreover, the theory is too complex to strip away the kind of details required for a logical reconstruction. Hence, we need to go directly to the processes involved in constructing the theory in order to understand its structure and the way it relates to the phenomena in question.

But physics is not the only place where this type of top down strategy has proved successful. The development of population genetics followed a similar path, and for reasons that are reminiscent of those operating in the electroweak case—difficulties building a theory from phenomena that appeared incompatible. In the final section let me provide a brief account of how this strategy, albeit with a twist in the form of an analogy, was responsible for establishing the theory of modern population genetics; a theory that brought together Mendelism and Darwinism in a context where unification seemed both quantitatively and qualitatively impossible.

---

[12]I should also mention here the importance of symmetry and the eightfold way in predicting the existence of particles.

5 Top Down and Sideways: From Physics to Biology via Mathematics

While it is commonly known that econometrics borrows techniques and modelling frameworks from physics, models such as the kinetic exchange models of markets, percolation models, models with self-organizing criticality as well as other models developed for earthquake prediction, it is less well known that population genetics also has its origins in physics. Although it was thought to be rather controversial at the time, before the advent these "top down" strategies based on mathematical techniques from statistical physics, there was no apparent way to incorporate Mendelian genetics into the Darwinian framework. The reasons for this will be explained below but first let me briefly describe the context in which the problem arose and why the challenges seemed insurmountable.

In a series of papers over a twelve-year period Karl Pearson (1904 etc.) provided a statistical foundation for Darwinism that extended Galton's law of ancestral hereditary beyond immediate parentage. The goal was to identify the theory of heredity with a theory of correlation rather than causation, in keeping with Pearson's commitment to positivism. The problem the plagued Darwinism at the time was a lack of any explanation of variation. The correlation coefficient in brothers was around .54 (amount of variance due to ancestry), leaving 46% of the variance to be accounted for in some other way. Pearson tried to incorporate a version of Mendelism (for 3 or 4 genes) but was unsuccessful, primarily because there were no methods available within biometrical statistics to correctly model the requisite information. The reason for this was largely the result of Pearson's own views about the nature of Mendelian assumptions that needed to be accounted for. He believed that biology differed from physics in the sheer number of variables one needed to incorporate in any single case of inheritance. Because physics dealt with inanimate objects one didn't need to pay particular attention to individual features. By contrast, there were certain properties of a population that needed to be specified in order to arrive at a proper statistical description and a proper theory of heredity.

Pearson assumed that for each Mendelian factor (gene) one needed to know a number of specific details: which allelomorph was dominant and the extent to which dominance occurred; the relative magnitudes of the effects produced by different factors; whether the factors were dimorphic or polymorphic; extent to which they were coupled and in what proportion the allelomorphs occurred in the general population. Other more general considerations included the effects of homogamy (preferential mating) as opposed to random mating as well as selection versus environmental effects. Pearson claimed all of these needed to be treated separately if one was to construct a theory of the genetic basis of the inheritance of particular traits. Consequently, it became immediately obvious that an analysis involving a large number of genes was virtually impossible using biometrical techniques. There were simply too many variables that needed to be taken account of. So, Pearson's difficulties with Mendelism were essentially two-fold. Not only did Mendelism require us to go beyond a theory of correlation which he saw as the proper foundation for science, but the statistical methods available to him were simply not powerful enough to encode all the information necessary for describing a Mendelian population.

The solution to the problem came via a mathematical reconceptualization of the notion of a population. Instead of treating genes as biological individuals, populations were re-described using the methods of statistical mechanics and the velocity distribution law. All of this came about at the hands of R. A. Fisher (1918, 1922) who reasoned as follows: Just as a sufficiently large number of independent molecules would exhibit a stable distribution of velocities, a sufficiently large number of Mendelian factors or genes in a population will enable one to establish general conclusions about the presence of particular traits. Unlike Pearson, Fisher did not assume that different Mendelian genes were of equal importance, so all dominant genes did not have a like effect. He also assumed an infinite population with random mating as well as the independence of the different genes. Because genes were sufficiently numerous some small quantities could be neglected and by assuming an infinite number it was possible to ignore individual peculiarities and obtain a statistical aggregate that had relatively few constants. Contra Pearson, biology is like physics after all!

Underwriting these results is, of course, the central limit theorem which, if the population is large enough, guarantees the true value for a specific trait. Fisher also introduced a number of other mathematical techniques including the analysis of variance which was used to determine how much variance was due to dominance, environmental causes and additive genetic effects. He then went on to specify the conditions under which variance could be maintained.

Once these issues about variation were resolved Fisher investigated how gene frequencies would change under selection and environmental pressures. To do this he introduced further stochastic considerations and examined the survival of individual genes by means of a branching process analysed by functional iteration. He then set up a chain binomial model (often used in epidemiology) and analysed it by a diffusion approximation. Fisher's "mathematical population" allowed him to conclude that selection acting on a single gene (rather than mutation, random extinction, epistasis, etc.) was the primary determinant in the evolutionary process. This top-down mathematical approach enabled him to "measure" selection effects that couldn't be observed in natural populations. The reason for this is that genotype or allele frequencies are easily measured but their change is not. Most naturally occurring genetic variants have a time scale of change that is on the order of tens of thousands to millions of years, making them impossible to observe. Similarly, fitness differences are likewise very small, less than 0.01%, also making them impossible to measure directly. So, by abstracting away from the specific features of a population and making general assumptions based on statistical and mathematical considerations Fisher was able to forge a synthesis of Mendelism and Darwinian selection in a way that would have otherwise been impossible.

A particularly interesting feature of this case for the purposes of "building theories" is Pearson's use of a bottom up approach, focusing on individual genes as the basis for theory construction. By insisting on the primacy of a genetic description that faithfully represented all the relevant information he was unable to construct a theory that allowed him to incorporate Mendelian genes into an account of Darwinian selection. As we saw above, this is especially true when we have two groups or types of phenomena we wish to unify into one coherent theory as in the electroweak case where we had two very different types of particles/forces with very different characteristics. In both of these contexts the top-down strategy provided a way of introducing specific structural constraints that could form the basis for a new theory.

What these structural constraints do is inform the way we conceptualise or reconceptualise the phenomena under investigation. Instead of focusing on specific features of the phenomena themselves, the mathematics provides a framework for constructing new approaches to characterizing the phenomena and rendering them compatible. In Fisher's case the top down strategy began with an analogy from statistical mechanics and the velocity distribution law, with the mathematical features functioning as the driving force for modelling populations of genes. What this produced was a conceptual shift in how to conceive of biological populations, a shift that involved a radically different account of how genes should be described in order to underwrite the kind of statistical analysis required for explaining variation. Although much of the methodology was borrowed from statistical physics the top down "theory building" involved in synthesizing natural selection and Mendelism involved using the methodology in new and innovative ways. In both of the top down examples I have considered, the electroweak theory and the new science of population genetics, there was a clear explanatory foundation that provided a new way of understanding the phenomena in question, an understanding that came not from the phenomena themselves but rather from the constraints imposed in constructing the theory.[13]

6   Conclusions—Where to Go Next

While it is certainly true that the top down strategy and its accompanying symmetry principles has dominated the history of high energy physics, its ongoing development and the search for physics beyond the standard model has embraced a rather different approach and methodology. Due to the lack of experimental evidence for supersymmetry and explanations for dark matter we find a move away from top down analyses to a more bottom up approach that, while consistent with certain features of the standard model, doesn't advance a full blown physical model. Rather, the "strategy" involves looking at available data to see which models might furnish a very basic description. This approach has become known as the method of "simplified models" and involves characterizing the basic properties of signal data from particle interactions in a way that allows comparison to any model, not just a

---

[13]For a longer discussion of this case see Morrison (2002).

spec_fic one. (Alwall and Schuster 2009). The process begins with an effective Lagrangian and a minimal set of parameters that typically include particle masses and production cross sections. The goal is to reproduce kinematics and multiplicities of observed particles with the simplified model fits. These are then used as a representation of the data which in turn is compared to a full model by simulating both in a detector simulation.

This is a marked departure from the strategies that we have seen historically in the development of modern physics. The reason for this is the fact that there is very little consensus on what future physics should or will look like, what is required to complete the standard model, and whether the standard model itself will survive as a theory. It faces outstanding problems such as the hierarchy problem, the number of free parameters, unexplained neutrino masses, dark matter and energy, to name a few. The theory has unverified predictions, phenomena it can't explain, as well as a host of other fundamental theoretical problems. In each case there seems to be no suitable strategies for resolving the issues. Hence, what we have is a level of genu_ne uncertainty with no obvious direction to take. Consequently, the simplified model approach is seen less as a strategy for theory building than a strategy for going forward in some very minimal sense.

What the development of this new approach involving simplified models indicates is that theory building is first and foremost the construction of strategies responsive to theoretical problems; strategies that take account of the phenomena under investigation, the goals, and how those goals might be accomplished. For that there is often no blueprint, nor do philosophical reconstructions via first order logic and model theory aid in either the process of construction or retrospective understand_ng about the theory's structure. Hence, we are left with the challenge of determining what benefits are yielded by philosophical accounts of theory structure, given that they seem unable to encode the multi-faceted activities that constitute "theory building" even in the more straightforward cases.[14]

References

Alwall J., & Schuster, P. (2009). Simplified models for a first characterisation of new physics at the LHC. *Physical Review D, 79*, 075020.

da Costa, N. C. A., & French, S. (2003). *Science and partial truth: A unitary approach to models and scientific reasoning*. Oxford: Oxford University Press.

Fisher, R. A. (1918). The correlation between relatives on the supposition of Mendelian inheritance. *Transactions of the Royal Society of Edinburgh, 52*, 399–433.

Fisher, R. A. (1922). On the dominance ratio. *Proceedings of the Royal Society of Edinburgh, 42*, 321–341.

Galison, P. (1987). *How experiments end*. Chicago: University of Chicago Press.

Glashow, S. (1961). Partial symmetries of weak interactions. *Nuclear Physics, 22*, 579–588.

[14]I would like to thank the Social Sciences and Humanities Research Council of Canada for research support and the editors for helpful comments and suggestions for improvements.

Giere, R. (1988). *Explaining science: A cognitive approach.* Chicago: University of Chicago Press.

Giere, R. (2004). How models are used to represent reality. *Philosophy of Science, 71,* 742–752.

Higgs, P. (1964a). Broken symmetries, massless particles and gauge fields. *Physics Letters, 12,* 132–133.

Higgs, P. (1964b). Broken symmetries and masses of gauge bosons. *Physical Review Letters, 13,* 508–509.

Lagrange, J. L. (1788). *Mécanique analytique.* Paris: Mallet-Bachelier.

Maxwell, J. C. (1873). *Treatise on electricity and magnetism.* Vols. 2. (Reprint from 1954 New York, Dover: Oxford: Clarendon Press).

Maxwell, J. C. (1965). *The scientific papers of James Clerk Maxwell.* 2 vols. Edited W. D. Niven. New York: Dover.

Morrison, M. (1999). Models as autonomous agents. In M. Morgan & M. Morrsion (Eds.), *Models as mediations: Essays in the philosophy of the natural and social sciences* (pp. 38–65). Cambridge, UK: Cambridge University Press.

Morrison, M. (2000). *Unifying scientific theories: Physical concepts and mathematical structures.* New York: Cambridge University Press.

Morrison, M. (2002). Modelling populations: Pearson and fisher on mendelism and biometry. *British Journal for Philosophy the of Science, 53,* 39–68.

Morrison, M. (2007). Where have all the theories gone? *Philosophy of Science, 74,* 195–228.

Morrison, M. (2008). Fictions, representation and reality. In Mauricio Suarez (Ed.), *Fictions in science: Philosophical essays on modelling and idealization* (pp. 110–138). London: Routledge.

Pickering, A. (1984). *Constructing quarks.* Chicago: University of Chicago Press.

Schwinger, J. (1957). A theory of fundamental interactions. *Annals of Physics, 2,* 407–434.

Suppe, Frederick. (1989). *The semantic conception of theories and scientific realism.* Urbana: University of Illinois Press.

Suppes, P. (1961). A Comparison of the Meaning and Use of Models in the Mathematical and Empirical Sciences. In H. Freudenthal (Ed.), *The Concept and Role of the Model in Mathematics and Natural and Social Sciences* (pp. 163–177). The Netherlands, Dordrecht: Reidel.

Suppes, P. (1962). Models of data. In E. Nagel, P. Suppes & A. Tarski (Eds.), *Logic, methodology and philosophy of science: Proceedings of the 1960 International Congress* (pp. 252–261). Stanford, CA: Stanford University Press.

Suppes, P. (1967). What Is a scientific theory? In S. Morgenbesser (Ed.), *Philosophy of science today* (pp. 55–67). New York: Basic Books.

Suppes, P. (2002). *Representation and invariance of scientific structures (Stanford.* CSLI): CA.

Tarski, A. (1953). *Undecidable theories.* Amsterdam: North Holland.

van Fraassen, B. (1980). *The scientific image.* Oxford: Oxford University Press.

van Fraassen, B. (1989). *Laws and symmetries.* Oxford: Oxford University Press.

van Fraassen, B. (2008). *Scientific representation: Paradoxes of perspective.* Oxford: Oxford University Press.

Weinberg, S. (1967). A model of leptons. *Physical Review Letters, 19,* 1264–1266.

Richer Than Reduction

David Danks

Abstract There are numerous routes for scientific discovery, many of which involve the use of information from other scientific theories. In particular, searching for possible reductions is widely recognized as one guiding principle for scientific discovery or innovation. However, reduction is only one kind of intertheoretic relation; scientific theories, claims, and proposals can be related in more, and more complex, ways. This chapter proposes that much scientific discovery proceeds through the use of *constraints* implied by those intertheoretic relationships. The resulting framework is significantly more general than the common reduction-centric focus. As a result, it can explain more prosaic, everyday cases of scientific discovery, as well as scientists' opportunistic use of many different kinds of scientific information. I illustrate the framework using three case studies from cognitive science, and conclude by exploring the potential limits of analyses of scientific discovery via constraints.

Keywords Scientific discovery · Intertheoretic constraints · Intertheoretic relations · Reduction · Cognitive science

1 Routes to Discovery

The diverse paths and techniques for scientific discovery, invention, and construction form perhaps the most heterogeneous part of science. There are many ways and methods, whether structured or intuitive, to develop a novel scientific theory or concept. In fact, people have sometimes thought that scientific discovery does not—perhaps, could not—exhibit any systematic patterns at all. While this latter pessimism is arguably unwarranted, the skeptics are correct that there is great diversity in routes to scientific discovery. At one extreme, a relatively minimal type

D. Danks (✉)
Departments of Philosophy and Psychology, Carnegie Mellon University,
Pittsburgh, PA, USA
e-mail: ddanks@cmu.edu

© Springer International Publishing AG 2018
D. Danks and E. Ippoliti (eds.), *Building Theories*, Studies in Applied Philosophy,
Epistemology and Rational Ethics 41, https://doi.org/10.1007/978-3-319-72787-5_3

of discovery occurs when the scientist starts with an existing theory, and then adjusts its parameters in light of new data. For example, a novel experiment might reveal the importance of a previously unconsidered causal factor. A more speculative type of scientific discovery depends on analogical reasoning, as that can lead the scientist to consider entirely new classes or types of theories. Alternately, various abductive or inductive strategies can point towards scientific theories, models, or concepts that have not previously been considered. And of course, there might be no explicit or conscious "method" at all in a case of scientific discovery; it might, from the perspective of the scientist herself, be the result of unexplainable inspiration.

This paper explores a particular set of methods for scientific discovery—those that use constraints from other scientific theories. Scientific innovation and discovery is often driven by consideration of *other* (folk and scientific) theories and models, where the resulting constraints can be both structural and substantive. Past discussions of this constraint-based scientific discovery have almost always centered on reductionism or reductionist commitments as a discovery strategy (Bechtel and Richardson 2000; Schouten and de Jong 2012; Wimsatt 1980). More specifically, there are two principal ways to use reductionism as a method for scientific discovery and innovation. First, suppose that one has a theory T_H that captures the higher-level (in some relevant sense) phenomena or structure. Reductionism, as an overarching meta-scientific commitment, implies that there must be some lower-level theory T_L—in fact, potentially many such theories if there are many lower levels—such that T_H reduces to T_L. (For the moment, I leave aside the question of the meaning of 'reduces to'.) Scientific discovery at the T_L-level can thus be guided by our knowledge of T_H: the higher-level theory can provide substantial information about features of T_L (e.g., the relevant inputs and outputs), and thereby significantly reduce the possibility space. For example, the search for underlying causal mechanisms is frequently guided in exactly this way by a higher-level theory about the structure of the system or the functional roles of various components (e.g., Darden 2002; Darden and Craver 2002). Of course, this use of reductionism does not eliminate the need for discovery; even though T_H might reduce the space of possible T_L's, it will rarely uniquely determine one particular T_L. Thus, we will still need to use one or more strategies from the previous paragraph, such as adjustment in light of novel data. Nonetheless, we can use reductionist commitments as a substantive "downward guide" to scientific discovery, and thereby greatly simplify the task.

A second way to use reductionism as a discovery strategy starts with a lower-level T_L that specifies particular components of the system (perhaps mechanisms in a strong sense, perhaps not). We can then seek to discover a T_H that captures the functional roles or higher-level regularities and relations of the system, and that reduces to T_L. For example, we might have a robust scientific theory about some set of regulatory mechanisms within a cell, and then aim to find a higher-level theory that captures the patterns that result from interactions of these mechanisms in particular environments. More generally, T_H will typically incorporate elements of T_L as particular realizations or implementation specifications of the T_H-components.

This lower-level information significantly constrains the possible functional, computational, or causal roles for elements of T_H, precisely because we require that T_H reduce to T_L. Although T_L might sometimes uniquely determine T_H at some levels (e.g., if T_H is the asymptotic behavior of dynamical model T_L), the discovery situation will typically be more complex: the proper T_H may depend on our explanatory goals, or specific initial conditions, or aspects of the background context. This second use of reductionism and reductionist commitments does not obviate the need for scientific discovery, but nonetheless provides guiding "upward constraints" that can significantly speed or improve that discovery.

Regardless of which strategy we pursue, the exact constraints will depend on both the details of the scientific case, and also the particular account of 'reduction' that one employs. For example, syntactic theories of 'reduction' (e.g., Dizadji-Bahmani et al. 2010; Nagel 1961) will emphasize discovery through manipulations of the symbolic representations of the theories. In contrast, causal theories of 'reduction' (e.g., Churchland 1985; Hooker 1981a, b) will focus on discovery of similar causal roles or capacities across the theories. However, *all* theories of 'reduction' agree that the relevant relation involves a very close connection between the two theories. Thus, scientific discovery via reductionism inevitably results in a new scientific theory that is tightly coupled with the pre-existing theory—either discovery of a T_H that reduces to the existing T_L, or discovery of a T_L to which the existing T_H can reduce. This tight connection between old and new theories provides much of the power of reductionism as a discovery strategy (when it is successful). For a given T_H, there will often be a relatively small class of lower-level realizations or implementations that actually exhibit the precise higher-level phenomena. For a given T_L, information about the relevant initial or background conditions will often almost determine the higher-level T_H. And we gain enormous further benefits if we can discover a suitable $<T_H, T_L>$ pair, as we can use each to refine the other, combine them into integrated multi-level models, and thereby establish cross-level, cross-theory, and cross-disciplinary connections.

However, although there can be significant benefits from requiring a reductionist connection between T_H and T_L (regardless of direction of discovery), such a connection comes with a significant cost: the required tight couplings are usually very difficult to establish. First, all extant theories of 'reduction' require that both T_H and T_L be full scientific theories, even though scientists frequently work with vaguer or more uncertain not-quite-theories (e.g., observation of a correlation between two factors, or knowledge that some manipulation produces a probabilistic change in a target variable). Second, reductionist discovery must involve levels that are an appropriate distance from one another, as reductions are very hard to establish across large "level gaps." Third, the requirements for a full reduction are often quite stringent, and so we might not be able to establish the appropriate connections between T_H and T_L (though searching for those connections could potentially be useful for discovery purposes). Fourth, for a given T_L, there might simply not be an appropriate T_H at our desired level, as we might not be able to abstract away or modularize the implementation details in T_L. Fifth, for a given T_H, the relevant distinctions or objects might not be respected in T_L (e.g., cognitive symbols might not be directly found in neural models), and so T_H could actually be a misleading guide for scientific discovery.

Reductionism and reductionist commitments are very powerful guides for scientific discovery, but also very limited. If we look more broadly, we can find many cases in which information from other scientific theories has been used for scientific discovery, but where those uses simply cannot be understood in terms of the search for reductions. Reduction is, however, only one intertheoretic relation of many, and so we might suspect that scientific discovery via reductionist commitments is only one way to employ information from other scientific theories. Perhaps we can have a more general, more useful model of scientific discover by considering alternative intertheoretic relations. This chapter aims to provide such an account via the use of intertheoretic constraints generated by those relations; reductive constraints are simply one special case. To that end, Sect. 2 provides a more general account of the notion of 'intertheoretic constraint', with a particular eye towards their use in discovery. Section 3 then uses that account to explicate several cases of scientific discovery in the cognitive sciences. Those case studies might well seem banal and ordinary, but that is part of the point: everyday scientific discovery is largely a matter of trying to fit together disparate puzzle pieces, and scientists employ many different constraints and relations—not just reduction—to find the next piece of the puzzle.

2 Discovery via Constraints

There are many different intertheoretic relations, involving multiple theoretical virtues. Reduction is one salient relation, and it holds when there is a tight coupling —syntactic, semantic, causal, functional, or other—between two theories. Autonomy is a different intertheoretic relation that obtains when features of theory T_A are essentially independent of T_B. For example, macroeconomics is thought to be explanatorily autonomous from quantum mechanics. More controversially, psychology has been claimed to be ontologically autonomous from neuroscience (Fodor 1974, 1997). These two relations of reduction and autonomy clearly fall at the extremes; theories can be related to one another in more subtle and fine-grained ways, as we will see in Sect. 3. Importantly, these intertheoretic relations imply intertheoretic constraints (though perhaps an empty set of constraints, as in the case of autonomy). For example, if T_H reduces to T_L, then if T_L is true, then T_H must also be true.[1] Moreover, this constraint (or its contrapositive: given a reduction, if T_H is false, then T_L must be false) does much of the work when reductionism is used as a guide for scientific discovery, which suggests that perhaps much scientific discovery proceeds through the use of intertheoretic constraints of all sorts, not just those grounded in reductions.

---

[1] Readers who are skeptical about notions of 'truth' with regards to scientific theories should instead substitute 'accurate' or 'approximately true' or whatever notion they prefer.

A general account of intertheoretic constraints should include reduction and autonomy as special case intertheoretic relations, but should also apply more generally, though that requires some complications.[2] At its most abstract, a scientific theory S (or model, or claim, or …) *constrains* another theory T relative to some theoretical virtue V just when the extent to which S has V is relevant in some way to the extent to which T has V. That is, an intertheoretic constraint exists between S and T if S's status with respect to V (e.g., truth, simplicity, predictive accuracy, explanatory power, etc.) matters in some way for T's status with respect to the same V. For example, the existence of a reduction relation between T_H and T_L yields (at least, on most accounts of 'reduction') the constraint that T_L's truth implies T_H's truth. That is, the truth of T_L is directly relevant to whether T_H is true. Crucially, though, a reduction implies this tight constraint only for some theoretical virtues (e.g., truth). The existence of a reduction relation does not, for example, necessarily imply any constraint with respect to explanatory power, as T_H could reduce to T_L but provide explanations with different scope and generalizability. Moreover, although I have been using the word 'theory' in this paragraph, this account of 'constraint' does not actually require S and T to be full-blown theories. Relevance can arise between scientific claims, data descriptions, partially specified models, and other kinds of not-quite-theories, and thus constraints based in those intertheoretic relevance relations can obtain between them. Of course, the specific relation underlying particular constraints could have more stringent requirements of the relata (e.g., a reduction requires theories), but that is not intrinsic to intertheoretic constraints more generally.

This high-level characterization of 'intertheoretic constraint' is qualitative and vague in certain key respects (e.g., what does it mean for S's theoretical virtues to be "relevant" to T's virtues?), but is already sufficiently precise to highlight some notable features (see also Danks 2014). Perhaps most importantly, this account implies that constraints are objective, not subjective: the constraint obtains if S is actually relevant for T, regardless of whether any scientists realize that it is relevant. In fact, a common scientific activity is the discovery of novel intertheoretic relations and constraints that were previously unknown (but were present all along). A less-obvious implication is that constraints are, on this account, comparative in both relata: whether S constrains T with respect to V depends not only on S and T themselves (and their relations), but also on the alternatives to S and T. This property of constraints might be surprising, but follows immediately from the focus on relevance, as whether one theory or model is relevant to another will depend on what we take to be the serious alternatives. For example, suppose that T is a particular associationist model of human (psychological) causal learning that uses prediction errors in learning (e.g., Rescorla and Wagner 1972), and S is neural evidence that prediction errors are computed in the brain. Is S relevant for whether T is the correct theory? If the only alternatives to T are other models that use

---

[2]For space reasons, I only summarize my account of intertheoretic constraints here. More details and discussion can be found in Danks 2013 or Chap. 2 of Danks 2014.

prediction errors (e.g., other standard associationist models, such as Pearce 1987), then the answer is "no," as S does not provide information that distinguishes between them. However, if the alternatives to T include models that do not directly employ prediction errors (e.g., more rationalist models, such as Griffiths and Tenenbaum 2005), then the answer is potentially "yes," as S might rule out (or make less plausible) some of these alternatives to T. More generally, relevance (of all different types) can depend on what else might have been the case, and so the alternatives to S and T matter.[3]

The use of these intertheoretic constraints in scientific discovery is relatively direct and immediate. Suppose that I am trying to discover a new scientific theory, model, or other account of phenomena P (in domain D and at level L) for purposes or goals G. For this discovery task, the first step is to list possibly-relevant theories and models $S_1, \ldots, S_n$ (and their corresponding sets of competitors $\mathbf{S}_1, \ldots, \mathbf{S}_n$). These S's are my scientific beliefs and knowledge that could perhaps imply constraints that are relevant to our theory of P (at level L for goal G). They might be about other phenomena in D, or characterized at a different level, or offered to fulfill a different function, but still potentially related. I also presumably have some ideas about what kind of theory or model is desired, even if only a vague sense. That is, we can assume that I have some set $\mathbf{T}$ of possible "targets," where $\mathbf{T}$ will frequently be infinite, or involve a number of unknown parameters, or otherwise be very broad.

Given these components, the use of intertheoretic constraints for scientific discovery is straightforward, at least in the abstract: (a) for each pair of S_i and $\mathbf{S}_i$, we determine the G-constraints (i.e., the constraints that are relevant for the scientific goal) that they imply for $\mathbf{T}$; (b) aggregate the G-constraints together, perhaps deriving further implied constraints; and (c) compute the resulting impacts on $\mathbf{T}$ (e.g., ruling out certain possibilities, or making others more likely). In some special cases, this process will result in only one T_j at the end, in which case the constraints were fully sufficient for our discovery problem. More typically, this process will reduce the possibility space, but not fully determine T. We can then look for additional S's (since it will not always be obvious a priori which scientific beliefs are potentially relevant), or try to discover additional constraints implied by the current S's (since we cannot use a constraint if we do not know about it), or turn to one of the other types of discovery strategy outlined at the beginning of this chapter (e.g., collecting novel data to further refine or specify the scientific theory or claim).

Scientific discovery via reductionism can easily be understood in terms of this schema. Consider the "bottom-up" strategy in which we know T_L and are trying to discover T_H. In this case, T_L largely sets the domain and phenomena, and other factors (perhaps extra-scientific) determine the level and set of possible target T_H's.

---

[3]As an aside, notice that this alternative-dependence implies that the particular constraints that scientists entertain can depend on contingent historical facts about the science (that influence the set of alternatives considered), even though the existence and nature of those constraints are not history-dependent.

This discovery problem is truth-centric,[4] and so we are concerned with truth-constraints: given that T_L is true, how does this constrain the possible truth of elements of $\mathbf{T_H}$? The notion of a truth-constraint is significantly more complicated than one might initially suspect (see Danks 2014 for details), but it is relatively simple if we require that the target T_H be reducible to T_L: any candidate T_H that is inconsistent with T_L in the relevant domain can be eliminated. That is, we get exactly the constraint that is used by reductionists in scientific discovery. And a similar analysis can be given for "top-down" reductionist discovery in which we know T_H and are trying to discover T_L. Thus, the much-discussed reductionist strategies are simply special cases of this more general account of the use of intertheoretic constraints for scientific discovery.

This picture of "scientific discovery via intertheoretic constraints" is similar to, but (at least) generalizes and extends, the constraint-inclusion model of scientific problem-solving (Nickles 1978, 1981). The constraint-inclusion framework for scientific discovery (and confirmation) contends that scientific problems, not theories, are the relevant units of inquiry, and that the goal of inquiry is a satisfactory answer, not truth (Laudan 1981; Nickles 1988). "Constraints" then provide critical information about what would count as a satisfactory answer to a problem: any such answer must satisfy the relevant constraints (Nickles 1978). Scientific problems are not *defined* by constraints, but they form a major part of the characterization of problems, and provide one way to gain understanding about the structure of a scientific problem. As Nickles (1981) puts it: "The more constraints on the problem solution we know, and the more sharply they are formulated, the more sharply and completely we can formulate the problem, and the better we understand it." (p. 88).

There are several shared features of the constraint-inclusion model and the framework proposed in this section: (a) problems, questions, and goals are central, not simple truth; (b) constraints play an important role in the scientific discovery process; and (c) the existence of constraints does not depend on contingent scientific history or human psychology, though our awareness of them might depend on these factors. At the same time, though, we employ somewhat different understandings of 'constraint'. Most notably, the constraint-inclusion model focuses on relatively "hard" or quasi-logical constraints, where these are derived for a particular problem. For example, a "reductive" constraint C on problem solutions specifies that the solution (whatever it may be) must be re-representable as specified by C (Nickles 1978); this type of constraint thus focuses on the mathematical relations between syntactically characterized scientific theories. Of course, there are many kinds of constraints in the constraint-inclusion model, but in general, "every single constraint, by definition of 'constraint', rules out *some* conceivable solution as inadmissible." (Nickles 1981, p. 109; emphasis in original) In contrast, my constraints need only influence plausibility without definitively ruling anything in or out. A constraint can be useful even if nothing is *inadmissible* as a result of it.

---

[4]Again, readers should freely substitute their preferred term for 'truth', such as 'accuracy' or 'approximate truth'.

In addition, my account does not identify constraints with problems, but rather provides an account of how they arise from intertheoretic relations. The "discovery via constraints" model proposed here thus generalizes the constraint-inclusion model by allowing for "soft" constraints, and also provides an account of the source of problem-specific constraints in terms of the potentially relevant theories (and their intertheoretic relations).

Of course, one might object that my account is too high-level and abstract to be useful, precisely because it attempts to cover a wide range of cases and constraints. In general, there is only a limited amount that can be said if we restrict ourselves to talking in terms of letters—D's, G's, S's, and so forth—rather than specific domains, phenomena, and discovery problems. For example, we need to know the relevant goal(s), as the very same S might truth-constrain T, but not explanation-constrain T. Thus, the trajectory of scientific discovery for one goal can be quite different than for another goal.[5] The details make a critical difference, and it is hard to evaluate this account without considering its applicability to particular cases of scientific discovery, and so we now examine some particular instances of scientific discovery.

3   Case Studies of Constraint-Driven Discovery

This section considers three case studies from cognitive science, each of which shows an instance of constraint-based scientific discovery, and that collectively show how scientific discovery can be an iterative process in which the outputs of one episode can be the inputs or constraints of the next. Although all three examples are drawn from cognitive science, I suggest that the lessons apply across many scientific disciplines. I focus on these examples only because I know them best, not because there is anything special or distinctive about them (at least, with respect to the use of intertheoretic constraints for discovery). In fact, as noted earlier, these case studies should hopefully seem somewhat anodyne, as one claim of this chapter is that "discovery via constraints" is a completely normal and regular scientific activity.

---

[5]This goal-dependence does not necessarily imply some sort of goal-dependent pragmatism or perspectivism (though I do also endorse that; see, e.g., Danks 2015). Rather, this dependence is just a generalization of the old reductionist observation that two theories could stand in a reduction relation without thereby constraining one another's explanations in any interesting or informative way (e.g., Putnam 1975).

3.1 Representations of Causal Knowledge

People have a great deal of causal knowledge about the world: we know which switches cause the lights to turn on; we know ways to alleviate pain; we might understand the causes of the functioning of a car engine; and so on. Causal knowledge is arguably one of the key guides throughout our cognition (Sloman 2005), and the first case study focuses on this phenomenon P of causal knowledge, within the domain D and level L of cognitive psychology/science. In particular, consider the discovery problem of finding a theory (or not-quite-theory) T that describes the structure of these cognitive representations. There are many different plausible candidate theories, as our causal knowledge might be structured as: lists of pairwise associations (e.g., Shanks 1995); stimulus $\rightarrow$ response or environment $\rightarrow$ action mappings (e.g., Timberlake 2001); causal graphical models (e.g., Danks 2014; Griffiths and Tenenbaum 2005); or in some other way.[6]

The first step in scientific discovery via constraints is to provide the different pairs of S_i and $\mathbf{S_i}$—that is, the potentially relevant scientific claims or theories, as well as their relevant, plausible alternatives. For the particular phenomenon of causal knowledge, there are an enormous number of potentially relevant scientific claims; for simplicity, we consider only two. First, there is substantial empirical evidence that people understand (perhaps implicitly) many of their actions as having a relatively uncaused (i.e., self-generated) component (Hagmayer and Sloman 2009), and this emerges at a very young age (Rovee and Rovee 1969). This understanding is arguably part of the reason that we have experiences of free will: we see our actions as not caused solely by the environment around us, but rather attribute some of the causation to ourselves. There are many alternatives to this claim S_1 (i.e., other members of $\mathbf{S_1}$); for example, people might understand their choices as entirely determined by environmental conditions, or by their own prior cognitive or emotional state.

A second, potentially relevant scientific claim S_2 is that people's decisions are appropriately responsive to indirect information about the state of the world. Obviously, we adjust our decisions so that they are appropriately tuned to the world. The relevant feature of human decision-making here is that we can use information that is not immediately relevant in order to make inferences about those factors that are directly relevant. For example, I might not bother to flip a light switch if I see my neighbor's lights are off, as I might thereby infer that the power is out in my neighborhood. That is, we employ disparate pieces of information to shape our decisions in order to maximize our chances of achieving a desired outcome. Alternative scientific possibilities to this S_2 are that people's decisions might depend only on local or immediate factors, or even be truly random in important ways.

---

[6]As a matter of historical interest, these were the three main types of theories of causal structure representation being proposed in cognitive science in the early 2000s.

Now consider the scientific discovery problem of the nature of our cognitive representations of causal structure. If S_1 holds (rather than some other possibility in $\mathbf{S_1}$), then our representations must enable us to derive predictions of the outcomes of exogenous actions. In particular, the representations should honor the basic asymmetry of intervention for causal relations (Hausman 1998): exogenous actions to change the cause C (probabilistically) change the effect E, but exogenous changes in E do not lead to changes in C. Thus, our cognitive representations cannot be composed solely of lists of associations, as those are symmetric in nature. On the other side, if S_2 holds, then our representations of causal structure must be relatively integrated or unified, since we can use disparate pieces of information to shape or constrain our choices. Thus, they cannot consist solely in environment $\rightarrow$ action mappings, as those are not "tunable" in the appropriate way.[7] If we think back to our original set $\mathbf{T}$ of possibilities, we find that only causal graphical models can satisfy the constraints implied by both S_1 and S_2. And as a matter of historical fact, causal graphical models are currently the dominant theory of cognitive representations of causal structure knowledge, in large part because they are the only representations that can explain diverse reasoning, inference, and decision-making abilities such as S_1 and S_2 (Danks 2014).

In this case study, scientific discovery occurred partly through understanding how our prior scientific commitments and beliefs constrained the possibilities. We did not need to perform a new experiment, or engage in analogical reasoning. More importantly for this chapter, the process looks nothing like "discovery via reduction." There is simply no possibility of a reduction relation between "causal representations are structured as causal graphical models" and either S_1 or S_2. None of these claims rise to the level of a full-fledged theory (as required for a reduction). More importantly, these not-quite-theories are not accounts of the same phenomena at different levels, and so a reduction would simply be inappropriate. In order to make sense of this example, we need to see scientific discovery about P (= the structure of our causal knowledge) as shaped and informed by other scientific claims that are relevant because they impose constraints, not because they are involved in a reduction.

3.2   Concepts Based on Causal Structure

For the second case study, we turn to the phenomenon of conceptual representation in our cognitive psychology: that is, we plausibly want to discover the structure of our everyday concepts, such as DOG, STUDENT, or APPLE, though with the recognition that there might be multiple types of concepts depending on the particular

---

[7]One might object that they *could* be tunable, if we understood "environment" in an appropriately broad and rich way. The problem is that this move makes the mappings essentially unlearnable, as every experience now involves a unique, never-before-seen environment.

domain or even individual (e.g., Barsalou 2008; Machery 2009). Many different theories of conceptual structure have been proposed over the years (Murphy 2004), and so we have a rich set $\mathbf{T}$ of theoretical possibilities, and a correspondingly difficult scientific discovery problem. One natural S_1 is the empirical finding that people frequently (and often spontaneously) group together different individuals on the basis of their shared or similar causal structure (Carey 1985; Keil 1989). The contrast class $\mathbf{S_1}$ here includes, for example, the claim that perceptual similarity always determines grouping. And given this S_1, we can sensibly include S_2 about causal structure: namely, people's representations of causal knowledge are structured like causal graphical models. These S_1 and S_2 constrain the space of theories of conceptual representations: at least some of our concepts are (likely) structured as causal graphical models (Rehder 2003a, b). Moreover, S_1 provides us with a relatively precise characterization of when our concepts will have that structure.

One might object that this example is not really a case of scientific discovery, but rather is "simple" scientific reasoning. However, this characterization is overly simplistic and dismissive. As an historical matter, the story that I provided in the previous paragraph largely captures the scientific history: causal graphical models were only proposed as a possible model of some concepts once people combined the information in S_1 and S_2. The causal model theory of concepts was "discovered" largely by thinking through the implications of these constraints. More generally, this objection assumes a sharp distinction between scientific reasoning and scientific discovery, but part of the point of these case studies is precisely that there is no bright line to be drawn. Scientific practice partly consists in trying to put various pieces together into a relatively integrated account. That integration can involve both discovery (e.g., proposing an entirely new theory of conceptual representation in terms of causal graphical models) and reasoning (e.g., showing the relevance of empirical findings that are not directly about the nature of conceptual representations).

This case study also demonstrates the dynamic nature of these processes in two different ways. First, notice that S_2 here is T from the previous case study. The product of some scientific discovery will itself usually imply constraints on other $\mathbf{T}$'s, though those might not immediately be recognized by the scientists. These connections provide one way in which a single empirical finding can have wide-ranging "ripple effects": the impact of an empirical finding is not necessarily limited to the immediately relevant scientific question or problem, as the answer to that question can imply constraints that help answer a second question, which can thereby imply constraints for a third question, and so on.[8] Second, this "discovery via constraints" is dynamic in nature because it leads to a new theory with novel empirical predictions that can subsequently be tested and explored (e.g., Hadjichristidis et al. 2004; Rehder 2009; Rehder and Kim 2010). And those experiments and observations provide additional, novel S_i claims that further

---

[8]A framework for characterizing and modeling this dynamics represents another extension of the constraint-inclusion model of Laudan, Nickles, and others.

constrain our theory of conceptual representations, either in detailed structure or in the scope of a particular theory. Scientific discovery and reasoning do not proceed in a discrete, staged manner, but rather involve a complex dynamic between using constraints to develop new theoretical ideas, and using ideas to find novel constraints.

3.3 Goals and Learning

The third case study looks a bit more like a case of "traditional" scientific discovery than the prior two. Consider the general question of the role of goals—more generally, beliefs about future tasks—on what and how we learn from the environment. Arguably, almost all major theories of (high-level) learning in cognitive psychology assume that goal or future task information only influence the domain from which we learn, but do not further influence the method or dynamics of learning. For example, essentially all theories of concept learning assume that I have domain knowledge about which features are potentially relevant to the new concept, but that the goal and (beliefs about) future tasks do not otherwise influence my concept learning. That is, given the same domain and same stimuli, learning is (on all of these theories) predicted to have the same dynamics. However, this dominant assumption has been called into question, and a new theory was discovered or proposed, in large measure by considering constraints implied by other scientific commitments.

The first theoretical claim S_1 that is potentially relevant to this problem is that much of our learning depends partly on attention. If we do not attend to a factor, then we typically learn less about it (e.g., Desimone and Duncan 1995; Huang and Pashler 2007), though some learning can occur even when we do not consciously attend to the items (DeSchepper and Treisman 1996). We do not need to make any particularly strong theoretical commitments about the nature of attention here. Rather, S_1 simply expresses the fact that attention and learning are sometimes closely connected. The relevant contrast class S_1 here includes the possibilities that attention does not directly modulate learning, or that attention is merely a necessary condition for learning (i.e., a "gate" on learning) rather than influencing it in a more nuanced fashion.

The second theoretical claim S_2 is that attention allocation depends partly on one's current task or goal. That is, my current task influences the particular way that I allocate my attention across my perceptual or cognitive field. For example, the current task or goal helps to determine which dimensions of objects are salient, and so which dimensions or objects are subsequently ignored as I perform that task (Maruff et al. 1999; Tipper et al. 1994). More colloquially, people pay much less attention to things that do not matter for their tasks, though they do not necessarily completely ignore those features of the stimuli or environment. As with S_1, this claim is likely not particularly surprising or controversial, though the human mind could have functioned differently (e.g., selection of task-relevant factors might have involved only cognitive mechanisms, rather than lower-level attentional processes).

Both S_1 and S_2 are widely (though not universally) endorsed in cognitive psychology, and both imply constraints on whether goals might influence the dynamics of learning. For concreteness, consider only two theoretical claims in **T**: (a) "goals only determine domain of learning input," labeled $T_{current}$ since it is the assumption of most current learning theories; and (b) "goals influence learning dynamics," labeled T_{new} since it is a novel theory (in this domain). Now consider the constraints implied by S_1 and S_2 for the two possible, though different, tasks of "learning for goal A" or "learning for goal B" (e.g., "learning to *predict* a system's behavior" vs. "learning to *control* a system's behavior"). By S_2, we should expect differential attention allocation; by S_1, we should expect this differential attention to translate into differential learning. That is, S_1 and S_2 jointly raise the plausibility of T_{new} and decrease the plausibility of $T_{current}$, even though none of these theoretical claims stands in any particular reductive relation with one another. Their intertheoretic relationships are more complicated, but equally able to support scientific discovery. In fact, this case study was historically a true case of discovery: although the analysis here evaluates T_{new} and $T_{current}$ as contemporaneous competitors, T_{new} was originally invented and proposed (in Danks 2014) only after recognizing that the constraints implied by S_1 and S_2 were in significant tension with $T_{current}$, and so a new theory was needed. Moreover, subsequent experimental results spoke strongly in favor of T_{new} (Wellen and Danks 2014; see also Hagmayer et al. 2010).

In sum, these three case studies provide three different ways in which intertheoretic constraints can be used to suggest or "discover" new scientific ideas. In contrast with "discovery via reduction," this account in terms of "discovery via constraints" can explain how disparate theories, as well as claims and other not-quite-theories, can inform and guide our scientific investigations. One might be concerned by the relatively prosaic and banal nature of these case studies, as we often think about scientific discovery as something grand or transformative. However, scientific discovery is also an everyday phenomenon, as particular scientists discover novel ways to synthesize or unify disparate scientific pieces. This type of everyday scientific thinking requires explanation and clarification just as much as the grand discoveries of Newton, Einstein, or others. And while reduction might sometimes be the basis of everyday scientific discovery, the more typical case is to use multidimensional intertheoretic constraints in order to add new pieces to our scientific puzzles.

4 Constraints All the Way Up?

The previous section focused on small-scale cases of scientific discovery via constraints, largely to provide enough detail to help demonstrate the mechanics of the framework. These smaller cases also clearly demonstrate that constraints are doing the relevant work, rather than full-blooded reductions. At the same time, one might wonder whether the "discovery via constraints" approach might be useful for understanding much larger-scale scientific discovery.[9] So, I close with some speculative thoughts about whether even scientific "revolutions" (using the term in a very broad way) could be understood in terms of discovery via constraints. At first glance, it is not obvious how constraints might be playing a role, particularly given the many stories about the crucial role of creativity in inventing or discovering scientific theories with wide scope. These stories highlight the role of **T** as a "free parameter" in the present account: I provided no explanation or account about how or why particular theoretical possibilities are included in **T**, even though discovery is (in a certain sense) limited to the elements of that set, and creativity or intuitive insight might be one way to generate elements of **T**. On the "discovery via constraints" view, creativity in theoretical innovation can thus have a very large impact, even though it is not directly modeled or explained.[10] We can only consider the impact of various constraints on a theoretical idea if we recognize the idea as possible or worth considering, and imagination or creativity might help explain why some possibility is included in **T**.

In many cases of scientific revolutions, this creative innovation is an important part of the overall story, but a single creative act is almost never the full story of any scientific revolution. Constraints arguably play a large role in the dynamics of scientific change that can result *after* the initial innovation. In many scientific "revolutions," there is a significant initial shift in approach or "paradigm" (again, using terms broadly) that is followed by significant work to put the empirical and theoretical pieces back together inside the new framework. The initial creative idea alone typically predicts and explains many fewer phenomena than were captured using the prior scientific theory and paradigm. Completion of the scientific revolution thus depends on finding auxiliary theories, background conditions, special cases, and other additional theories and not-quite-theories that generate explanations and predictions. Discovery via constraints will frequently play a significant role in these discoveries: these additional scientific elements can be discovered by trying to integrate constraints from the initial innovation, as well as constraints from prior empirical data and other posits that "survive" the revolution. For example, the Copernican revolution that shifted astronomy to a heliocentric view of the solar system started with a creative innovation, but then required substantial work to determine the appropriate constants, parameters, structures, and so forth.

---

[9]Thanks to Donald Gillies for encouraging me to consider this possibility, even after I had initially dismissed it.

[10]That being said, creativity could perhaps be modeled as discovery via constraints in the following way: suppose creativity results, as some have suggested (e.g., Simonton 1999), from profligate, unguided idea generation, followed by thoughtful pruning of the outputs. This pruning process could potentially be based on the use of constraints, and so we have the beginnings of a picture in which *all* discovery is based on constraints. Of course, this proposal does not explain the "idea generator," and much more work would need to be done before we have a full story in terms of constraints. Nonetheless, it is suggestive that even the "singular creative act" might be captured by this framework.

A dichotomy is sometimes drawn between periods of "revolutionary" and "normal" science, but a scientific revolution typically requires many steps of normal science along the way, and those can all (I argue) be fruitfully understood in terms of discovery via constraints. Discovery via constraints might (or might not) help us understand creativity or true innovation, but much of the rest of the process of large-scale scientific change can potentially be helpfully modeled as discovery via constraints.

In general, the process of scientific discovery often employs constraints from other scientific ideas, claims, theories, and not-quite-theories. These constraints result from the complex, multidimensional intertheoretic relationships that obtain between these pieces and the to-be-discovered scientific claim. Reduction is one salient intertheoretic relationship, and a source of particularly powerful constraints when it obtains. The corresponding "discovery via reduction" can thus also be particularly powerful, but only when the reduction relation obtains. In actual scientific practice, and particularly in everyday science, reductions are rarely forthcoming. Instead, scientific discovery proceeds through the opportunistic use of less powerful, but more widespread, constraints grounded in weaker intertheoretic relationships. "Scientific discovery via intertheoretic constraints" includes "discovery via reduction" as a special case. More importantly, it provides us with a richer, more nuanced understanding of some ways in which scientists develop novel ideas and theories.

Acknowledgements The ideas in this paper were initially presented at the "Building Theories: Hypotheses & Heuristics in Science" conference at Sapienza University. Thanks to the audience at the conference for their comments and criticisms, particular Emiliano Ippoliti, Lindley Darden, Donald Gillies, and Margie Morrison. Thanks also to two anonymous reviewers for valuable feedback on an earlier draft.

References

Barsalou, L. W. (2008). Grounded cognition. *Annual Review of Psychology, 59,* 617–645.

Bechtel, W., & Richardson, R. C. (2000). *Discovering complexity: Decomposition and localization as strategies in scientific research.* Princeton, NJ: Princeton University Press.

Carey, S. (1985). *Conceptual change in childhood.* Cambridge, MA: MIT Press.

Churchland, P. M. (1985). Reduction, qualia, and the direct introspection of brain states. *Journal of Philosophy, 82,* 1–22.

Danks, D. (2013). Moving from levels & reduction to dimensions & constraints. In M. Knauff, M. Pauen, N. Sebanz, & I. Wachsmuth (Eds.), *Proceedings of the 35th Annual Conference of the Cognitive Science Society* (pp. 2124–2129). Austin, TX: Cognitive Science Society.

Danks, D. (2014). *Unifying the mind: Cognitive representations as graphical models.* Cambridge, MA: The MIT Press.

Danks, D. (2015). Goal-dependence in (scientific) ontology. *Synthese, 192,* 3601–3616.

Darden, L. (2002). Strategies for discovering mechanisms: Schema instantiation, modular subassembly, forward/backward chaining. *Philosophy of Science, 69*(S3), S354–S365.

Darden, L., & Craver, C. (2002). Strategies in the interfield discovery of the mechanism of protein synthesis. *Studies in History and Philosophy of Science Part C: Studies in History and Philosophy of Biological and Biomedical Sciences, 33*(1), 1–28.

DeSchepper, B., & Treisman, A. (1996). Visual memory for novel shapes: Implicit coding without attention. *Journal of Experimental Psychology. Learning, Memory, and Cognition, 22,* 27–47.

Desimone, R., & Duncan, J. (1995). Neural mechanisms of selective visual attention. *Annual Review of Neuroscience, 18,* 193–222.

Dizadji-Bahmani, F., Frigg, R., & Hartmann, S. (2010). Who's afraid of Nagelian reduction? *Erkenntnis, 73,* 393–412.

Fodor, J. A. (1974). Special sciences: Or the disunity of science as a working hypothesis. *Synthese, 28,* 97–115.

Fodor, J. A. (1997). Special sciences: Still autonomous after all these years. *Nous, 31,* 149–163.

Griffiths, T. L., & Tenenbaum, J. B. (2005). Structure and strength in causal induction. *Cognitive Psychology, 51*(4), 334–384.

Hadjichristidis, C., Sloman, S., Stevenson, R., & Over, D. (2004). Feature centrality and property induction. *Cognitive Science, 28,* 45–74.

Hagmayer, Y., & Sloman, S. A. (2009). Decision makers conceive of their choices as interventions. *Journal of Experimental Psychology: General, 138,* 22–38.

Hagmayer, Y., Meder, B., Osman, M., Mangold, S., & Lagnado, D. A. (2010). Spontaneous causal learning while controlling a dynamic system. *The Open Psychology Journal, 3,* 145–162.

Hausman, D. M. (1998). *Causal asymmetries.* Cambridge: Cambridge University Press.

Hooker, C. A. (1981a). Towards a general theory of reduction, part I: Historical and scientific setting. *Dialogue, 20,* 38–59.

Hooker, C. A. (1981b). Towards a general theory of reduction, part II: Identity in reduction. *Dialogue, 20,* 201–236.

Huang, L., & Pashler, H. (2007). Working memory and the guidance of visual attention: Consonance-driven orienting. *Psychonomic Bulletin & Review, 14,* 148–153.

Keil, F. C. (1989). *Concepts, kinds, and cognitive development.* Cambridge, MA: MIT Press.

Laudan, L. (1981). A problem solving approach to scientific progress. In I. Hacking (Ed.), *Scientific revolutions.* Oxford: Oxford University Press.

Machery, E. (2009). *Doing without concepts.* Oxford: Oxford University Press.

Maruff, P., Danckert, J., Camplin, G., & Currie, J. (1999). Behavioural goals constrain the selection of visual information. *Psychological Science, 10,* 522–525.

Murphy, G. L. (2004). *The big book of concepts.* Cambridge, MA: Bradford.

Nagel, E. (1961). *The structure of science: Problems in the logic of scientific explanation.* New York: Harcourt.

Nickles, T. (1978). Scientific problems and constraints. In *PSA: Proceedings of the Biennial Meeting of the Philosophy of Science Association* (pp. 134–148). Chicago: University of Chicago Press.

Nickles, T. (1981). What is a problem that we may solve it? *Synthese, 47,* 85–118.

Nickles, T. (1988). Questioning and problems in philosophy of science: Problem-solving versus directly truth-seeking epistemologies. In M. Meyer (Ed.), *Questions and questioning* (pp. 43–67). Berlin: Walter de Gruyter.

Pearce, J. M. (1987). A model for stimulus generalization in Pavlovian conditioning. *Psychological Review, 94*(1), 61–73.

Putnam, H. (1975). Philosophy and our mental life. In *Mind, language, and reality: Philosophical papers* (Vol. 2, pp. 291–303). Cambridge, MA: Cambridge University Press.

Rehder, B. (2003a). A causal-model theory of conceptual representation and categorization. *Journal of Experimental Psychology. Learning, Memory, & Cognition, 29,* 1141–1159.

Rehder, B. (2003b). Categorization as causal reasoning. *Cognitive Science, 27,* 709–748.

Rehder, B. (2009). Causal-based property generalization. *Cognitive Science, 33,* 301–344.

Rehder, B., & Kim, S. (2010). Causal status and coherence in causal-based categorization. *Journal of Experimental Psychology. Learning, Memory, & Cognition, 36*(5), 1171–1206.

Rescorla, R. A., & Wagner, A. R. (1972). A theory of Pavlovian conditioning: Variations in the effectiveness of reinforcement and nonreinforcement. In A. H. Black & W. F. Prokasy (Eds.), *Classical conditioning II: Current research and theory* (pp. 64–99). New York: Appleton-Century-Crofts.

Rovee, C. K., & Rovee, D. T. (1969). Conjugate reinforcement of infant exploratory behavior. *Journal of Experimental Child Psychology, 8,* 33–39.

Schouten, M. K. D., & de Jong, H. L. (Eds.). (2012). *The matter of the mind: Philosophical essays on psychology, neuroscience and reduction.* London: Wiley-Blackwell.

Shanks, D. R. (1995). Is human learning rational? *The Quarterly Journal of Experimental Psychology, 48A,* 257–279.

Simonton, D. K. (1999). *Origins of genius: Darwinian perspectives on creativity.* New York: Oxford University Press.

Sloman, S. A. (2005). *Causal models: How people think about the world and its alternatives.* Oxford: Oxford University Press.

Timberlake, W. (2001). Integrating niche-related and general process approaches in the study of learning. *Behavioural Processes, 54,* 79–94.

Tipper, S. P., Weaver, B., & Houghton, G. (1994). Behavioural goals determine inhibitory mechanisms of selective attention. *Quarterly Journal of Experimental Psychology: Human Experimental Psychology, 47,* 809–840.

Wellen, S., & Danks, D. (2014). Learning with a purpose: The influence of goals. In P. Bello, M. Guarini, M. McShane, & B. Scassellati (Eds.), *Proceedings of the 36th Annual Conference of the Cognitive Science Society* (pp. 1766–1771). Austin, TX: Cognitive Science Society.

Wimsatt, W. C. (1980). Reductionistic research strategies and their biases in the units of selection controversy. In T. Nickles (Ed.), *Scientific discovery: Case studies* (pp. 213–259). D. Reidel Publishing.

Theory Building as Problem Solving

Carlo Cellucci

Abstract After giving arguments against the claim that the so-called Big Data revolution has made theory building obsolete, the paper discusses the shortcomings of two views according to which there is no rational approach to theory building: the hypothetico-deductive view and the semantic view of theories. As an alternative, the paper proposes the analytic view of theories, illustrating it with some examples of theory building by Kepler, Newton, Darwin, and Bohr. Finally, the paper examines some aspects of the view of theory building as problem solving.

Keywords Theory building · Big data revolution · Analytic view of theories
Novelty of non-deductive rules · Plausibility · Problem solving

1 Big Data Revolution Versus Theory Building

Dealing with theory building today may seem anachronistic because of the so-called 'Big Data revolution', the view that theory is dead at the hands of Big Data analysis. But, has the Big Data revolution really made theory building obsolete?

The best known formulation of the Big Data revolution is by Anderson who states that, "faced with massive data," the old "approach to science—hypothesize, model, test is becoming obsolete" (Anderson 2008). We are at "the end of theory," since "the data deluge makes the scientific method obsolete" (ibid.). The "new availability of huge amounts of data, along with the statistical tools to crunch these numbers, offers a whole new way of understanding the world. Correlation supersedes causation, and science can advance even without coherent models, unified theories" (ibid.). Indeed, "correlation is enough," so "we can stop looking for models. We can analyze the data without hypotheses about what it might show. We can throw the numbers into the biggest computing clusters the world has ever seen

C. Cellucci (✉)
Sapienza University of Rome, Via Carlo Fea 2, 00161 Roma, Italy
e-mail: carlo.cellucci@uniroma1.it

© Springer International Publishing AG 2018
D. Danks and E. Ippoliti (eds.), *Building Theories*, Studies in Applied Philosophy, Epistemology and Rational Ethics 41, https://doi.org/10.1007/978-3-319-72787-5_4

and let statistical algorithms find pattern where science cannot" (ibid.). For example, "Google's founding philosophy is that we don't know why this page is better than that one: If the statistics of incoming links say it is, that's good enough. No semantic or causal analysis is required" (ibid.). So, "there's no reason to cling to our old ways. It's time to ask: What can science learn from Google?" (ibid.).

These claims, however, are unjustified. The claim that we can analyze the data without hypotheses about what it might show, is based on the assumption that the data are completely independent of theories. This assumption is favoured by the fact that the term 'data' comes from the Latin verb *dare* which means 'to give', which suggests that the data are raw elements that are given by phenomena. But it is not so, because observation is always selective, every choice of the data is a reflection of an, often unstated, set of assumptions and hypotheses about what we want and expect from the data. The data are not simple elements that are abstracted from the world in neutral and objective ways. There is always a 'viewpoint' preceding observation and experiment, namely a theory or hypothesis which guides observation and experiment, and generally data-finding. The data do not speak for themselves, but acquire meaning only when they are interpreted, and interpreting them requires a theory through which to observe them, and extract information from them. So, the Big Data revolution has not made theory building obsolete.

In particular, systems of data analysis are designed to capture certain kinds of data, the algorithms used for that purpose are based on some theory, and have been refined through testing. Thus, a statistical strategy of identifying patterns within data is based on previous findings and theories. Then, it illusory to think that statistical strategies may automatically discover insights without presupposing any theory or testing. This is admitted even by some supporters of the Big Data revolution, such as Berman who states that, "for Big Data projects, holding a prior theory or model is almost always necessary; otherwise, the scientist is overwhelmed by the options" (Berman 2013, 147).

Moreover, it is incorrect to say that correlation is enough, therefore we can stop looking for models. Calude and Longo (2016, 4) "document the danger of allowing the search of correlations in big data to subsume and replace the scientific approach". This "is mathematically wrong" (ibid., 15). For, "the overwhelming majority of correlations are spurious. In other words, there will be regularities, but, by construction, most of the time (almost always, in the mathematical sense), these regularities cannot be used to reliably predict and act" (ibid., 16). Indeed, "the bigger the database which one mines for correlations, the higher is the chance to find recurrent regularities and the higher is the risk of committing such fallacies" (ibid., 15). Therefore, "the more data, the more arbitrary, meaningless and useless (for future action) correlations will be found in them. Thus, paradoxically, the more information we have, the more difficult is to extract meaning from it" (ibid., 6). This shows that, as "no theory can be so good to supplant the need for data and testing," so "big data analytics cannot replace science" (ibid., 16).

Supporters of the Big Data revolution often portray Bacon as a precursor. They claim that "Bacon proclaimed that science could discover truths about nature only by empirical testing of all the possible explanations for all the observed phenomena" (Siegfried 2013). For example, science could discover the nature of heat only by recording "all observed facts about all manner of heat-related phenomena" and then performing "experiments to eliminate incorrect explanations" (ibid.). Thus, "Bacon was a fan of Big Data. With today's massive computerized collections of massive amounts of data on everything," at last "Bacon's dreams have been realized" (ibid.).

But these claims are unjustified. Bacon was not a fan of Big Data. According to him, "experience, when it wanders in its own track" unguided by theory, is "mere groping in the dark" (Bacon 1961–1986, I, 203). Without theory, human beings "wander and stray with no settled course, but only take counsel from things as they fall out," and hence "make little progress" (ibid., I, 180). Therefore, human beings must not only "seek and procure abundance of experiments," but must also develop theories "for carrying on and advancing experience" (ibid., I, 203).

It is somewhat ironical that, at the same time as Anderson proclaimed the Big Data revolution, the financial crisis of 2007–08 occurred. Financial analysts who had thrown the numbers into the biggest computing clusters the world had ever seen and had let statistical algorithms find pattern, failed to foresee the worst financial crisis since the Great Depression of the 1930s. While not invalidating the Big Data revolution, this raises serious doubts about it.

2 Theory Building and Rationality

Although the Big Data revolution has not made theory building obsolete, today theory building is not a well-developed subject in its own right. This is due to the fact that a crucial part of theory building is the process of discovery, and most scientists and philosophers think that there is no rational approach to discovery. To them, discovery appears as a mysterious process that somehow happens, and happens through intuition.

Thus Einstein states that "there is no logical path" to the basic laws of physics, "only intuition, resting on sympathetic understanding of experience, can reach them" (Einstein 2010, 226).

Popper states that "there is no such thing as a logical method of having new ideas, or a logical reconstruction of this process," every "discovery contains 'an irrational element', or 'a creative intuition', in Bergson's sense" (Popper 2005, 8).

The lack of a rational approach to the process of discovery, and hence to theory building, characterises two main views of scientific theories, the 'syntactic', or 'hypothetico-deductive' view, and the 'semantic', or 'model-theoretic' view. Neither of them provides a rational account of theory building.

3 The Syntactic or Hypothetico-Deductive View of Theories

According to the 'syntactic', or 'hypothetico-deductive' view, a theory consists, on the one hand, of "an uninterpreted deductive system, usually thought of as an axiomatized calculus C, whose postulates correspond to the basic principles of the theory and provide implicit definitions for its constitutive terms," and, on the other hand, of "a set R of statements that assign empirical meaning to the terms and the sentences of C by linking them to potential observational or experimental findings and thus interpreting them, ultimately, in terms of some observational vocabulary" (Hempel 2001, 49). Such statements R are referred to as 'rules of correspondence' or 'rules of interpretation' (ibid.).

Theories cannot be obtained by "any process of systematic inference" (Hempel 1966, 15). They "are not derived from observed facts, but invented in order to account for them" (ibid., 15). They "are invented by an exercise of creative imagination" (Hempel 2001, 32). Their discovery "requires inventive ingenuity; it calls for imaginative, insightful guessing" (Hempel 1966, 17). That is, it calls for intuition. Thus, according to the hypothetico-deductive of theories, theories cannot be obtained by any process of systematic inference, they are the result of an act of intuition.

The syntactic view is very widespread. In particular, both Einstein and Popper support it.

Thus Einstein states that "the intuitive grasp of the essentials of a large complex of facts leads the scientist" to "a basic law," then "from the basic law (system of axioms) he derives his conclusion as completely as possible in a purely logically deductive manner" (Einstein 2002, 108). The conclusions deduced from the basic law are then "compared to experience and in this manner provide criteria for the justification of the assumed basic law. Basic law (axioms) and conclusion together form what is called a 'theory'" (ibid.).

Popper states that, from a basic law obtained through creative intuition, "conclusions are drawn by means of logical deduction" and compared with experience, so the basic law is tested "by way of empirical applications of the conclusion which can be derived from it" (Popper 2005, 9). This is the scientific method, which is then the method of "deductive testing" (ibid.).

However, the syntactic view is inadequate. By Gödel's first incompleteness theorem, for any consistent, sufficiently strong, formal system, there is a sentence of the system which is true but cannot be deduced from the axioms of the system. So, by Gödel's result, there will be laws of a theory which cannot be deduced from the postulates of the theory.

Moreover, the syntactic view is unable to provide a rational account of theory building. Saying that theories are invented by an exercise of creative imagination is an irrational explanation.

Furthermore, the syntactic view is unable to account for theory change, the process by which one theory comes to be replaced by another. For, according to the syntactic view, a theory has no rational connection with the preceding one, except that it agrees with more observational and experimental data than the preceding one.

4   The Semantic or Model-Theoretic View of Theories

According to the 'semantic', or 'model-theoretic' view, a theory is "identified with its class of models" (van Fraassen 1989, 222). Thus "to present a theory is to specify a family of structures, its models" (van Fraassen 1980, 64). Such family of structures is specified "directly, without paying any attention to questions of axiomatizability, in any special language" (van Fraassen 1989, 222). A "model is a mathematical structure" (van Fraassen 2008, 376, Footnote 18). More precisely, "a model is a structure plus a function that interprets the sentences in that structure" (van Fraassen 1985, 301). If "a theory is advocated then the claim made is that these models can be used to represent the phenomena, and to represent them accurately," where we say that "a model can (be used to) represent a given phenomenon accurately only if it has a substructure isomorphic to that phenomenon" (van Fraassen 2008, 309).

Theories cannot be obtained by any process of systematic inference, "all those successes of science which so many people have thought must have been produced by induction or abduction" were "initially good guesses under fortunate circum-stances," then they "were made effective by means of the precise formulation and disciplined teasing out of their implications through logic and mathematics" (van Fraassen 2000, 275). Indeed, "if our pursuit of knowledge" is "to be successful, we must be lucky—we have no way to constrain such fortune" (ibid., 273).

The semantic view became popular a few decades ago, but is inadequate. A model is a structure and hence a mathematical object, while a phenomenon is not a mathematical object. Van Fraassen himself states: "If the target," that is, the phenomenon, "is not a mathematical object, then we do not have a well-defined range for the function, so how can we speak of an embedding or isomorphism or homomorphism or whatever between that target and some mathematical object?" (van Fraassen 2008, 241). Van Fraassen's answer is that we compare the model not with the phenomenon but rather with the data model, that is, our representation of the phenomenon. The data model "is itself a mathematical structure. So there is indeed a 'matching' of structures involved; but is a 'matching' of two mathematical structures, namely the theoretical model and the data model" (ibid., 252).

This answer, however, is inadequate, because the data model is a mathematical object, while the phenomenon is not a mathematical object. This raises the question of the matching of the data model and the phenomenon. Thus, van Fraassen's answer just pushes the problem back one step. Besides, even a fiction can have a model, in the sense of a mathematical structure. Therefore, it is not models that can make a distinction between fictions and reality.

Moreover, the semantic view is unable to provide an account of theory building. Saying that theories are good guesses under fortunate circumstances is a non-explanation, it completely evades the issue.

Furthermore, the semantic view entails that scientific theories, being families of structures, are static things. But scientific theories undergo development, and the semantic view has no alternative than treating their development as a progression of successive families of models. Then the issue arises how the transition from a theory to the next one in the progression comes about. The semantic view has nothing to say about this, because it does not account for the process of theory building, which is essential to explain the development of theories and theory change. Therefore, it cannot account for the dynamic character of scientific theories.

5   The Analytic View of Theories

It is somewhat ironical that, while contemporary views of scientific theories are unable to give accounts of theory building, the two philosophical giants of antiquity, Plato and Aristotle, gave such accounts. This is not the place to describe their accounts, but it seems right and proper to mention that Plato and Aristotle gave them. (On their accounts, see Cellucci 2013, Chaps. 4 and 5, respectively.) In particular, it seems right and proper to mention Plato's account, because a suitable account of theory building, the 'analytic view of theories', can be given by modifying Plato's original account.

According to the analytic view of theories, a scientific theory is an open set of problems about the world and hypotheses that permit to solve them. An open set, because the hypotheses are not given once for all but new hypotheses can always be introduced, or the existing ones can always be modified. Theory building consists in starting from problems, arriving at hypotheses and deducing solution to problems from them. Hypotheses are arrived at through non-deductive rules—such as induction, analogy, metaphor, and so on (see Cellucci 2013, Chaps. 20 and 21)— and must be plausible, that is, the arguments for a hypothesis must be stronger than the arguments against it, on the basis of experience. Solutions to problems are not absolutely certain but only plausible.

This amounts to saying that theory building is carried out by the analytic method. Indeed, the latter is the method according to which, to solve a problem, one looks for some hypothesis that is a sufficient condition for solving the problem, namely, such that a solution to the problem can be deduced from it. The hypothesis is arrived at from the problem, and possibly other data already available, by some non-deductive rule, and must be plausible. But the hypothesis is in its turn a problem that must be solved, and is solved in the same way. That is, one looks for another hypothesis that is a sufficient condition for solving the problem posed by the previous hypothesis, it is arrived at from the latter, and possibly other data already available, by some non-deductive rule, and must be plausible. And so on.

ad infinitum. (For more on the analytic method, see Cellucci 2013, Chap. 4; 2017, Chap. 12).

Being carried out by the analytic method, theory building does not come to an end, it is an ongoing process. Hypotheses are subject to be modified or replaced when they become implausible as new data emerge. The modified or new hypotheses are obtained through an analysis of the reasons why the former hypotheses have become implausible.

The analytic view is supported by Gödel's first incompleteness theorem because, according to it, no system of hypotheses can solve all the problems of a given field. The hypotheses are bound to be replaced sooner or later with other more general ones through a potentially infinite process, since every system of hypotheses is incomplete and needs to appeal to other systems to bridge its gaps.

The analytic view is also supported by Gödel's second incompleteness theorem because, according to it, no solution to a problem is absolutely certain but only plausible.

Moreover, the analytic view is able to provide a rational account of theory building, in terms of hypotheses obtained by non-deductive rules and validated by their plausibility.

Furthermore, the analytic view is able to account for theory change because, it establishes a rational connection between subsequent theories. On the basis of it, the hypotheses of the new theory can be formulated through an analysis of the reasons why the hypotheses of the preceding theory have become implausible.

6 Kepler's Theory Building

An example of theory building in accordance with the analytic view is Kepler's theory of the motion of planets.

The problem Kepler wanted to solve by his theory was to explain what is the moving power of the planets, how this moving power acts on them, and why the planets further away from the Sun move slower than those closer to the Sun. His theory was based on the hypothesis that the moving power of the planets is in the Sun, that such moving power acts at a distance, and that it impels each planet more strongly in proportion to how near it is to the Sun.

Kepler arrived at this hypothesis through an analogy between the light emanating from the Sun that illuminates the planets, and a power emanating from the Sun that causes the planets to move. His starting point was that the light that illuminates the planets emanates from the Sun, it acts at a distance, and gets weaker with distance. From this, by analogy, Kepler inferred that the power that causes the planets to move emanates from the Sun, it acts at a distance and gets weaker with distance. (Generally, an inference by analogy is one by which, if a is similar to b in certain respects and a has a certain property, then b too will have that very same property; for more on this, see Cellucci 2013, Chap. 20).

Indeed, Kepler states: "Let us suppose" that "motion is dispensed by the Sun in the same proportion as light" (Kepler 1981, 201). Then, since Sun's light acts at a distance and "does not exist in the intermediate space between the source and the illuminable, this is equally true of the motive power" (Kepler 1992, 383). And, as "Sun's light also grows thinner with distance from the Sun," similarly "this moving cause grows weaker with distance" (Kepler 1981, 201).

Thus Kepler arrived at the hypothesis upon which his theory was based through an inference by analogy.

7 Newton's Theory Building

Another example of theory building in accordance with the analytic view is Newton's theory of universal gravitation.

The problem Newton wanted to solve by his theory was to explain the structure of the system of the world. His theory was based on the hypothesis that gravity exists in all bodies universally, and the gravitation toward each of the particles of a body is inversely as the square of the distance of places from those particles.

Newton arrived at this hypothesis through an induction. His starting point was that gravity exists in all planets, and the gravity toward any one planet is inversely as the square of the distance of places from the center of the planet. From this, by induction, Newton inferred that gravity exists in all bodies universally and the gravitation toward each of the particles of a body is inversely as the square of the distance of places from those particles. (Generally, an inference by induction is one by which, if a number of things of a certain kind have a certain property, all things of that kind will have that property; for more on this, see Cellucci 2013, Chap. 20).

Indeed, Newton states: "We have already proved that all planets gravitate toward one another and also that the gravity toward any one planet, taken by itself, is inversely as the square of the distance of places from the center of the planet" (Newton 1999, 810). From this we may infer that "gravity exists in all bodies universally" (ibid.). And "the gravitation toward each of the individual equal particles of a body is inversely as the square of the distance of places from those particles" (ibid., 811).

Thus Newton arrived at the hypothesis upon which his theory was based through an inference by induction. (On Newton's views about theory building, see Cellucci 2013, Chap. 8).

8 Darwin's Theory Building

Another example of theory building in accordance with the analytic view is Darwin's theory of evolution by natural selection.

The problem Darwin wanted to solve by his theory was to explain the characteristics of existing living things, and how these characteristics came to be. His theory was based on two hypotheses: (1) Natural selection produced different species of animals and plants. (2) As more individuals of any species are produced than can possibly survive, there must be a struggle for existence.

Darwin arrived at these two hypotheses through an analogy and an induction. Indeed, Darwin arrived at hypothesis (1) through an analogy between artificial selection and natural selection. His starting point was that breeders used artificial selection to produce different breeds of animals and plants. From this, by analogy, Darwin inferred that nature used natural selection to produce different species of animals and plants. On the other hand, Darwin arrived at hypothesis (2) through an induction. His starting point was Malthus' observation that, as more human beings are produced than can possibly survive, there must be a struggle for existence. From this, by induction, Darwin inferred that, as more individuals of any species are produced than can possibly survive, there must be a struggle for existence.

Indeed, Darwin states that (1) he "came to the conclusion that selection was the principle of change from the study of domesticated productions" (Darwin 1903, I, 118). Namely, "from what artificial selection had done for domestic animals" (Darwin 2009b, II, 118). And (2) he came to the conclusion that, in any species, "as more individuals are produced than can possibly survive, there must in every case be a struggle for existence," from "the doctrine of Malthus applied with manifold force to the whole animal and vegetable kingdoms" (Darwin 2009a, 50).

Thus Darwin arrived at the hypotheses upon which his theory was based through an inference by analogy and an inference by induction.

9 Bohr's Theory Building

Another example of theory building in accordance with the analytic view is Bohr's theory of the structure of the atom.

The problem Bohr wanted to solve by his theory was to explain the structure of the atom. His theory was based on the hypothesis that atoms consist of a nucleus surrounded by a cluster of electrons which rotate around the nucleus in fixed quantized orbits.

Bohr arrived at this hypothesis through a metaphor: the atom behaves as if it were a minuscule quantized solar system, thus some properties of a quantized solar system can be transferred to the atom. (Metaphor is an inference by which a thing of a certain kind, called the primary subject, behaves as if it were a thing of another kind, called the secondary subject. In Bohr's case, the primary subject was the atom, and the secondary subject was a quantized solar system. While analogy expresses similarity, metaphor does not express similarity, it creates similarity; for more on this, see Cellucci 2013, Chap. 21).

Indeed, Bohr states that the atom behaves as if it were a minuscule quantized solar system, consisting "of a positively charged nucleus surrounded by a cluster of electrons" (Bohr 1913, 476). The "states of the electron system" can be pictured "as planetary motions obeying Keplerian laws" (Bohr 1963, 85). Specifically, the "electrons are arranged at equal angular intervals in coaxial rings rotating round the nucleus" and "the angular momentum of every electron round the centre of its orbit is equal to the universal value $h/2\pi$, where h is Planck's constant" (Bohr 1913, 477).

Thus Bohr arrived at the hypothesis upon which his theory was based through an inference by metaphor.

10 Novelty

In the process of theory building according to the analytic view, non-deductive rules play an essential role, because hypotheses are arrived at through them, not through deductive rules. This depends on the fact that non-deductive rules can produce new knowledge, while deductive rules cannot produce new knowledge.

This claim may seem problematic, because some people maintain that deductive rules can produce new knowledge. Thus, Prawitz states that "in mathematics one solves problems by deductive proofs from premises held to be true" (Prawitz 2014, 75). For instance, Wiles and Taylor inferred Fermat's Last Theorem "deductively from initial premises that were agreed by mathematicians to express known truths" (ibid., 89). Generally, all "mathematical knowledge is obtained by deduction from truths already known" (ibid., 87).

This, however, conflicts with the fact that, for example, when Cantor demonstrated that for any cardinal number there is a greater cardinal number, he did not deduce this from truths already known, since it was impossible to demonstrate it within the bounds of traditional mathematics. Demonstrating it required formulating new concepts and a new theory of the infinite. Therefore, not all mathematical demonstrations are deductions from truths already known.

It could be objected that, since the use of deductive rules requires labour, deductive rules can produce new knowledge. Thus Frege argues that, although the conclusion of a deduction is "in a way contained covertly in the whole set" of premises "taken together," this "does not absolve us from the labour of actually extracting them and setting them out in their own right" (Frege 1960, 23). Indeed, "what we shall be able to infer from" the premises, "cannot be inspected in advance; here, we are not simply taking out of the box again what we have just put into it. The conclusions we draw" from the premises may "extend our knowledge, and ought therefore, on Kant's view, to be regarded as synthetic" (ibid., 100–101). They are contained in the premises, "but as plants are contained in their seeds, not as beams are contained in a house" (ibid., 101).

This objection, however, is invalid because there is an algorithmic method for enumerating all deductions from given premises, and hence all the conclusions which can be deduced from given premises. Then, as Turing states, we can "imagine that all proofs take the form of a search through this enumeration for the theorem for which a proof is desired" (Turing 2004a, 193). Although in practice "we do not really want to make proofs by hunting through enumerations for them" since this is a long method, nevertheless the usual procedure for finding a proof "is always theoretically, though not practically, replaceable by the longer method" (Turing 2004b, 212). So "ingenuity is replaced by patience" (Turing 2004a, 193). Given enough time and space, the algorithmic method will enumerate all proofs, and, if a conclusion can be proved, the algorithmic method will sooner or later find a proof of it. So, contrary to Frege's claim, extracting conclusions from the premisses is a purely mechanical task, it can be performed by a computer and hence requires no labour, only some electric power.

Moreover, saying that conclusions are contained in the premisses as plants are contained in their seeds is misleading. For, plants can develop from seeds only absorbing water from the soil and harvesting energy from the sun, so, using something which is not contained in the seeds. On the contrary, conclusions can be deduced from premisses without using anything not contained in the premisses.

In relation to Frege's arguments, a distinction must be made between objective novelty and psychological novelty. The conclusions of deductions may be psychologically surprising and hence may have psychological novelty, because we are incapable of making even comparatively short deductions without the help of processes external to us. But this does not mean that the conclusions of deductions extend our knowledge, and hence have objective novelty. Frege's claim that they extend our knowledge is a form of psychologism, since it mistakes psychological novelty for objective novelty.

11 Plausibility

In the process of theory building according to the analytic view, the concept of plausibility plays the central role in the validation of theories. Indeed, the hypotheses on which a theory is based can be accepted only if they are plausible, where, as already stated, a hypothesis is said to be plausible if the arguments for the hypothesis are stronger than the arguments against it, on the basis of experience.

Solutions to problems are not certain but only plausible. This depends on the fact that the process of theory building according to the analytic view is based on heuristic reasoning, and heuristic reasoning cannot guarantee certainty. On the other hand, heuristic reasoning is essential in theory building, because only heuristic reasoning can produce novelty. As Pólya states, "if you take a heuristic conclusion as certain, you may be fooled and disappointed; but if you neglect heuristic conclusions altogether you will make no progress at all" (Pólya 1971, 181).

That solutions to problems are not certain but only plausible does not mean that 'plausible' can be identified with 'probable'. Pólya claims that one can "use the calculus of probability to render more precise our views on plausible reasoning" (Pólya 1954, II, 116). For "the calculus of plausibilities obeys the same rules as the calculus of probabilities" (Pólya 1941, 457).

But it is not so. Plausibility involves a comparison between the arguments for the hypothesis and the arguments against it, so it is not a mathematical concept. Conversely, probability is a mathematical concept. This is made quite clear by Kant, who states that, on the one hand, "plausibility is concerned with whether, in the cognition, there are more grounds for the thing than against it," on the other hand, "there is a mathematics of probability" (Kant 1992, 331). (On Kant's distinction between plausibility and probability, see Capozzi 2013, Chap. 7, Sect. 5, and Chap. 15).

In fact, there are hypotheses which are plausible but, in terms of the classical concept of probability, have zero probability. On the other hand, there are hypotheses which are not plausible but, again in terms of the classical concept of probability, have a non-zero probability. The same holds on other concepts of probability (see Cellucci 2013, Chap. 20).

Rather than being related to 'probable', 'plausible' is related to Aristotle's *endoxon*. For, Aristotle states that, in order to determine whether a hypothesis is *endoxon*, we must "examine the arguments for it and the arguments against it" (Aristotle, *Topica*, Θ 14, 163 a 37–b 1). In fact, Striker states: "'plausible': I use this word to translate the Greek *endoxon*, literally, 'what enjoys a certain fame or reputation'" (Striker 2009, 77). However 'plausible' is not the same as *endoxon* (see Cellucci 2017, Chap. 9).

A hypothesis which is plausible at one stage may become implausible at a later stage, or vice versa, because new data may always emerge which change the balance between the arguments for a hypothesis and the arguments against it. For example, the hypothesis on which Bohr's theory of the structure of the atom was based—that atoms consist of a nucleus surrounded by a cluster of electrons which rotate around the nucleus in fixed quantized orbits—was plausible when Bohr first stated it but became implausible later on, for example when new data showed that it could not explain the spectral lines of atoms with more than one electron.

12 Theory Building and Problem Solving

The crucial step in theory building in accordance with the analytic view is the discovery of solutions to problems.

The question how solutions to problems are discovered often receives incongruous answers, such as Pólya's answer that "the first rule of discovery is to have brains and good luck. The second rule of discovery is to sit tight and wait till you get a bright idea" (Pólya 1971, 172). This answer is incongruous because it is of the same kind as that of Molière's Bachelierus: "Mihi a docto doctore domandatur causam et rationem quare opium facit dormire. A quoi respondeo, Quia est in eo virtus dormitiva, cujus est natura sensus assoupire [I am asked by a learned doctor for the cause and reason why opium makes one sleep. To which I reply, Because there is in it a dormitive virtue, whose nature is to make the senses drowsy]" (Molière, *The Imaginary Invalid*, Act III, Interlude III). Indeed, Pólya's answer amount to saying that the reason why the mind discovers solutions to problems is that there is in it a discoveritive virtue, whose nature is to make the mind inventive.

Pólya's answer is all the more incongruous because he admits that there is a method of discovery. Indeed, he states that there are procedures which are "typically useful in solving problems" and are "practiced by every sane person sufficiently interested in his problems" (Pólya 1971, 172). The best of such procedures is "the method of analysis, or method of 'working backwards'" (ibid., 225).

Indeed, a reasonable answer to the question, 'How can solutions to problems be discovered?' is that they can be discovered by the analytic method. The latter has been recognized from the antiquity as the main method for problem solving.

Since the crucial step in theory building in accordance with the analytic view is the discovery of solutions to problems, theory building is a variety of problem solving, and specifically, problem solving by the analytic method.

13   A Critical Remark on an Alternative Problem Solving Approach

The view that theory building is problem solving by the analytic method is apparently related to Laudan's view that "science is essentially a problem-solving activity" (Laudan 1977, 11). We "do not have any way of knowing for sure (or even with some confidence) that science is true," or "that it is getting closer to the truth" (ibid., 127). Therefore, we cannot say that the aim of science is truth or approximation to truth. Such aims "are utopian, in the literal sense that we can never know whether they are being achieved" (ibid.). Rather, "science fundamentally aims at the solution of problems" (ibid., 4–5). While a criterion of truth will never be found, "we can determine whether a given theory does or does not solve a particular problem" (ibid., 127).

So far so good. (On the claims that the aim of science cannot be truth and that a criterion of truth will never be found, see Cellucci 2017, Chap. 8). However, in addition to stating that science fundamentally aims at the solution of problems, Laudan also states that "the case has yet to be made that the rules governing the techniques whereby theories are invented (if such rules there be) are the sorts of things that philosophers should claim any interest in" (Laudan 1981, 191). Discovery cannot be a concern of philosophy, because "a theory is an artifact," and "the investigation of the mode of manufacture of artifacts" is "not normally viewed as philosophical activity. And quite rightly, for the techniques appropriate to such investigations are those of the empirical sciences, such as psychology, anthropology, and physiology" (ibid., 190–191).

Thus Laudan ends up accepting Frege's view that philosophy cannot be concerned "with the way in which" new results "are discovered" (Frege 1960, 23). The question of discovery is a merely subjective, psychological one, because it "may have to be answered differently for different persons" (Frege 1967, 5).

Like Frege, then, Laudan claims that discovery is a matter that can be a subject only for empirical sciences such as psychology.

This view, however, is unjustified because, as already mentioned, from the antiquity a method of discovery is known and has been recognized as the main method for problem solving, that is, the analytic method, in which there is nothing subjective or psychological.

14   The Big Data Revolution Revisited

In the light of the analytic view of theories, it is possible to revisit certain claims currently made about the Big Data revolution.

A significant representative of these claims is Malle's assertion that the Big Data revolution "requires a complete break from Cartesian logic. It calls for the non-scientific part of human thought: inductive reasoning" (Malle 2013, n.p.). The "analysis of huge amounts of data mainly focuses on finding correlations," such correlations will be found by inductive algorithms, and "induction allows algorithms to reproduce observed phenomena by generalizing beyond their scope," but "without trying to make models out of them" (ibid.). An "algorithm designed within an inductive approach," is not "intended to test an existing hypothesis," and yet it may achieve "a specified aim" (ibid.). This "is far from Cartesian principles," since Cartesian logic is based on deduction, and induction and deduction are opposed to each other, indeed, inductive thinking is "the exact opposite of deductive thinking" (ibid.). So "either we reason in the context of a deductive approach, or else we firmly choose an inductive approach" (ibid.). While "deductive reasoning always comes to an end, inductive reasoning generally produces no finished status. The results of inferences are likely to alter the inferences already made. It is possible to continue the reasoning indefinitely" (ibid.). And yet, inductive reasoning "produces an imperfect but useful knowledge" (ibid.).

In terms of the analytic view of theories, these claims appear to be completely unjustified.

Indeed, it is unjustified to say that inductive reasoning is the non-scientific part of human thought. According to the analytic view of theories, inductive reasoning is a main constituent of theory building. In the latter, induction plays an important role as one of the main non-deductive rules by which hypotheses are arrived at, so inductive reasoning is an important tool for scientific discovery.

Moreover, it is unjustified to say that inductive thinking is the exact opposite of deductive thinking, so either we reason in the context of a deductive approach, or else we firmly choose an inductive approach. According to the analytic view of theories, induction and deduction work side by side in theory building. For, in the analytic method, every step upward by which a hypothesis is arrived at from a problem through a non-deductive rule, is accompanied by a step downward by which a solution to the problem is deduced from the hypothesis.

In addition, it is unjustified to say that, while deductive reasoning always comes to an end, inductive reasoning generally produces no finished status, the results of inferences are likely to alter the inferences already made, and it is possible to continue the reasoning indefinitely. According to the analytic view, like deductive reasoning, the non-deductive reasoning through which a hypothesis is arrived at from a problem always comes to an end. What does not come to an end is theory building as a whole, which is an ongoing process. But it is so because, beyond every hypothesis, one can always find a deeper hypothesis by some non-deductive rule. The deeper hypothesis thus obtained is likely to alter the justification of the hypotheses found by the non-deductive inferences already made and it is possible to continue this process indefinitely, but only because this leads to a more comprehensive justification. The process in question produces an imperfect but useful knowledge because it brings about ever more plausible hypotheses, which result into an unlimited deepening of the problem.

15 The Purpose of Theory Building

In the light of the analytic view of theories, we may give an answer to the question: What is the purpose of theory building?

The purpose is twofold: first, to solve problems by finding hypotheses which are the key to the discovery of solutions; second, to justify solutions by showing that they follow from hypotheses which are plausible. Both purposes are essential to knowledge. The first purpose is essential to the extension of knowledge, the second purpose is essential to the validation of knowledge.

Neither purpose, however, can be achieved with certainty. For, non-deductive rules do not absolutely guarantee to produce hypotheses which are the key to the discovery of solutions. And even when they produce such hypotheses, they do not guarantee to justify the solutions absolutely, since the solutions will be only plausible, hence not absolutely certain.

On the other hand, this is no limitation, because there is no source of knowledge capable of guaranteeing truth or certainty. Plausible knowledge is the best we can achieve in all fields, including mathematics where absolutely certain knowledge cannot be reached by Gödel's second incompleteness theorem (see Cellucci 2017, Chap. 20). As Plato states, "certain knowledge is either impossible or extremely difficult to come by in this life," we can only "adopt the best and least refutable of human hypotheses, and embarking upon it as upon a raft, run the risk of sailing the sea of life" (Plato, *Phaedo*, 85 c 3–d 2).

Acknowledgements I wish to thank Reuben Hersh and an anonymous reviewer for their comments and suggestions.

References

Anderson, C. (2008). The end of theory: The data deluge makes the scientific method obsolete. *Wired Magazine*, 23 June. https://www.wired.com/2008/06/pb-theory/.

Bacon, F. (1961–1986). *Works*. Stuttgart–Bad Cannstatt: Frommann Holzboog.

Berman, J. (2013). *Principles of big data: Preparing, sharing, and analyzing complex information*. Amsterdam: Elsevier.

Bohr, N. (1913). On the constitution of atoms and molecules. *Philosophical Magazine 26*, 1–25, 476–502, 857–875.

Bohr, N. (1963). *Essays 1958–1962 on atomic physics and human knowledge*. New York: Interscience Publishers.

Caluce, C. S., & Longo, G. (2016). The deluge of spurious correlations in big data. *Foundations of Science*. https://doi.org/10.1007/s10699-016-9489-4.

Capozzi, M. (2013). *Kant e la logica I*. Naples: Bibliopolis.

Cellucci, C. (2013). *Rethinking logic: Logic in relation to mathematics, evolution, and method*. Dordrecht: Springer.

Cellucci, C. (2017). *Rethinking knowledge: The heuristic view*. Dordrecht: Springer.

Darwin, C. (1903). *More letters of Charles Darwin*. London: Murray.

Darwin, C. (2009a). *The origin of species by means of natural selection*. Cambridge: Cambridge University Press.

Darwin, C. (2009b). *The life and letters of Charles Darwin, including an autobiographical chapter*. Cambridge: Cambridge University Press.

Einstein, A. (2002). Induction and deduction in physics. In *Collected papers* (Vol. 7, pp. 108–109). Princeton: Princeton University Press.

Einstein, A. (2010). *Ideas and opinions*. New York: Crown.

Frege, G. (1960). *The foundations of arithmetic: A logico-mathematical enquiry into the concept of number*. New York: Harper.

Frege, G. (1967). *Begriffsschrift*, a formula language, modeled upon that of arithmetic, for pure thought. In J. van Heijenoort (Ed.), *From Frege to Gödel: A source book in mathematical logic, 1879–1931*. Cambridge: Harvard University Press.

Hempel, C. G. (1966). *Philosophy of natural science*. Englewood Cliffs: Prentice-Hall.

Hempel, C. G. (2001). *The philosophy of Carl G. Hempel: Studies in science, explanation, and rationality*. Oxford: Oxford University Press.

Kant I. (1992). *Lectures on logic*. Cambridge: Cambridge University Press.

Kepler, J. (1981). *Mysterium Cosmographicum*. New York: Abaris Books.

Kepler, J. (1992). *New astronomy*. Cambridge: Cambridge University Press.

Laudan, L. (1977). *Progress and its problems*. Berkeley: University of California Press.

Laudan, L. (1981). *Science and hypothesis: Historical essays on scientific methodology*. Dordrecht: Springer.

Malle, J.-P. (2013). Big data: Farewell to Cartesian thinking? *Paris Innovation Review*, 15 March. http://parisinnovationreview.com/articles-en/big-data-farewell-to-cartesian-thinking.

Newton, I. (1999). *The Principia: Mathematical principles of natural philosophy*. Berkeley: University of California Press.

Prawitz, D. (2014). The status of mathematical knowledge. In E. Ippoliti & C. Cozzo (Eds.), *From a heuristic point of view* (pp. 73–90). Newcastle upon Tyne: Cambridge Scholars Publishing.

Pólya, G. (1941). Heuristic reasoning and the theory of probability. *The American Mathematical Monthly, 48*, 450–465.

Pólya, G. (1954). *Mathematics and plausible reasoning*. Princeton: Princeton University Press.

Pólya, G. (1971). *How to solve it: A new aspect of mathematical method*. Princeton: Princeton University Press.

Popper, K. R. (2005). *The logic of scientific discovery*. London: Routledge.

Siegfried, T. (2013). Rise of big data underscores need for theory. *Science News*, 3 December. https://www.sciencenews.org/blog/context/rise-big-data-underscores-need-theory.

Striker, G. (2009). Commentary. In Aristotle, *Prior analytics*, Book I (pp. 67–246). Oxford: Oxford University Press.

Turing, A. M. (2004a). Systems of logic based on ordinals. In B. J. Copeland (Ed.), *The essential Turing* (pp. 146–204). Oxford: Oxford University Press.

Turing, A. M. (2004b). Letter to Max Newman, 21 April 1940. In B. J. Copeland (Ed.), *The essential Turing* (pp. 211–213). Oxford: Oxford University Press.

van Fraassen, B. (1980). *The scientific image*. Oxford: Oxford University Press.

van Fraassen, B. (1985). Empiricism in the philosophy of science. In P. Churchland & C. Hooker (Eds.), *Images of science: Essays on realism and empiricism* (pp. 245–308). Chicago: The University of Chicago Press.

van Fraassen, B. (1989). *Laws and symmetry*. Oxford: Oxford University Press.

van Fraassen, B. (2000). The false hopes of traditional epistemology. *Philosophy and Phenomenological Research, 60,* 253–280.

van Fraassen, B. (2008). *Scientific representation: Paradoxes of perspective*. Oxford: Oxford University Press.

Part II
Theory Building in Action

Discovering Cures in Medicine

Donald Gillies

Abstract The paper begins by suggesting that the discovery of a theory involves not just the formulation of a theory, but some degree of justification of the theory as well. This will be illustrated by the example of the discovery of the special theory of relativity by Einstein. The paper then goes on to apply this approach to medicine. The discovery of a cure involves first the formulation of a theory to the effect that such and such procedure will result in the disappearance of a disease or condition without unacceptable harm to the patient. The discovery is not complete until a theory of this form (a cure theory) has been confirmed empirically. The final section of the paper will illustrate this general view of discovery by two case histories. The first is the discovery of the statins. The second concerns the drug thalidomide.

Keywords Discovery · Justification · Einstein · Cures in medicine
Statins · Thalidomide

1 The Distinction Between Discovery and Justification

This paper is about discovering cures in medicine, and I will begin by saying something about discovery in general. In the writings of philosophers of science about discovery, there are two favourite themes. The first concerns the mysterious notion of serendipity. I will say something about this in Sect. 5. The second is the distinction between discovery and justification. Many philosophers of science think that this distinction has a fundamental importance. So let us begin by considering a classic statement of the distinction, which is to be found in Popper's (1934) *The Logic of Scientific Discovery*. Popper writes (1934, p. 31):

D. Gillies (✉)
University College London, London, UK
e-mail: donald.gillies@ucl.ac.uk

© Springer International Publishing AG 2018
D. Danks and E. Ippoliti (eds.), *Building Theories*, Studies in Applied Philosophy, Epistemology and Rational Ethics 41, https://doi.org/10.1007/978-3-319-72787-5_5

> ... the work of the scientist consists in putting forward and testing theories.

> The initial stage, the act of conceiving or inventing a theory, seems to me neither to call for logical analysis nor to be susceptible of it. The question how it happens that a new idea occurs to a man[1] - whether it is a musical theme, a dramatic conflict, or a scientific theory – may be of great interest to empirical psychology; but it is irrelevant to the logical analysis of scientific knowledge. This latter is concerned not with *questions of fact* (Kant's *quid facti?*), but only with questions of *justification* or *validity* (Kant's *quid juris?*). Its questions are of the following kind. Can a statement be justified? And if so, how? Is it testable? Is it logically dependent on certain other statements or does it perhaps contradict them?

So for Popper discovery and justification of scientific theories are treated by two different disciplines—discovery by empirical psychology, and justification by the logical analysis of scientific knowledge, i.e. by philosophy of science as it was understood in the days of the Vienna Circle.

Popper adds some emphasis to this position in the following passage (1934, p. 32):

> ... my view of the matter, for what it is worth, is that there is no such thing as a logical method of having new ideas, or a logical reconstruction of this process. My view may be expressed by saying that every discovery contains 'an irrational element', or 'a creative intuition' in Bergson's sense.

These passages from Popper contain some slightly curious features. It is surprising to find, in the second passage, Popper citing Bergson with approval. It is perhaps still more surprising to find Popper in the first passage emphasizing justification (Kant's *quid juris?*). Later in his life Popper was to declare that he opposed all forms of justificationism. Here, for example, is a passage from Popper's *Realism and the Aim of Science* (1983, p. 19):

> I assert ... that we cannot give any positive justification or any positive reason for our theories and beliefs. ... the belief that we can give such reasons ... is ... one that can be shown to be without merit. ... my solution to the central problem of justificationism ... is ... *unambiguously negative* ...

So Popper seems to have moved away from the discovery/justification distinction which he made in his 1934. I too once held that this distinction was a valid one, but have recently also moved away from it, though for reasons different from those of Popper. In fact I have been influenced here by Carlo Cellucci's account of discovery in mathematics (see Cellucci 2013).[2] Cellucci focuses on discovery in mathematics, while here I want to focus on discovery in the natural sciences and medicine. In both the natural sciences and medicine, I am assuming that the justification of a theory is given by the degree to which it is empirically confirmed or corroborated.[3] Instead of discovery and justification, therefore, I will speak of discovery and confirmation.

---

[1] 1934 was earlier than the rise of feminism.

[2] I benefited from an email correspondence with Carlo Cellucci just before writing this paper.

[3] In this paper I will use confirmation and corroboration as synonyms, though Popper uses them with different meanings.

Popper speaks of 'conceiving or inventing' a scientific theory. I will instead use the expression 'formulating a scientific theory'. My main point is that formulating a scientific theory is not the same as discovering a scientific theory. Once such a theory has been formulated, it is compared to the evidence available and also to new evidence which may be collected in order to assess it. Only if the theory is confirmed sufficiently by such evidence in order to persuade the majority of the relevant scientific community to accept it, can the theory be said to have been discovered. If the theory is instead strongly disconfirmed by the evidence, we do not say that it has been discovered, even though it has been formulated. So discovery of a scientific theory involves both formulation *and* subsequent confirmation. Thus discovery and confirmation overlap rather than being separate.

I still think, however, that there is a distinction between discovery and confirmation, and that we cannot reduce confirmation to discovery. This is because, once a theory has been discovered, scientists continue to compare it to empirical evidence, and such further comparisons may result in either confirmation or disconfirmation, even though this stage of confirmation or disconfirmation has nothing to do with the discovery of the theory. For example, Newton's theory was formulated by Newton and published by him in 1687. By the early 18th century, Newton's theory had been sufficiently strongly confirmed empirically to be accepted by the majority of the scientific community. By that stage, therefore, we can say that the discovery of the theory was complete. More than a century later in 1846, the discovery of Neptune gave additional confirmation to Newton's theory, but this confirmation has nothing to do with the original discovery of the theory which had occurred long before.

This completes my account of the relation between discovery and justification (confirmation) for scientific theories. In the next section, I will illustrate it with the well-known example of Einstein's discovery of special relativity. Then in the rest of the paper, I will apply this approach to the discovery of cures in medicine.

2   Example of Einstein's Discovery of Special Relativity

Most textbooks of special relativity begin by describing the Michelson-Morley experiment of 1887. Although this was important in providing evidence for Einstein's theory of special relativity, it was certainly not the stimulus, which produced it. The null result of the experiment was explained by the Lorentz-Fitzgerald contraction, and 18 years elapsed from 1887 until 1905 when Einstein published the special theory of relativity. The stimulus for special relativity was a series of remarkable discoveries which occurred in the decade before 1905.[4] In 1896 Becquerel discovered that uranium emits radiation. This phenomenon of radioactivity was investigated by the Curies from 1898, and they discovered radium. Rutherford in 1899 began his

---

[4]In this account of the discovery of special relativity, I have taken much information from Arthur I. Miller's classic (1981). My account is given in more detail in Gillies (2016).

investigations of radioactivity, and distinguished between α- and β-rays. Meantime J. J. Thomson had made the most important discovery of all, that of the electron. This particle seemed to be the bearer of electric charge, but its mass was soon estimated to be more than 1500 times less than that of the lightest particle hitherto known. In 1900 Becquerel connected J. J. Thomson's discovery with that of radioactivity by showing that β-rays were streams of electrons. The novelty of these discoveries should be emphasized. They had shown the existence of hitherto unknown entities, and the laws governing these entities were completely unknown.

It was natural then that the physics community should from 1900 put forward radically new ideas to explain these radically new phenomena. The first such idea turned out to be incorrect, but had many followers initially. This was the electromagnetic world picture which was proposed by Wien in 1900. In the 19th century, mechanics had been regarded as fundamental. Maxwell tried to explain electricity and magnetism in terms of mechanical models. Wien's idea was that one should do the opposite and explain mechanics in terms of electromagnetism. The hope was that matter would be ultimately reduced to electrical charges and the aether.

Applying these ideas to the electron, it seemed possible to explain at least part of its mass as originating from the electromagnetic field. This *electromagnetic mass* turned out to have rather curious properties. First of all it varied with the particle's velocity. Secondly it had a different value when the particle's velocity was in the same direction as its acceleration (*longitudinal mass*) from its value when the velocity was perpendicular to its acceleration (*transverse mass*). The electromagnetic world-picture naturally suggested that the mass of the electron might be entirely electromagnetic in character, and it was this hypothesis which an experimenter in Göttingen (Walter Kaufmann) set out to investigate.

Kaufmann borrowed from the Curies a radium chloride source only 0.2 mm thick. The β-rays, i.e. streams of electrons, from this source were found to move at velocities exceeding 0.9c, where c = the velocity of light (Miller 1981, p. 48). Thus Kaufmann was investigating a phenomenon where relativistic effects might become noticeable. He concluded in 1902, as a result of his experiments, that the electron's mass was purely electromagnetic in character. This was obviously a great success for the electromagnetic world picture.

A colleague of Kaufmann's at Göttingen was a theoretical physicist (Max Abraham) who was an adherent of the electromagnetic world picture. In the years 1902-3 he developed a theory in which the electron was a rigid charged sphere, whose mass was entirely electromagnetic in character. He calculated the longitudinal and transverse mass of such an electron. Indeed Abraham was the first to introduce the terms longitudinal and transverse mass.

In his 1905 paper, Einstein adopted an approach quite different from Abraham's. Instead of trying to reduce mechanics to electrodynamics, Einstein introduced a new mechanics, which differed from Newtonian mechanics, and was based on two postulates. The first of these was the principle of relativity, and the second the principle of the constancy of the velocity of light. From his new mechanics, Einstein derived formulas for the longitudinal and transverse masses, not just of the electron, but of any moving body. These formulas, however, differed from

Abraham's but agreed with the formulas which Lorentz had produced in 1904 from the hypothesis of a deformable electron. Kaufmann therefore decided to carry out another set of experiments to distinguish between the theory of Abraham on the one hand and those of Einstein and Lorentz on the other. He published his results in 1906 and these agreed with Abraham's formula for the transverse mass of the electron, but disagreed with the formula of Einstein and Lorentz.

Let us pause to analyse the situation in terms of our scheme regarding the discovery of scientific theories. Both Abraham and Einstein had formulated scientific theories. Abraham's was the theory of the rigid electron, and Einstein's the theory of special relativity. These theories had then been tested experimentally. Abraham's had been confirmed, and Einstein's disconfirmed. It should be added, however, that Kaufmann's experiments were not the only relevant evidence available. In the period 1902-4 some further aether drift experiments had been carried out using quite different set-ups from that of Michelson-Morley and achieving higher levels of accuracy. They had all produced a null result. Rayleigh had carried out an experiment in 1902 which achieved an accuracy of one part in 10^{10}. This experiment was repeated in 1904 by Brace who achieved an accuracy of one part in 10^{13}. Another quite different experiment was carried out in 1903 by Trouton and Noble, again with a null result (Miller 1981, p. 68). Abraham's theory, however, predicted a positive result within the limits of accuracy of Rayleigh and Brace. Einstein's theory naturally predicted a null result in all cases. At this stage, therefore, the evidence was mixed. Some of the evidence confirmed Abraham and disconfirmed Einstein, while other parts of the evidence disconfirmed Abraham and confirmed Einstein. I think it is fair to say that nothing had been discovered at that stage.

The evidence was soon, however, going to swing in favour of Einstein. In 1908, Bucherer carried out further experiments on electrons moving with high velocities, which avoided some of the flaws in Kaufmann's experiments. This time the results agreed with Einstein's formula for transverse mass, and disagreed with Abraham's. Within a decade or so, Einstein's theory of special relativity had come to be accepted by nearly all physicists, whereas Abraham's theory of the rigid spherical electron was largely forgotten.

We can therefore sum up as follows. Both Abraham and Einstein formulated scientific theories. However, Abraham's theory was disconfirmed by the evidence, whereas Einstein's was confirmed by the evidence and came to be accepted by the physics community. It is standard to say that Einstein discovered special relativity, but no one would say that Abraham discovered that the electron was a rigid sphere. This example shows clearly that the discovery of a scientific theory does not consists just of the formulation of that theory, but requires that the theory be sufficiently well confirmed by the evidence for it to be accepted. So discovery involves confirmation (justification) and the concepts of discovery and confirmation (justification) overlap rather than being separate.

3   Application of This Approach to the Discovery of Cures

We naturally think of cures as being cures of a disease, but I will here extend the notion a little by including cures of conditions as well as diseases. A condition is something like high blood pressure, which, although it may not be exactly a disease itself, is a causal factor for a disease. High blood pressure is a causal factor for heart disease, so that those with high blood pressure have an above average probability of developing heart disease. For this reason, it is considered wise to treat this condition, and try to lower high blood pressure either by changes in diet or by the use of drugs. As a result of such a treatment the patient will have a lower probability of going on to have heart disease.

We can distinguish three types of cure, namely (i) surgery, (ii) life style changes, particularly in diet and exercise, and (iii) drug treatments. Of course very often all three types of cure are combined. In this paper, however, I will focus on drug treatments. Now the discovery of a drug treatment must begin by the discovery of a substance, which will constitute the drug. For example, Fleming discovered a particular substance (penicillin), which was not known before. However, the discovery of the substance does not in itself constitute the discovery of a cure. Although Fleming discovered penicillin, he did not discover that it could be used successfully as an antibiotic for many bacterial infections. In fact, though he originally suggested in his 1929 paper on penicillin that this might be the case, he came to doubt whether penicillin could be a successful antibiotic. It was Florey and his team at Oxford, who showed later on that penicillin was an excellent antibiotic.[5]

The discovery that a particular substance (s) can successfully cure a disease (D) involves first the discovery of a possibly curative substance s, and then the development of what could be called a cure theory (CT), which shows how s can be used to cure D. A cure theory for s and D, or CT(s, D) has the following form:

CT(s, D): The substance s given in such and such a manner and quantities to patients of such and such a type will cure, or at least ameliorate, the disease, or condition D, *without unacceptable harm to the patient*.

I will now make a few comments on the last clause in italics, which is obviously of crucial importance. It is often possible to eliminate a disease using a drug, which, however, causes damage to the patient which is as bad, if not worse, than the disease itself. Ideally, of course, one would like drugs to act '*without any harm to the patient*', but sometimes it is necessary to tolerate quite a harmful side effect in order to eliminate a very serious disease. Chemotherapy is an example of this.

---

[5]For a fuller account of the two phases in the discovery of penicillin as an antibiotic, see Gillies (2006).

Still one always wants to ensure that the benefit to the patient of a drug in terms of cure outweighs the damage which the drug does. This is what is meant by the phrase 'without unacceptable harm to the patient'.[6]

Discovering the cure of a disease D by a drug treatment is thus equivalent to discovering a cure theory CT(s, D). Using our earlier analysis, this involves first formulating a cure theory, and confirming this theory to a level which makes it acceptable to the majority of medical practitioners. Sometimes a cure theory CT(s, D) may be disconfirmed rather than confirmed, but it can then turn out that a different cure theory CT(s, D') is confirmed. To put it another way, it may be that a substance s is not a successful cure for the disease or condition for which it was first introduced, but it may turn out to be successful cure for a completely different disease or condition. As we shall see in Sect. 5, this was the case for thalidomide. In the next Sect. 4, I will use the 'cure theory' framework to analyse the discovery of a less problematic drug treatment, namely the use of statins to lower cholesterol levels. This has proved to be one of the safest and most successful drug treatments to be introduced in the last thirty or so years.

4 Case History (I) Statins

Statins are a treatment for a condition, namely high levels of blood cholesterol. Unfortunately two different systems of units are used to measure the level of blood cholesterol. In the USA it is measured in milligrams per decilitre (mg/dL), whereas in Canada and the UK it is measured in millimoles per litre (mmol/L). One can convert the Canadian measure to that of the USA by multiplying by 38.6598.

Naturally there are debates about what constitutes a high level of blood cholesterol, but it is generally accepted nowadays that a level greater than 240 mg/dL (6.19 mmol/L) is too high. The problem with a high level of blood cholesterol is that it is generally held to be a causal factor for atherosclerosis (or 'hardening of the arteries'). Atherosclerosis consists in the formation of plaques in the walls of the arteries. These plaques are indeed largely composed of cholesterol. Atherosclerosis in turn often leads to one or more of a number of unpleasant diseases such as angina, strokes, deterioration of blood flow to the legs, or heart attack. A heart attack is usually caused by a blood clot forming at the surface of an artery affected by advanced atherosclerosis. In 25–35% of cases, the first attack is fatal. Quite a number of statistics have been collected, and they show that the higher an individual's blood cholesterol level, the greater the probability of that individual developing atherosclerosis and one or more of its characteristic consequences. For example, a study of healthy middle aged Americans showed that those with a blood

---

[6]The phrase 'without unacceptable harm to the patient' was introduced as a result of a comment by an anonymous referee. Previously I had used the phrase 'without significant harm to the patient', but, as the referee pointed out, this is contradicted by the case of chemotherapy. I am very grateful for this good criticism.

cholesterol level of over 260 mg/dL had 4 times the probability of having a heart attack compared with those whose blood cholesterol level was under 200 mg/dL (Keys and Keys 1963, p. 18). So those who have high levels of blood cholesterol are well-advised to take steps to reduce it.

There are basically two ways of reducing the level of blood cholesterol, namely (i) changes in diet, and (ii) taking a cholesterol reducing drug. The relations between diet and the level of blood cholesterol have been fairly well worked out. Some of the cholesterol in our body is directly absorbed from the food we eat. However, among the foods normally eaten, only egg yolks are particularly rich in cholesterol. So it is sensible to limit the number of eggs consumed in order to lower blood cholesterol. However, the main agent in raising blood cholesterol is saturated fat, the principal sources of which are meat and dairy products, such as butter, cream and cheese. In fact typically around 7 times more cholesterol is made in the body from dietary saturated fats than is absorbed as cholesterol from food. So to lower the level of blood cholesterol someone simply needs to eat less meat and dairy products. The traditional Japanese diets and Mediterranean diets had low levels of meat and dairy products, and sure enough epidemiological studies of those who ate such diets showed that they had low blood cholesterol levels and suffered very few heart attacks. The contemporary Western diet which is based on the Northern European dietary tradition, with the addition of a good deal of fast food, is of course high in saturated fats, and atherosclerosis and heart attacks are common among those who eat it. Regular consumption of cheeseburgers is a sure way of raising blood cholesterol.

It would thus seem simple for people to reduce their blood cholesterol. All they need to do is switch from a typical Western diet to some combination of traditional Japanese or Mediterranean diets. However, this is easier said than done. Many people are very conservative about what they eat, and reluctant to change their diet. This tendency is reinforced by fast food companies, who regularly advertise their products, such as cheeseburgers. Moreover, although the effect of diet on blood cholesterol level is always in the same direction, its magnitude varies a lot from person to person. Some people can eat quite large amounts of saturated fat without their blood cholesterol level rising very much, while others have high blood cholesterol levels even though they eat only small quantities of saturated fat. An extreme case of this latter group is constituted by those who suffer from a genetic disorder known as familial hypercholesterolemia. Due to the lack of an important gene, such people fail to develop the usual system for regulating blood cholesterol; whatever diet they eat, they have very high blood cholesterol levels, and, without treatment, can develop angina and have heart attacks while still children. For these reasons then, pharmaceutical companies started the search for drugs, which would reduce the level of blood cholesterol. This search resulted in the discovery of the statins.

Before going on to describe this discovery, I would like to make a technical point. So far, I have talked rather loosely of the level of blood cholesterol. This was what the original researchers in this field measured. Later, however, it was discovered that blood cholesterol is of two types, namely LDL and HDL cholesterol.

LDL (Low Density Lipoprotein) cholesterol is the bad cholesterol. It is the type of cholesterol which leads to atherosclerosis. HDL (High Density Lipoprotein) is good cholesterol. It is actually protective against atherosclerosis. So one should aim to reduce the level of LDL cholesterol, but not that of HDL cholesterol. It is normally considered that LDL cholesterol should be below 160 mg/dL (4.12 mmol/L).

Fleming had discovered penicillin when a petri dish containing colonies of staphylococci had become contaminated with a mould, later identified as *penicillium notatum*. Fleming noticed that the staphylococcus colonies had not developed near the mould, and concluded that the mould was producing a substance which inhibited this pathogenic bacterium. This substance he later named penicillin.

Following this example, many pharmaceutical companies started to conduct systematic investigations of substances produced by moulds and fungi to see whether they could find more effective antibiotics in this manner. In 1971, Akira Endo was conducting an investigation of this sort for the Japanese pharmaceutical company Sankyo in Tokyo.[7] He reasoned that it might be worth testing the products of fungi not just for antibiotic properties, but for their ability to reduce cholesterol levels. Cholesterol levels in the body are increased or reduced through the action of the enzyme HMG-CoA reductase, and Endo thought that there might be HMG-CoA inhibitors of microbial origin, because (Endo 1992, p. 1570): "certain microbes would produce such compounds as a weapon in their fight against other microbes that required sterols or other isoprenoids for growth. Inhibition of HMG-CoA reductase would thus be lethal to these microbes." So Endo had a strong heuristic guiding his search.

There is a certain irony in the fact that the most important cholesterol reducing drugs were discovered by a Japanese, since, because of the Japanese diet, high cholesterol was not a serious problem in Japan. However, Endo had worked for several years from 1965 in the USA on the problems of cholesterol synthesis. He knew a fast and cheap method of assaying for a substance's ability to inhibit cholesterol synthesis from labelled acetate in a cell-free system. Accordingly he and his colleague Dr. Masao Kuroda started testing fungal broths in 1971. Over a two year period, they tested over 6000 microbial strains, and then finally hit on a very powerful inhibitor of cholesterol synthesis, produced by the mould *Penicillium citrinum*. Endo called this substance ML-236B, but it was later named mevastatin— the first of the statins. Once again the *Penicillia* had produced something of great benefit to humanity.

By this stage, Endo had clearly formulated the hypothesis that ML-236B might be a useful cholesterol reducing drug for humans, but he cannot yet be said to have discovered that it was such a drug. There were many hurdles still to cross, and one of them might well have prevented a genuine discovery being made. So far it had been shown that mevastatin inhibited cholesterol synthesis in vitro, but would it also do so in vivo? Unfortunately when Endo tried out mevastatin on the usual

---

[7]This account of the discovery of the statins is based on Endo (1992), Brown and Goldstein (2004) and Steinberg (2007, pp. 176–193).

experimental animals (rats and mice), it turned out that it did not lower blood cholesterol levels (Endo 1992, p. 1573):

> Unexpectedly, the feeding of rats with a diet supplemented with 0.1% mevastatin for 7 days caused no changes in plasma cholesterol levels .... Plasma cholesterol was not lowered even when the agent was given to the animals at a dose as high as 500 mg/kg for 5 weeks. Furthermore, mevastatin was ineffective in mice, producing no detectable effects on plasma lipids at 500 mg/kg for 5 weeks.

Many researchers might have given up at that point, but the persistent Endo decided to try mevastatin out on other animals. Knowing that egg yolks are very high in cholesterol (a fact which was mentioned earlier), Endo reasoned that the level of cholesterol synthesis in hens which actively producing eggs would be high, so that it would be good to try out mevastatin on such hens. This time the experiment was a success (Endo 1992, p. 1574):

> We fed hens a commercial diet supplemented with 0.1% mevastatin for 30 days. As expected, plasma cholesterol was reduced by as much as 50%, while body weight, diet consumption, and egg production were not significantly changed throughout the experiments ....

> The success in the experiments in hens opened up an opportunity to conduct experiments in dogs and monkeys. In dogs, mevastatin reduced plasma cholesterol by 30% at a dose of 20 mg/kg and as much as 44% at 50 mg/kg .... β-Lipoprotein (LDL) was markedly reduced by mevastatin while α-lipoprotein (HDL) was not lowered but, rather, increased slightly.

Similar results were obtained with monkeys. Mevastatin appeared to reduce the bad (LDL) cholesterol significantly, while slightly increasing the good (HDL) cholesterol. If the results of these animal experiments applied to humans, then mevastatin, if it were safe, would become a most important drug. However, its effectiveness and safety for humans still remained to be confirmed before the discovery was complete.

In 1977 Endo gave a paper on drugs affecting lipid metabolism at a major conference in Philadelphia, but his presentation was poorly attended and seemed to arouse no interest. Luckily, however, two leading American researchers in the field (Brown and Goldstein) did realise the importance of Endo's work. They invited him to their laboratory in Dallas, and the three of them went on to write a joint paper.

Curiously another research group working for the pharmaceutical company Beecham's in England had isolated mevastatin from another strain of penicillium mould slightly before Endo. They were primarily interested in the possible anti-gonococcal properties of the molecule, and, when these proved not very strong, they abandoned their research on the compound. Like Endo, however, they tried out the compound, which they called compactin, on rats to see whether it reduced cholesterol levels, and like him, found that it did not. They were also in touch with Brown and Goldstein, and sent them 500 mg of compactin. However, according to Brown and Goldstein (2004, pp. 13–14):

> ... they warned us informally that compactin was not a useful hypocholesterolemic agent. They had given the drug to rats, and it didn't lower the cholesterol at all!

It seems that some persistence in the face of apparently negative evidence is important in drug discovery.[8]

But why did mevastatin not reduce the blood cholesterol levels in rats? Brown and Goldstein suggest (2004, p. 14) that this was because rats have most of their cholesterol in the form of HDL and very little LDL. At all events rats seem to thrive much better than humans on a diet rich in cholesterol and saturated fats. If only human beings could become more like rats, a diet of contemporary fast food would suit them perfectly!

Brown and Goldstein stimulated an American pharmaceutical company (Merck) to take an interest in statins, and they discovered their own statin (lovastatin) in 1979. Curiously Endo in Japan had isolated exactly the same compound though from a different mould. Lovastatin differs from mevastatin only because one hydrogen atom in mevastatin is replaced by a methyl group CH_3. This new agent was slightly more active in inhibiting HMG-CoA reductase.

By early 1980, then, the stage seemed set for the commercial development of statins, but now another set back occurred. It was reported that Sankyo had shown in a trial that mevastatin produced toxic effects in dogs. Some mystery surrounds this episode because the full details of the experiment were never made public. This what Endo says about the matter (1992, p. 1575):

> In mid 1980 ... mevastatin had been found to produce toxic effects in some dogs at higher doses in a long-term toxicity study. In this experiment, mevastatin was given to the animals at doses of 25, 100, and 200 mg/kg per day for 104 weeks. Although details of the experiment have not been reported, the purported toxicity was apparently due to the accumulated toxicity of the drug. It should be noted that mevastatin is effective in humans at as low as 0.2 mg/kg or less ...; thus a dose of 200 mg/kg given to dogs is 1000 times higher than the effective dose in man."

Sankyo abandoned work on mevastatin and lovastatin, and Merck was doubtful about whether to do the same. Eventually, however, the problem of treating patients suffering from familial hypercholesterolemia provided the stimulus for Merck to continue. Clinical trials carried out in Japan in 1979 had indicated that mevastatin was an effective treatment for some patients suffering from familial hypercholesterolemia. As no other treatment was available for such patients, it was argued that it was justified to carry out clinical trials of lovastatin with them. The FDA gave its permission in 1987, and the clinical trials showed that lovastatin was indeed an effective treatment and had no severe side effects. Statins have since then been prescribed to millions of patients and they have proved to be among the very safest drugs. So, Sankyo's result concerning toxicity in dogs had given rise to unfounded alarm.

The cure theory for statins states that when given in appropriate doses, they reduce the blood cholesterol level without causing unacceptable harm to the patient. As we have seen, this cure theory seemed to be initially disconfirmed by two negative results. However, further tests in 1987 showed that the cure theory was

---

[8]This principle is also illustrated by the case of Fleming and penicillin. See Gillies (2006).

sound after all. On our analysis therefore, the statins were not discovered in 1973 when Endo isolated the first statin, but only in 1987 when the cure theory for statins was sufficiently well confirmed to convince the medical community.

That concludes my account of the discovery of statins, and, in the next section, I will turn to the case of thalidomide where events took a very different course. Initially the cure theory for the first use of thalidomide seemed to be strongly confirmed by the evidence, but then it was disconfirmed in a striking and horrible manner.

5   Case History (II) Thalidomide

Thalidomide was originally proposed as cure for a condition rather than a disease.[9] This condition consisted of anxiety and insomnia. It was quite widespread in the 1950s—perhaps not surprisingly. The Second World War had ended a few years before in 1945, and many had had, during the War, the kind of horrifying and traumatic experiences which leave lasting effects. The War had ended with the explosions of two atomic bombs, and had been succeeded by the Cold War, which, in the nuclear age, seemed to threaten the continued existence of mankind. Moreover, the Cold War was not entirely cold, and the Korean War raged from 1950 to 1953.

Given this general background, it is not surprising that the use of sedatives was very common. In Britain it was estimated that a million people took sedatives, and in the USA that one person in seven took sedatives (Brynner and Stephens 2001, p. 4). These sedatives were barbiturates, and an overdose could be fatal. There were many tragic accidents in which children took several of their parents' sleeping pills, thinking them to be sweets, and died as a result. The pharmaceutical industry was looking for a sedative, which could be taken in large quantities without ill effects. In 1954, a German pharmaceutical company, Chemie Grünenthal, came up with what they thought was the answer, a drug which they named thalidomide.

Thalidomide was indeed a powerful sedative and effective sleeping pill for humans. Moreover it appeared in animal trials to be completely non-toxic. In such trials, animals are fed a chemical to determine the dosage at which half of the tested animals die; this is called LD_{50}. In the case of thalidomide, they could not find a dose large enough to kill rats. This was most unusual. Moreover, the drug appeared to be non-toxic for the further animals tested, namely mice, guinea pigs, rabbits, cats, and dogs (Brynner and Stephens 2001, p. 9).

---

[9]This account of thalidomide is largely based on Brynner and Stephens 2001 book: *Dark Remedy*. This is an excellent book, which I strongly recommend. It was written by a historian with a M.A. in Philosophy (Rock Brynner), and a Professor of Anatomy and Embryology who had spent twenty-five years researching into thalidomide (Trent Stephens). The book contains a highly informative account of all aspects of the thalidomide case—scientific, methodological, sociological, political, and historical—including many aspects, which I will not be discussing in this paper.

Grünenthal concluded that thalidomide was much safer than the barbiturates currently used as sedatives. Indeed, thalidomide was believed to be so safe that it was released as an over the counter drug in Germany on 1 October 1957. The advertising campaign for thalidomide stressed that thalidomide was much safer than the other sedatives currently available. Even a determined suicide could not take enough of it to die, and tragic accidents with children would no longer be possible (Brynner and Stephens 2001, pp. 14–15). The cure theory for thalidomide was that it was an effective sedative and sleeping pill which could be used safely by anyone. There was some, though not enough, empirical support for this theory when thalidomide was released, but it was soon to refuted in a tragic fashion.

Thalidomide, as we now know, produces horrifying birth defects in babies when it is taken by pregnant women. It also has another less grim but still very unpleasant side effect when taken either by men or women. This is peripheral neuropathy, that is to say damage to peripheral nerves rather than those in the brain. Nerve damage occurred in between 5 and 20% of those who took thalidomide for several months. It usually affected the feet and lower part of the legs, producing numbness, pain, loss of balance, and difficulty in walking (Brynner and Stephens 2001, pp. 24–25). Sometimes the hands were affected as well. Unfortunately the damage proved to be irreversible when the patient ceased to take the drug.

When thalidomide's ability to cause birth defects and peripheral neuropathy came to light, thalidomide was withdrawn in Germany in November 1961, and, though there were unfortunate delays, in the rest of the world by the end of 1962. During the 4 to 5 years when it was on the market, thalidomide produced around 40,000 cases of peripheral neuropathy, and between 8000 and 12,000 deformed babies of whom about 5000 survived beyond childhood (Brynner and Stephens, 2001, p. 37).

The first question which the thalidomide case raises is whether the disaster could have been avoided by testing the drug more severely before putting it on the market. As we have seen, some animal trials were conducted on the drug, but not very many. However, it is not clear that more animal trials would have brought the problems of thalidomide to light. The difficulty is that thalidomide in small doses does not cause birth defects in any animals other than primates. After the disaster, birth defects produced by thalidomide were shown in rabbits, but only with doses 150 times greater than the therapeutic dose. Moreover, it was difficult to show this effect. Dr. Helen Taussig, a leading American researcher, said that she was unable to obtain abnormalities in baby rabbits using thalidomide, because the massive doses needed produced abortions (Brynner and Stephens 2001, p. 13).

More consideration of the evidence of mechanism would, however, have helped to avoid the disaster. This is another illustration of the importance of evidence of mechanism in assessing drug treatments.[10] Brynner and Stephens (2001, pp. 12–13) write:

---

[10]On this topic, see Clarke et al. (2014).

... it had been known since 1955 that any substance with a molecular weight of less than
1000 could cross the placenta and enter the fetal blood. The molecular weight of
thalidomide is 258. ... it had been demonstrated in 1948 that the dye known as trypan blue
could cause birth defects in rat embryos, whereas the mother rats exhibited no symptoms.

Evidence of this kind should have raised doubts in the minds of the regulators
whose job it was to approve thalidomide. Indeed perhaps it did in some cases.
Although thalidomide was approved, not only in West Germany and many other
countries (the British Commonwealth, Italy, Japan), it was not approved in some
countries, notably East Germany and the USA.

A major disaster with thalidomide was narrowly avoided in the USA. A leading
American pharmaceutical company (Richardson-Merrell) was preparing to launch
thalidomide in March 1961. 10 million tablets were already manufactured, when
they submitted an application to the American regulatory authority (the FDA) on
8 September 1960. It was expected that the application would go through with no
problems, but, unexpectedly, difficulties were raised by the FDA officer assigned to
the case (Dr. Frances Kelsey). Kelsey delayed granting approval until 29 November
1961 when Grünenthal withdrew the drug in West Germany. After that, it was
obvious that thalidomide would not be approved in the USA. In August of the next
year, Kennedy presented Dr. Kelsey with the President's Award for Distinguished
Federal Civilian Service (Brynner and Stephens 2001, p. 55). Obviously it was
richly deserved.

Now the interesting fact here is that, before joining the FDA, Dr. Kelsey had
carried out research with her husband on malaria. Among other things they had
examined the effects of quinine on pregnant rabbits, and discovered that quinine is
toxic to the fetus, but not to the mother (Brynner and Stephens 2001, p. 45).

After the thalidomide disaster, it would be natural to suppose that thalidomide
would be banned forever. Instead something very surprising occurred. It was discovered that thalidomide was a cure for some terrible diseases for which previously
there had been no remedy. The man who made this discovery was Dr. Jacob
Sheskin. Dr. Sheskin was Jewish, but had managed to survive the holocaust in the
ghetto of Vilna. After the war, he emigrated to Venezuela, and, as leprosy was still
common there, he specialised in that disease. Later he resettled in Israel, where he
became director of the Jerusalem hospital for leprosy.

Leprosy is caused by the *Mycobacterium leprae*, which is similar to the
Mycobacterium tuberculosis, and, like the tubercle bacillus, is hard to treat with
antibiotics. In the 1960s, leprosy could be controlled, though not completely cured,
using some antibiotics called sulphones. These are, however, useless against a
severe inflammatory complication of the disease, which occurs in about 60% of the
worst forms of the disease. This complication is called *Erythema Nodosum
Laprosum* (or ENL) by Brynner and Stephens (2001, p. 122), but is referred to by
Sheskin himself as Lepra Reaction (1975). The symptoms are large, persistent, and
very painful weeping boils all over the body; severe inflammation of the eyes that
often leads to blindness; and severe pains in the joints and abdomen, as well as
headaches. The pain is very intense, despite the injections of morphine and other

pain killers several times a day which were made in the early 1960s. Patients can neither eat nor sleep, and become emaciated.

In 1964, a critically ill patient with this condition was sent to Dr. Sheskin by the University of Marseilles. His condition is described by Brynner and Stephens, as follows (2001, p. 122):

> The man had been bedridden for nineteen months, and by this time, on the verge of death, he was almost deranged from the unremitting pain that had denied him sleep for weeks; doctors in France had tried every existing sedative, but nothing he had been given helped for more than an hour.

Sheskin wondered whether any sedative, which had not yet been tried, could be given to this patient. Thalidomide had been withdrawn by this time, but there was still a bottle of 20 tablets in the hospital. Sheskin knew that thalidomide was a powerful sedative and sleeping pill. Of course he also knew that it was a banned drug, but, in the circumstances, there seemed no harm in giving it a try. He therefore gave the patient two thalidomide tablets.

The results seemed almost miraculous. The patient slept soundly for twenty hours, and, upon waking, was well enough to get out of bed without assistance. Moreover with further doses of thalidomide his sores began to heal. This continued as long as thalidomide was given. Six other patients in the hospital were then treated with similarly excellent results (Brynner and Stephens 2001, p. 123).

Sheskin was scientifically minded, and he realised that, to convince his peers, he would have to perform some control trials with thalidomide. He therefore returned to Venezuela, where thalidomide was still available, and carried out a trial with it on 173 of the patients whom he had treated earlier. The result of this trial was published in 1965. Sheskin went on to organise further trials in the next decade, and sums up the results in his 1975 as follows (p. 575):

> A double-blind study performed in 173 leprosy reactions indicated therapeutic effectiveness in 92% of the trials in which thalidomide was given.

> A survey which I performed of the use of thalidomide in lepra reaction of lepromatous leprosy showed successful results in 99% of the cases. The study involved 62 therapeutic centers in 5 continents, covering 4552 patients of different ages, sexes, and races, living in different climates and having different ways of life and dietary habits. All signs and symptoms of lepra reaction of lepramatous leprosy showed improvement during the first 24-48 hours, and total remission was completed during the second week of treatment.

Evidence of the mechanism by which thalidomide cures ENL was then found by Dr. Gilla Kaplan. One factor in the immune system is known as Tumour Necrosis Factor-alpha or TNF-α. This is useful in suppressing tumours. However, in ENL, as in many inflammatory conditions, the immune system over-reacts, and becomes itself the cause of the symptoms. Kaplan showed that patients with ENL had very high levels of TNF-α in their blood and lesions. She went on to show that thalidomide could reduce TNF-α levels by as much as 70% in vitro, and tests on ENL patients showed that a similar reduction occurred in vivo as well. Thus by the early 1990s, the mechanism of action of thalidomide in curing ENL has been established. As in other cases of discovery, the discovery of the cure for the lepra reaction in lepromatous leprosy involved confirmation of the cure theory.

Sheskin's discovery is often referred to as a serendipitous one. In Gillies (2014, 2015), I discuss serendipity in some detail, and argue in favour of adopting the following definition. Serendipity consists in "looking for one thing and finding another". Sheskin's discovery is not strictly serendipity if we adopt this definition, because Sheskin was looking for sedative, which would make his patient sleep, and he found what he was looking for. However, his discovery is an example of what I call in my 2014 *additional serendipity*. In additional serendipity, the researcher does discover what he or she was looking for, but, in addition, discovers something else unexpected. Sheskin did discover what he was looking for, namely a sedative which would make his patient sleep, but, in addition, he discovered something else unexpected, namely that the drug cured his patient's symptoms.

Despite this striking results regarding the curative properties of thalidomide in what was an otherwise incurable condition, the FDA were understandably reluctant to approve the drug. However, in 1998, the FDA did approve thalidomide for the treatment of ENL in leprosy. The approval was given only under very strict conditions relating to possible side effects. First of all patients had to be monitored for signs of peripheral neuropathy. Secondly patients had to follow a strict protocol to ensure that they did not become pregnant, while taking the drug. The precautions were more complicated than might at first be thought. Research had shown that thalidomide produced birth defects when taken early on in pregnancy, between the twentieth and thirty-sixth day after conception, and many women might not realise they were pregnant at this stage. Moreover, thalidomide could affect the semen. So men had to take precautions as well as women.

In fact there was a further breakthrough in the treatment of leprosy in the 1980s. Improved antibiotics for the disease have been developed, and, most importantly, a cocktail of several different antibiotics is now given, as is done in the treatment of tuberculosis. With this multi-drug therapy, leprosy has become curable in most cases. The need for thalidomide in the treatment of leprosy is thus reduced.

Brynner and Stephens remark (2001, p. 142) on: "an unusual feature of U.S. drug regulations. Once a medication has been approved for treating one condition, doctors may then legally prescribe it *for any other condition*." In fact thalidomide has been used in the treatment of no less than 130 different conditions. One possible reason for this is that thalidomide, when it is metabolized by the liver, produces over a 100 breakdown products. Some of these may be responsible for reducing TNF-α levels, others for producing birth defects, and others still for damaging the peripheral nerves.

This is illustrated by the use of thalidomide to treat multiple myeloma, a cancer of the bone marrow. Myeloma involves what is known as *angiogenesis*, which means the development of new blood vessels. Now thalidomide has been shown to be an antiangiogenic agent, and, though the matter is still uncertain, this may explain its effectiveness in the treatment of multiple myeloma, since other inhibitors of TNF-α do not have any effect on the disease (Brynner and Stephens 2001, p. 193).

6 Conclusions

The thalidomide case is a good illustration of many of the themes of this paper. Thalidomide shows that discovering the cure for a disease (or condition) is not just a matter of discovering a potentially curative substance s, but of discovering a cure theory CT(s, D) for that substance and a disease or condition D.

The original cure theory for thalidomide [CT(s, D)] was that it was a sedative safe for general use in treating anxiety and insomnia. CT(s, D) was refuted in a tragic fashion, but surprisingly a new cure theory was developed CT(s, D′), where the disease was D′ was now ENL in leprosy, and where the theory involved very rigorous conditions for using the drug, designed to avoid its known negative side effects.

Sheskin formulated the new cure theory for thalidomide, but he recognised that his discovery would only be complete if he obtained evidence in its favour, which was sufficient to convince the medical community. Sheskin himself produced statistical evidence from control trials of patients. Later on Kaplan produced evidence concerning the mechanism by which thalidomide cured the lepra reaction in lepromatous leprosy. This is a good illustration of how discovery involves not just formulation of a theory but its justification. In medicine justification takes the form of empirical confirmation of the theory.

Acknowledgements I am grateful to AHRC for supporting this research as a part of the project *Evaluating Evidence in Medicine* (AH/M005917/1). An earlier version of the paper was read at a workshop on 'New frontiers for evaluating evidence in medicine' organised as part of this project in University College London on 20 June 2016. I received many helpful comments on this occasion, and also when the paper was read in Rome on 16 June 2016 at the international conference on 'Building Theories. Hypotheses & Heuristics in Science.' I would also like to thank two anonymous referees whose comments led to several revisions of the paper.

References

Brown, M. S., & Goldstein, J. L. (2004). A tribute to Akira Endo, discoverer of a "Penicillin" for cholesterol. *Atherosclerosis Supplements, 5,* 13–16.

Brynner, R., & Stephens, T. (2001). *Dark remedy: The impact of thalidomide and its revival as a vital medicine.* New York: Basic Books.

Cellucci, C. (2013). *Rethinking logic: Logic in relation to mathematics, evolution, and method.* Berlin: Springer.

Clarke, B., Gillies, D., Illari, P., Russo, F., & Williamson, J. (2014). Mechanisms and the evidence hierarchy. *Topoi, 33*(2), 339–360.

Einstein, A. (1905). Zur Elektrodynamik bewegter Körper, *Annalen der Physik, 17,* 891–921. English edition: On the Electrodynamics of Moving Bodies' in Miller (1981) pp. 392–415.

Endo, A. (1992). The discovery and development of HMG-CoA reductase inhibitors. *Journal of Lipid Research, 33,* 1569–1582.

Fleming, A. (1929). On the antibacterial action of cultures of a penicillium, with special reference to their use in the isolation of *B. Influenzae, British Journal of Experimental Pathology, 10,* 226–236.

Gillies, D. (2006). Kuhn on discovery and the case of penicillin. In W. J. Gonzalez & J. Alcolea (Eds.), *Contemporary perspectives in philosphy and methodology of science* (pp. 47–63). Spain: Netbiblo.

Gillies, D. (2014). Serendipity and mathematical logic. In E. Ippoliti & C. Cozzo (Eds.), *From a heuristic point of view: Essays in honour of Carlo Cellucci* (pp. 23–39). Newcastle upon Tyne: Cambridge Scholars Publishing.

Gillies, D. (2015). Serendipity and chance in scientific discovery: Policy implications for global society. In D. Archibugi & A. Filipetti (Eds.), *The handbook of global science, technology, and innovation* (pp. 525–539). New Jersey: Wiley.

Gillies, D. (2016). Technological Origins of the Einsteinian Revolution. *Philosophy and Technology, 29*(2), 97–126.

Keys, A., & Keys, M. (1963). *Eat well and stay well.* New York: Doubleday.

Miller, A. I. (1981). *Albert Einstein's special theory of relativity. Emergence (1905) and early interpretation (1905–1911).* Boston: Addison-Wesley.

Popper, K. R. (1934). *The logic of scientific discovery.* 6th Revised Impression of English Translation, Hutchinson, 1972.

Popper, K. R. (1983). *Realism and the aim of science.* Paris: Hutchinson.

Sheskin, J. (1975). Thalidomide in lepra reaction. *International Journal of Dermatology, 14*(8), 575–576.

Steinberg, D. (2007). *The cholesterol wars: The skeptics versus the preponderance of evidence.* Cambridge: Academic Press.

The Product Guides the Process: Discovering Disease Mechanisms

Lindley Darden, Lipika R. Pal, Kunal Kundu and John Moult

Abstract The nature of the product to be discovered guides the reasoning to discover it. Biologists and medical researchers often search for mechanisms. The "new mechanistic philosophy of science" provides resources about the nature of biological mechanisms that aid the discovery of mechanisms. Here, we apply these resources to the discovery of mechanisms in medicine. A new diagrammatic representation of a disease mechanism chain indicates both what is known and, most significantly, what is not known at a given time, thereby guiding the researcher and collaborators in discovery. Mechanisms of genetic diseases provide the examples.

Keywords Discovery · Mechanism · Diagrams · Genetic disease

1 Introduction

While physicists often represent theories as sets of mathematical laws, biologists usually represent general knowledge with schematic representations of mechanisms. Biologists and medical researchers seek mechanisms because knowing a mechanism facilitates explanation, prediction, and control. A theme in work on discovering mechanisms is captured by the slogan—the product guides the process. The thesis is that characterizing a mechanism (the product) provides resources to guide the reasoning in its discovery (the process) (Craver and Darden 2013).

Recent philosophical analysis of mechanisms provides resources to aid the discovery of mechanisms. This work is being applied to the discovery of mechanisms in medicine. When the goal is to discover a disease mechanism, the nature of

L. Darden (✉)
Department of Philosophy, University of Maryland College Park, 1102A Skinner,
4300 Chapel Lane, College Park, MD 20742, USA
e-mail: darden@umd.edu

L. R. Pal · K. Kundu · J. Moult
Institute for Bioscience and Biotechnology Research, University of Maryland,
9600 Gudelsky Drive, Rockville, MD 20850, USA

© Springer International Publishing AG 2018
D. Danks and E. Ippoliti (eds.), *Building Theories*, Studies in Applied Philosophy,
Epistemology and Rational Ethics 41, https://doi.org/10.1007/978-3-319-72787-5_6

the product—the kind of disease mechanism—guides the process of searching for it. The kinds of products to be discussed here are representations of genetic disease mechanisms. In such diseases, genetic variants play a major role, together with environmental effects. The process is the reasoning to discover such mechanisms.

We develop a new graphical interface to aid medical researchers in hypothesizing and representing genetic disease mechanisms. We illustrate its use here in detailed diagrams of genetic mechanism schemas. The three examples are for a monogenic disease chain (cystic fibrosis), a cancer disease chain (affecting DNA mismatch repair), and one complex trait disease chain (one chain for one of many mutations for one of the loci associated with Crohn's disease).

This paper first summarizes recent work in mechanistic philosophy of science (e.g., Machamer, Darden and Craver 2000; Bechtel and Abrahamsen 2005; Glennan and Illari 2017). As philosophers have shown, diagrams of mechanism schemas play important roles in abstractly representing the product to be discovered and guiding the process of discovery (e.g., Craver and Darden 2013; Abrahamsen and Bechtel 2015). A key idea is to sketch both what is known and what is not known at a given time. Black boxes in the sketch indicate where to fill in missing mechanism components. The next section of this paper reviews the application of the mechanistic perspective in the philosophy of medicine (e.g., Thagard 1998, 1999, 2003; Darden 2012; Plutynski 2013). Then, we show how abstract mechanism chain diagrams serve to represent what is known or not known in genetic disease mechanisms. By depicting the state of knowledge about the genetic mechanism at a given time, the diagram perspicuously represents gaps in knowledge, namely, the sites of ignorance that researchers seek to remove. A set of heuristic questions provides guidance in filling the gaps. Three example diagrams of disease mechanism chains illustrate our new framework. We contrast our framework with two other graphical representation schemes. Finally, we propose future work, including plans for a web-based, graphical system that facilitates easy drawing and sharing of the individual mechanism chains, as well as discovery of interactions among them.

2   Mechanistic Philosophy of Science

Philosophers have been working for over twenty years to develop what is called the "new mechanistic philosophy of science" (Bechtel and Richardson 1993; Glennan 1996; Machamer, Darden and Craver 2000). This work calls attention to the importance of the search for mechanisms in biology and other disciplines, characterizes the nature of mechanisms, and compiles hindsight about the reasoning strategies used in the discovery of mechanisms (summarized in Craver and Darden 2013).

The discovery of a mechanism typically begins with a puzzling phenomenon. When the goal is to find what produces the phenomenon, then one searches for a mechanism. That decision rules out other parts of a large search space. One is not seeking merely a set of correlated variables. One is not seeking an economical equation that describes the phenomenon, although such an equation can provide a constraint in the search for a mechanism (Craver 2008; Bechtel and Abrahamsen 2013). One is not seeking a law from which a description of the phenomenon can be derived. One is not merely seeking a relation between one cause and the phenomenon as the effect, although such a relation provides clues about mechanism components (Darden 2013). Nor is one merely seeking to find a pathway, characterized by nodes and unlabeled links which do not depict the activities that drive the mechanism. Rather, one is attempting to construct a mechanism schema that describes how entities and activities are spatially and temporally organized together to produce the phenomenon.

Employing a specific characterization of a mechanism provides guidance in discovery. One oft-cited mechanism characterization is this: "Mechanisms are entities and activities organized such that they are productive of regular changes from start or set up to finish or termination conditions" (Machamer, Darden and Craver 2000, p. 3). The goal in mechanism discovery is to find the entities and activities, to describe how they are organized, and to show how that productively continuous organization produces the phenomenon of interest. This characterization directs one to ask: What are the set up and finish conditions? Is there a specific, triggering start condition? What is spatially next to what? What is the temporal order of the steps? What are the entities in the mechanism? What are their structures? What are the activities that drive the mechanism? What are their range and their rate? How does each step of the mechanism give rise to the next? What are the activity enabling properties that make possible the next step? What are the activity signatures (properties of an entity or group of entities in a subsequent step) that show the kinds of activities that operated in the previous step to produce them? How was each step driven by the previous one? What is the overall organization of the mechanism: does it proceed linearly or is the mechanism perhaps cyclic (with no clear start and stop), or is it organized with feedback loops, or does it have some other overall organizational motif? Where is it spatially located? In what context does the mechanism operate and how is it integrated with other mechanisms? These kinds of questions show how the nature of the product provides desiderata that guide the process of its discovery.

Mechanism schemas are representations of mechanisms. A "schema" (sometimes called a "model" of a mechanism) abstractly represents the structure of a target mechanism. Here is an example of a very abstract schema for the mechanism of protein synthesis: DNA $\rightarrow$ RNA $\rightarrow$ protein. Such schemas are often depicted in diagrams. William Bechtel and his collaborators (Sheredos et al. 2013; Abrahamsen and Bechtel 2015; Abrahamsen et al. 2017) discuss the many "visual heuristics" that diagrammatic representations of mechanism enable. They envisage biologists as reverse engineers, trying out various designs to spatially represent the interacting components of the mechanisms being discovered. The diagrams make salient specific aspects of the organization and operation of the mechanisms.

Schemas vary from one another along several dimensions: sketchy to sufficiently complete, abstract to specific, small to general scope of applicability, and possible to actual (Craver and Darden 2013, Chap. 3). A goal in discovering a mechanism is to convert an incomplete sketchy representation into an adequate one for the purpose at hand. Incomplete sketches indicate where black (unknown components) and grey (only functionally specified) boxes need to be filled in order to have a productively continuous schema in which it is clear how each step gives rise to the next. During the construction phase of discovery, moving from a sketch to a sufficiently complete schema allows one to work in a piecemeal fashion; one can work on one part of the mechanism at a time while leaving other parts as black or grey boxes. Because one is attempting to reveal the productive continuity of a mechanism from beginning to end, what one learns about one step of the mechanism places constraints on what likely has come before or what likely comes after a given step.

Abstraction comes in degrees and involves dropping details; specification involves adding details all the way to instantiation, with sufficient details to represent a productively continuous mechanism from beginning to end. A goal in discovery is to find a schema at a given degree of abstraction, from a very abstract type of schema with few specified components to a fully instantiated one for a particular case. For example, the schema DNA → RNA → protein is very abstract. Steps are condensed in this spare representation. However, any given step could be instantiated with specific details if needed for the project at hand. A more detailed schema would begin with a particular coding DNA sequence, show the transcription to complementary messenger RNA, and proceed through the well-known steps of reading the genetic code to order the amino acids in a particular protein.

The desired degree of abstraction depends on the purpose for which the mechanism is sought. Although degree of abstraction is an independent dimension from the scope of the domain to which the schema applies, more abstract schemas (if they have any instances at all) may have a wider scope of applicability. Hence, when the goal of the discovery process is to find a very generally applicable mechanism schema, it is likely to be represented at a high degree of abstraction, as in the above schema for protein synthesis.

The move from how possibly to how plausibly to how actually is driven by applying strategies for evaluation, such as experimental testing, and strategies for anomaly resolution, such as localizing faults and revising the schema. Ideally one wishes to find empirical evidence for each step in the mechanism (Craver and Darden 2013, Chaps. 6–9).

Consider the example of mechanisms connecting a gene mutation to a disease phenotype. One starts with the beginning point, e.g., a particular gene mutation, and a characterization of the disease phenotype (e.g., a set of symptoms). At the outset, between the gene/gene mutation and the phenotypic character is a black box. Having evidence of an association between a beginning point (the gene mutation) and the end point (the disease phenotype), the discovery task is to fill in the black box to some degree of detail. For example, if the goal is to replace an identified mutant gene during gene therapy, then it may be unnecessary to find all the intervening steps in the mechanism. A highly abstract schema may be sufficient to guide the work to find and replace the faulty gene. However, if the goal is to design a therapy to alter an entity or activity in a downstream mechanism site, then specific details become important: e.g., one may need to find the three-dimensional structure of a protein and identify its active site or locate the effect of an environmental factor.

A given gene to phenotype mechanism has entities of different size levels, beginning with the macromolecular DNA, proceeding through protein synthesis, which employs ribosomes (particles in the cytoplasm, composed of both proteins and ribosomal RNAs), and on to, in some cases, ever larger level cell organelle, membrane, cell, tissue and organ components. The appropriate size level depends on what the working entities are in the steps of the mechanism, on how the phenotype is characterized, and how much detail is needed for a given project. Hence, a single gene to phenotype mechanism likely has entities at many different size levels. (For more on the difference between size levels and mechanism levels, see Craver 2007, Chap. 5; Craver and Darden 2013, pp. 21–22)

The mechanism discovery process has at least four aspects: characterizing and recharacterizing the phenomenon, constructing a schema, evaluating the schema, and revising the schema (Darden 2006, Chap. 12; Craver and Darden 2013, Chaps. 4–9). These are often pursued in parallel and in interaction with one another. Strategies for mechanism schema construction are the most relevant here. One localizes where the mechanism operates. For gene to phenotype mechanisms, the mechanism starts with a DNA sequence; what the final stage is depends on the characterization of the phenotype. Thus, the overall structure of the mechanism to be discovered begins with DNA and ends with a phenotype. If a library of types of mechanism components is available, then those types of components become candidates to be specialized to construct steps of the target schema. For example, the module of protein synthesis is a likely module to use in an early step in a gene to phenotype mechanism. The strategy of forward/backward chaining allows the mechanism chain builder to reason forward from one step to the following step or backward to a likely previous step. Activity enabling properties in the previous step indicate possible types of mechanism modules to come. Activity signatures indicate what possibly came before, because once a specific kind of activity has operated to change the state of the next step it leaves specific traces (signatures). For example, a polarly charged DNA base is available to bond to its complementary base in the next step of the mechanism of DNA replication. Because hydrogen bonding leaves weakly bonded molecular components, when such a signature is detected, one can conclude that polar bonding occurred in a previous step (Darden 2002; Darden and Craver 2002; Craver and Darden 2013, Chap. 5).

With this synopsis of some key features of previous work on mechanisms in hand, we now turn to discovery of disease mechanisms. Here too, we argue, the nature of the product guides the reasoning process to find it. The product is a schema representing the steps in a target disease mechanism. The process is the reasoning by a chain builder to construct a diagram to represent the steps, and, while doing so, to fill black boxes to remove uncertainties.

3 Disease Mechanisms

In medicine, the following general types of mechanisms are of interest:

(i) The "normal" biological mechanism (noting that what is "normal" can nonetheless vary from person to person)

(ii) The general disease mechanism, which aids in finding sites for therapy and designing therapeutic treatments

(iii) The specific disease mechanism in an individual patient, which may aid choosing an effective therapy

(iv) The general mechanism of action of a drug or other therapeutic agent

(v) The specific mechanism of action of drug or therapy in an individual patient, given their genetic makeup and personal history

(vi) Possible mechanisms to account for side effects of therapies on other bodily mechanisms

Philosophers of medicine are participating in a lively debate about the role that knowledge of the mechanisms of the action of therapies (iii–vi above) should play in evidence based medicine. The debated issue is this: is evidence of the effectiveness of a therapy from randomized clinical trials sufficient to show the efficacy of a therapy, or is knowledge of its mechanism of action needed? (See, e.g., Russo and Williamson 2007; Howick 2011; Andersen 2012.) That is not our topic here. Those concerned with evidence for a therapy acknowledge that knowing the disease mechanism(s) (type ii and iii) can aid the rational design of therapies. That is one of our topics here.

The philosopher of science Paul Thagard analyzed reasoning in discovering disease mechanisms and possible therapeutic sites. Diseases are of different types, which he classified according to their causes (Thagard 1999). Thagard noted that one searches for different types of mechanisms if the disease is due to different types of causes. Thagard proposed different types of abstract mechanism schemas, based on the different types of diseases, including infectious disease, nutritional disease, and molecular genetic disease. The causes of diseases, he claimed, are most often identified by statistical and experimental means before researchers find the mechanism in which that cause participates. However, finding the cause and thereby classifying the kind of disease aids the search for the mechanism. In each type, finding where a normal mechanism is broken indicates sites for possible therapeutic intervention (Thagard 1998, 2003).

As to discovering such disease mechanisms, Thagard queried whether Darden's (2002) reasoning strategies for discovering mechanisms might be useful in medicine. This paper shows that they are: schema instantiation, modular subassembly, and forward/backward chaining are indeed relevant in disease mechanism discovery, as we will see below.

Several philosophers and historians of medicine have discussed cystic fibrosis. Cystic fibrosis (CF) is a monogenic, autosomal (i.e., not sex-linked), recessive (i.e., a patient must have two mutations, one inherited from each parent) disease.

The gene (labeled *CFTR*) is large: about 180,000 bases on the long arm of chromosome 7. The CFTR protein transports chloride ions across membranes in epithelial cells. Normal functioning aids in maintaining appropriate salt balance in those cells. Many different types of mutations in this single gene produce variants of the disease. Researchers have identified as many as 1324 different disease causing mutations in the *CFTR* gene (http://www.hgmd.cf.ac.uk/). These mutations produce a cluster of symptoms affecting the lungs, pancreas, liver, and other organs. Lung inflammation and frequent lung infections due to the build up of mucus in the lungs are the most serious problems for CF patients.

The mechanistic analysis applies well to this case. Since the discovery of the relevant gene in 1989, medical researchers have extensively studied the beginning steps in the disease mechanism and have targeted them for therapy (Darden 2013; Craver and Darden 2013, Chap. 11). Sadly, gene therapy has yet to work successfully to put a functioning copy of the large *CFTR* gene into the genome of cystic fibrosis patients (Lindee and Mueller 2011). Downstream stages of the mechanism have proved more promising targets, especially in mechanisms where the mutation produces a malformed protein whose function can be partially corrected with "chaperonin" molecules (Solomon 2015). Despite extensive study, black boxes remain in the later stages of the mechanism. Still unknown are all the details of exactly how defects in the chloride ion transport mechanism produces the final phenotypic symptoms of the disease in the lungs (Darden 2013). Also puzzling is why patients with the same genetic mutations nonetheless vary in the severity of their symptoms (Solomon 2015).

Cancer is another disease whose mechanisms have been discussed by philosophers. Thagard (2003) classifies cancer as a "disease of cells" due to genetic mutations. Hereditary and somatic mutations occur in oncogenes (genes that regulate cell division or survival) and tumor suppressor genes (suppress cell division). In contrast, the philosopher Anya Plutynski criticizes the view of cancer as merely a genetic disease (Plutynski, forthcoming). Cancer, she says, is a "complex process, due to many causes," not just to gene, chromosomal, and epigenetic changes but also "causes acting at the level of the cell and above" (Plutynski 2013, p. 466). Genetic mutations are often important difference makers in cancer etiology, but they are not the only ones. Genetic models, she argues, inappropriately "black box" environmental factors (Plutynski 2013, p. 474). Mechanism sketches for cancer should depict factors other than genes and indicate where such environmental factors should fill in black boxes.

The product and the discovery process for complex trait diseases, e.g., Crohn's disease, are much more complex than for monogenic diseases and more unknown than for the causes of cancer. The causes of this inflammatory bowel disease are hypothesized to include not only many genetic variants, but also interactions with the intestinal microbiome (the microbes in the gut), and tuning of the immune system through past exposures to invading microbes. So far, Crohn's is statistically associated with over 160 loci in the genome (de Lange et al. 2017). The gene products engage in complex interactions in producing the disease. Hence, Crohn's researchers need to find many mechanisms connecting gene variants to aspects of the disease phenotype. These will include roles of environmental factors, e.g., diet, and interactions with the gut microbiome. Then researchers need to find complex interactions among the products of the many mechanisms involved in order to explain the disease and find sites for therapies. This paper extends the mechanistic analysis beyond monogenic cases and cancer discussed in previous philosophical work to include these complex trait diseases.

Genome-Wide Association Studies (GWAS) provide data statistically associating genetic variants with disease risk (https://www.ebi.ac.uk/gwas/home). The presence of this association implies that either this variant SNP (single nucleotide polymorphism, a single base change in the DNA) or another variant nearby is involved in a disease mechanism. Many variants are just what may be called "proxy SNPs"; these are single base changes in the DNA that are somehow linked to parts of the genome associated with the disease, but do not themselves play roles in a disease mechanism. The question arises for each identified variant: Is there a disease mechanism that begins with the genetic variant and proceeds to the phenotype, characterized as disease risk?

Discovery of genetic disease mechanisms, we propose, is aided by an abstract diagrammatic representation for disease mechanisms. An abstract diagram sketches the overall structure of the product—the disease mechanism—and thereby aids the chair builder in the process—reasoning to its discovery.

4   Diagrammatic Representations for Genetic Disease Mechanisms

One analysis of understanding is that it involves the ability to manipulate a mental representation (Wilkenfeld 2013). An abstract diagrammatic representation facilitates the formation of an individual's understanding, guides the person in filling it in, and serves to convert a single person's visual mental representation into a publically accessible form. The proposed abstract general form for mechanism disease diagrams to be discussed below plays this role. It has several advantages as a representation of the product to guide the process of its discovery:

(a) It provides a general framework for integrating and expanding knowledge about disease mechanisms.
(b) It clearly delineates what is known and not known about the mechanism(s) of each disease.
(c) It provides a potential way of finding interactions when multiple mechanisms interact to produce or modulate the severity of a disease.
(d) It allows representation of interacting subsets of mechanisms [found in (c)] in individual patients.
(e) It facilitates identification of sites of potential therapeutic intervention.

Consider these abstract idealized mechanism diagrams of genetic variant to disease phenotype via disease mechanisms.

Figure 1 is a beginning point after a genetic variant is related to disease risk: does that variant mark the beginning of a mechanism? In contrast, if indeed a target mechanism is found, then an idealized general abstract diagram for it will have components such as in Fig. 2.

Figure 2 shows an idealized diagram of a genetic disease mechanism chain for a case where all the components of the chain are understood. It has no black boxes. The goal of chain building is to proceed from a figure such as Fig. 1, to progressively fill the black box, to draw a diagram such as Fig. 2 (or else conclude no mechanism likely exists). Figure 2 begins with a variant that affects the function of a gene. Rectangles represent altered substates, with in box text indicating how the substate is altered. Ovals, depicting mechanism modules of groups of activities and entities, label the arrows; text inside the oval names the module. The entities and activities of modules transform one substate to another. The chain proceeds through successive altered substates to a disease phenotype. Blue octagons indicate potential sites for therapeutic intervention. A white cloud entering the chain from below shows a possible role for environmental factors. All the boxes are glass boxes; one can look inside and see whatever details are relevant. Details in a box or oval may be telescoped (collapsed, as in a folded telescope) when the details are irrelevant. All the lines are green, indicating the chain builder's highest confidence level, based on evidence for those steps.

Fig. 1 The entire mechanism between a genetic variant and disease risk is a black box. The question mark queries whether a mechanism actually exists between the two

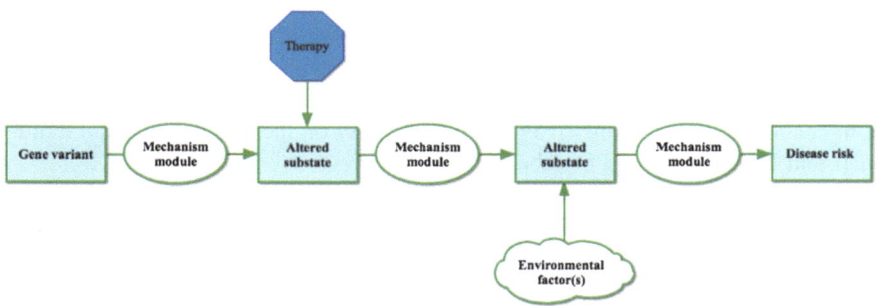

Fig. 2 Abstract genetic disease mechanism chain with no black boxes

Our diagrammatic framework enables us to suggest **a set of heuristic questions**. These serve to guide the chain builder in filling black boxes to remove ignorance and to reach a diagram that is complete enough for whatever is the purpose of the work.

Mechanism at all? The first step in removing ignorance is to inquire whether a mechanism exists at all. In Fig. 1, the chain begins with a genetic variant connected via a black box with a question mark to the disease phenotype. For a statistical association, the question mark asks whether there is a mechanism at all. To answer that question: try to fill the black box with a plausible mechanism. Given failure to find a possible mechanism, the chain builder will have to make a judgment call as to when to stop trying. Where a specific mutated gene is known, such as in monogenic diseases and some cancers, then the first box names the gene and its mutation. The chain builder can draw a green arrow to a black box with no question mark to indicate where additional specific mechanism components are expected and should be sought.

What kind of genetic variant begins the chain? Once the task becomes to fill the black box with a mechanism, the next question is what kind of variant begins the chain? Different kinds of variants likely require chains with different kinds of beginning steps. For example, a missense variant (a change in one DNA base that results in a changed amino acid in a protein) will proceed via protein synthesis. In contrast, a variant in a non-coding region of DNA that affects the binding of a regulatory protein will have earlier steps before the module of protein synthesis plays a role.

In addition to building the chain forward from the genetic variant, is it possible to begin at the end and build the chain backward? Black boxes show missing steps in need of elaboration. Because what comes before and what comes after are indicated in the diagram, the chain builder can reason forward from the previous step or backward from the subsequent one to conjecture what fills the box. Are there activity enabling properties in a step that indicate a likely module and a likely substate perturbation in the next step? Conversely, are there activity signatures (properties of an altered substate) that indicate what kind of activities operated in the previous module, earlier in the chain, that produced it?

Do environmental factors play a role? Is there a place where a white cloud representing an environmental factor should be added? What kind of substate change follows from its insertion?

Does the chain branch into subchains? Does the chain branch at a given step? If so, are the subchains mutually exclusive alternatives ("or" at the branch) or do both occur ("and" at a branch)? Is there uncertainty on the part of the chain builder such that branches should be labeled with "and/or"?

Other than branches in the chain, are there other nonlinear organizational motifs that need to be added? If no feedback or feed-forward loops are included, the question arises as to whether any should be? Are there other nonlinear organizational motifs to consider?

Is there a potential site for therapeutic intervention? Can types of therapies for types of steps be suggested, e.g., does a misfolded protein indicate that a chaperonin should be considered?

How strong is the evidence for each step? How confident is the chain builder in each step? As noted above, a black box with a question mark asks whether there is anything to be discovered at that point, either whether a mechanism as a whole exists or whether a branch of a chain exists. A black box without a question mark indicates a likely but currently unknown substate, mechanism module, or group of substates and mechanism modules. Green, pink and red colors of the lines indicate the confidence level of the chain builder in each specific perturbed substate and arrow/module. Just like black boxes, red and pink colors indicate where more work is needed to increase confidence, to convert red and pink to green. Evidence for particular steps includes the following: standard biological knowledge, one or more reliable published sources provide evidence for the step, and experimental evidence. (The diagrams below illustrate the use of all three kinds of evidence.)

These questions aid the chain builder in using the diagram to depict kinds of ignorance and direct resources to remove it.

5 Web-Based Graphical Interface for Chain Building

We are developing a web-based graphical interface to aid medical researchers in hypothesizing and representing genetic disease mechanisms. The interface is implemented using the conventions discussed above. We have used it to produce more detailed diagrams, such as those below. These three figures provide examples for a monogenic disease chain, a cancer disease chain, and one (of what will be many) complex trait disease chain.

Figure 3 shows a mechanism disease chain for cystic fibrosis (CF). It begins with the mutation, DeltaF508, in the cystic fibrosis transmembrane conductance regulator (*CFTR*) gene. This is the most common mutation among CF patients in the United States. Normally the CFTR protein inserts into epithelial cell membranes and transports chloride ions across the membrane. In this mutant form, three DNA

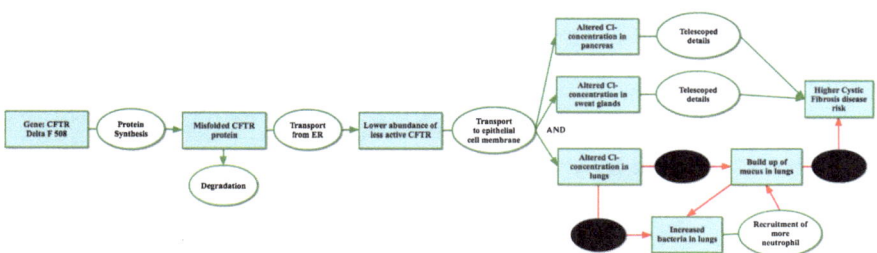

Fig. 3 A disease mechanism chain for cystic fibrosis, an example of a monogenic disease

bases are missing, resulting in one missing amino acid in the protein. The protein misfolds. The first branch in the chain indicates that some misfolded proteins are degraded but others are released from the endoplasmic reticulum (ER), where it is synthesized. This lower abundance of the protein and its misfolding results in altered concentrations of the misfolded protein in epithelial cells in the pancreas, sweat glands, and lungs, shown in the next three branches. This is a well-studied case so all the lines in the beginning of the mechanism chain are green. The black boxes and red arrows (toward the end in the lower branch of the chain) indicate the controversy that still surrounds exactly what contributes to the build up of thick mucus in the lungs. One hypothesis is that improper salt balance produces the mucus build up (shown in the top chain coming out of the lungs rectangle). Another hypothesis is that a contributing factor is the break down of the overexpressed immune cells, neutrophils, that are recruited to fight invading bacteria (shown in the loop in the bottom branch of the chain). (For more details, see Darden 2013.)

Figure 4 is an example of a disease mechanism chain for cancer. This is a hypothesized mechanism chain for a germline DNA variant in the human gene *MSH2*. The ID number identifies the particular variant. The DNA base change is expected to lead to nonsense mediated decay (NMD), decreasing the messenger RNA abundance by half, and as a consequence, also decreasing MSH2 protein abundance. As a result, all the complexes of this protein with other proteins will also be of reduced abundance, hence the "and" at the first branch. The Le Chatelier's Principle refers to a state in which, e.g., the concentration of a reactant changes in a system in equilibrium such that the equilibrium will shift so as to tend to counteract the effect. This activity lowers the abundance of macromolecular complexes in the next steps in all three branches. Then, the branches show the effects of less DNA mismatch repair of both short and longer mismatch regions in the DNA, as well as less apoptosis (programed cell death) triggered by recognition of drug induced DNA damage. Results include greater accumulation of somatic mutations, hence increased cancer risk. Greater microsatellite instability occurs, which may also somehow increase cancer risk, indicated by the black box with a question mark. The top branch of the chain shows the path to drug resistance. (For details, see the review, Li 2008.)

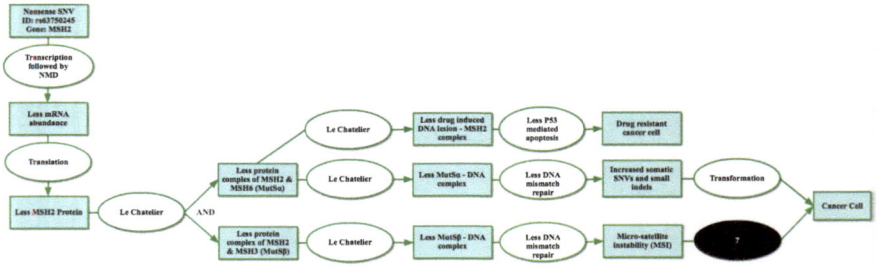

Fig. 4 Mechanism chain diagram for a cancer gene variant in the human gene *MSH2*

Figure 5 is an example of one of the hypothesized mechanism chains for Crohn's disease, originating in a locus containing a GWAS marker at the *MST1* gene, which codes for MSP (Macrophage Stimulating Protein). The mechanism begins at the perturbed DNA substate on the left, and progresses through protein, protein-protein complex, cell signaling, innate immune response, and gut barrier layer stages to disease risk. In this view, some parts of the chain, at the DNA, protein, and protein complex stages, are fully expanded, while others are partly telescoped (for example "cell signaling" and "innate immunity"). These telescoped steps have multiple substate perturbations and mechanism modules within them. (For details on research on this chain, see Gorlatova et al. 2011.) Black boxes, as well as pink and red lines, indicate uncertainty. The "or" at the first branch indicates two different ways that the chain might branch. The next "and/or" indicates that one or both of the branches may occur; the chain builder is not yet certain which is the case. This *MST1* chain represents just one of the many mechanisms involved in Crohn's disease. Much work remains to find additional chains and the ways they interact with each other.

This diagrammatic method clearly illustrates the way an abstract representation of the product to be discovered guides reasoning to its discovery through various stages. Admittedly, this diagrammatic representation abstracts away from many features of mechanisms discussed above. It is an open question whether any features of more fully represented mechanisms will need to be added. Note that the diagrams do not include structures of the proteins and protein complexes, although, if needed, it would be easy to add a link to the protein structure database. Also omitted are quantitative rates by which activities operate or quantitative measures of abundance of entities; this is a qualitative representation. Furthermore, the locations of the mechanism steps are not graphically shown, e.g., whether the steps occur in the nucleus, in the cytoplasm, within cell organelles, or elsewhere; however, when relevant, text in the altered substate box does indicate location, such as epithelial cells. It is an open question whether such general features of mechanisms (structures, rates, spatial locations) will need to be represented to fulfill the goals of adequately explaining the disease and locating sites for therapy. Should the need arise, the general philosophical analysis of mechanisms provides a storehouse of items that can be added to the simplified graphic in the future.

Our work contrasts with other graphical forms of representations; we discuss two here. One type is a directed acyclic graph (DAG) to represent causal chains. Philosophers of science are engaged in a lively debate about the adequacy or inadequacy of DAGs for representing normal biological mechanisms (e.g., Gebharter and

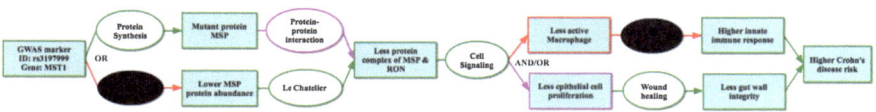

Fig. 5 Shows a chain for one variant in the *MST1* gene associated with increased risk of Crohn's disease

Kaiser 2014; Kaiser 2016; Weber 2016). From our perspective, causal graphs are impoverished in merely having unlabeled edges that represent generic cause relations. In contrast, our mechanism diagrams indicate the specific kind of activity or the group of entities and activities in a mechanism module that effect each particular instance of casual production.

Biologists have developed other graphical frameworks, but not (so far as we know) ones using analyses from the new mechanistic philosophy of science to specifically represent genetic disease mechanisms. Most represent normal molecular biological pathways. (For a list see, e.g., Jin et al. 2014.) One of the best developed the Kyoto Encyclopedia of Genes and Genomes. The KEGG Pathway database is a collection of manually drawn graphical diagrams. These represent molecular pathways for metabolism, genetic information processing, environmental information processing, other cellular processes, some human diseases, and drug resistance (Kanehisa et al. 2017). The disease diagrams are represented by perturbations in normal pathways.

The KEGG disease diagrams differ from our framework in numerous ways. Unlike our diagrams, KEGG depicts diseases in pathway wiring diagrams of groups of normal pathways with genes associated with a disease in color-coded rectangles. Furthermore, the focus is only on the early stages involving genes, proteins, and molecular interactions. In contrast, each of our diagrams begins with a specific single gene mutation and traces the changes resulting from that mutation through numerous other stages to the disease phenotype.

For example, in KEGG the Crohn's disease pathway is part of the pathway for inflammatory bowel diseases (IBD) in general. The IBD pathway depicts numerous genes and proteins (in rectangular boxes) in their normal pathways. A few genes known to be associated with diseases (not just Crohn's) are colored pink in contrast to normals, which are green. Light blue boxes indicate actual drug targets. Some anatomical details are depicted, such as a breach in the gut wall and a macrophage (an immune cell). Types of lines connecting genes and gene products indicate types of interactions, such as inhibition, activation, indirect effect, or dissociation. In contrast, our diagrams have ovals with text to label the arrows, thus showing the activities or mechanism modules that produce perturbed substates. Any new kind of activity easily fits within our framework whereas KEGG will need a new graphical symbol. Consequently, our framework is more easily extendable.

Also unlike ours, the KEGG pathway diagrams do not indicate confidence levels nor do they include black boxes to show ignorance. Our diagrams are thus better for directing the discovery process to produce the product of a genetic disease mechanism by filling black boxes, resolving uncertainties about branches, and increasing confidence levels.

6 Conclusion

This paper argues for the thesis that the product shapes the process: knowing what is to be discovered provides guidance as to how to discover it. Here the product is a schema to represent steps in a disease mechanism from gene variant to disease phenotype. Heuristic questions and abstract diagrams aid the reasoning process to discover a specific disease mechanism chain. By indicating black boxes and uncertainties, the chain builders represent their ignorance at a given time and show where to focus additional work. This new diagrammatic representational tool, grounded in philosophical analysis, aids in storing collective knowledge and guiding collective discovery.

Plans for future work include finding standardized ontology terms (Arp et al. 2015) for each stage of genetic disease mechanisms (e.g., Gene Ontology 2015). Such standardized terminology is especially important to facilitate finding interactions among the related chains. This standardization will also foster communication between groups of experts to complete the parts of the chains in their areas of expertise. An even longer-range goal is to apply this work in precision medicine. That goal requires finding specific interacting chains for individual patients (or groups of patients), given variability in their genes, environment, and lifestyle, so that personalized therapy can be designed and administered.

Acknowledgements LD thanks Emiliano Ippoliti and his colleagues at Sapienza University in Rome for the invitation to the Building Theories Workshop, and to them and all the participants for lively discussions of discovery heuristics. For helpful suggestions on earlier drafts, she thanks Carl Craver, Giamila Fantuzzi, Nancy Hall, and an anonymous reviewer, as well as Kal Kalewold and the other students in her graduate seminar on mechanisms and ontology. This work was supported in part by NIH (US National Institutes of Health) grants R01GM102810 and R01GM104436 to JM.

References

Abrahamsen, A. A., & Bechtel, W. (2015). Diagrams as tools for scientific reasoning. *Review of Psychology and Philosophy, 6*, 117–131.

Abrahamsen, A., Benjamin, S., & William, B. (2017). Explaining visually: Mechanism diagrams. In S. Glennan & P. Illari (Eds.), *The routledge handbook of mechanisms and mechanical philosophy* (pp. 238–254). New York: Routledge.

Andersen, H. (2012). Mechanisms: What are they evidence for in evidence-based medicine? *Journal of Evaluation in Clinical Practice, 18*(5), 992–999.

Arp, R., Smith, B., & Spear, A. D. (2015). *Building ontologies with basic formal ontology.* Cambridge, MA: MIT Press.

Bechtel, W., & Abrahamsen, A. (2005). Explanation: A mechanist alternative. In C. F. Craver & L. Darden (Eds.), *Special issue: Mechanisms in biology. Studies in History and Philosophy of Biological and Biomedical Sciences* (vol. 36, pp. 421–441).

Bechtel, W., & Abrahamsen, A. A. (2013). Thinking dynamically about biological mechanisms: Networks of coupled oscillators. *Foundations of Science, 18*, 707–723.

Bechtel, W., & Richardson, R. C. (1993). *Discovering complexity: Decomposition and localization as strategies in scientific research.* Princeton, NJ: Princeton University Press.

Craver, C. F. (2007). *Explaining the brain: Mechanisms and the mosaic unity of neuroscience.* New York: Oxford University Press.

Craver, C. F. (2008). Physical law and mechanistic explanation in the Hodgkin and Huxley model of the action potential. *Philosophy of Science, 75*(5), 1022–1033.

Craver, C. F., & Darden, L. (2013). *In search of mechanisms: Discoveries across the life sciences.* Chicago, IL: University of Chicago Press.

Darden, L. (2002). Strategies for discovering mechanisms: Schema instantiation, modular subassembly, forward/backward chaining. *Philosophy of Science, 69*(Proceedings), S354–S365.

Darden, L. (2006). *Reasoning in biological discoveries: Mechanisms, interfield relations, and anomaly resolution.* New York: Cambridge University Press.

Darden, L. (2013). Mechanisms versus causes in biology and medicine. In H. K. Chao, S. T. Chen, & R. L. Millstein (Eds.), *Mechanism and causality in biology and economics* (pp. 19–34). The Netherlands: Springer.

Darden, L., & Craver, C. F. (2002). Strategies in the interfield discovery of the mechanism of protein synthesis. *Studies in History and Philosophy of Biological and Biomedical Sciences, 33*, 1–28. Reprinted with corrections in Darden (2006, Chap. 3).

de Lange, K. M., et al. (2017). Genome-wide association study implicates immune activation of multiple integrin genes in inflammatory bowel disease. *Nature Genetics, 49*, 256–261. https://doi.org/10.1038/ng.3760.

Gebharter, A., & Kaiser, M. I. (2014). Causal graphs and biological mechanisms. In M. I. Kaiser, O. R. Scholz, D. Plenge, & A. Hüttemann (Eds.), *Explanation in the special sciences* (Vol. 367, pp. 55–85). Dordrecht: Synthese Library.

Gene Ontology Consortium. (2015). The gene ontology consortium: Going forward. *Nucleic Acids Research, 43*(database issue), D1049–D1056.

Glennan, S. S. (1996). Mechanisms and the nature of causation. *Erkenntnis, 44*, 49–71.

Glennan, S., & Illari, P. (Eds.). (2017). *Routledge handbook of mechanisms and mechanical philosophy.* New York: Routledge.

Gorlatova, N., Chao, K., Pal, L. R., Araj, R. H., Galkin, A., Turko, I., et al. (2011). Protein characterization of a candidate mechanism SNP for Crohn's disease: The macrophage stimulating protein R689C substitution. *PLOS ONE* (open access). http://dx.doi.org/10.1371/journal.pone.0027269.

Howick, J. (2011). Exposing the vanities—And a qualified defense—Of mechanistic reasoning in health care decision making. *Philosophy of Science, 78*, 926–940.

Jin, L., Xiao-Yu, Z., Wei-Yang, S., Xiao-Lei, Z., Man-Qiong, Y., Li-Zhen, H., et al. (2014). Pathway-based analysis tools for complex diseases: A review. *Genomics, Proteomics & Bioinformatics, 12*(5), 210–220. https://doi.org/10.1016/j.gpb.2014.10.002.

Kanehisa, M., Furumichi, M., Tanabe, M., Sata, Y., & Norishima, K. (2017). KEGG: New perspectives on genomes, pathways, diseases and drugs. *Nucleic Acids Research, 45*(D1), D353–D361. https://doi.org/10.1093/nar/gkw1092.

Kaiser, M. I. (2016). On the limits of causal modeling: Spatially-structured complex biological phenomena. *Philosophy of Science, 83*(5), 921–933.

Li, G. M. (2008). Mechanisms and functions of DNA mismatch repair. *Cell Research, 18*, 85–98.

Lindee, S., & Mueller, R. (2011). Is cystic fibrosis genetic medicine's canary? *Perspectives in Biology and Medicine, 54*(3), 316–331.

Machamer, P., Darden, L., & Craver, C. F. (2000). Thinking about mechanisms. *Philosophy of Science, 67*, 1–25.

Plutynski, A. (2013). Cancer and the goals of integration. *Studies in the History and Philosophy of Biological and Biomedical Sciences, 4*, 466–476.

Plutynski, A. (forthcoming). *Explaining Cancer: Finding Order in Disorder.*

Russo, F., & Williamson, J. (2007). Interpreting causality in the health sciences. *International Studies in the Philosophy of Science, 21*(2), 157–170.

Sheredos, B., Burston, D., Abrahamsen, A., & Bechtel, W. (2013). Why do biologists use so many diagrams? *Philosophy of Science, 80*(5), 931–944.

Solomon, M. (2015). *Making medical knowledge*. New York: Oxford University Press.

Thagard, P. (1998). Explaining disease: Causes, correlations, and mechanisms. *Minds and Machines, 8,* 61–78.

Thagard, P. (1999). *How scientists explain disease*. Princeton, NJ: Princeton University Press.

Thagard, P. (2003). Pathways to biomedical discovery. *Philosophy of Science, 70,* 235–254.

Weber, M. (2016). On the incompatibility of dynamical biological mechanisms and causal graphs. *Philosophy of Science, 83*(5), 959–971.

Wilkenfeld, D. (2013). Understanding as representation manipulability. *Synthese, 190,* 997–1016. https://doi.org/10.1007/s11229-011-0055-x.

"Take the Case of a Geometer…" Mathematical Analogies and Building Theories in Aristotle

Monica Ugaglia

Abstract In this paper the way of doing physics typical of mathematical physics is contrasted with the way of doing physics theorised, and practised, by Aristotle, which is not extraneous to mathematics but deals with it in a completely different manner: not as a demonstrative tool but as a reservoir of analogies. These two different uses are the tangible expression of two different underlying metaphysics of mathematics: two incommensurable metaphysics, which give rise to two incommensurable physics. In the first part of this paper this incommensurability is analysed, then Aristotle's way of using mathematics is clarified in relation to some controversial mathematical passages, and finally the relation between mathematics and the building of theories is discussed.

Keywords Aristotle · Mathematics · Problem-solving

1 Introduction

The contemporary way of doing science is grounded on the possibility of using mathematics as a proper demonstrative tool. In fact, the closer a branch of knowledge gets to a complete mathematisation, the more reliable it is considered. Hence the efforts of more and more disciplines to achieve the status of mathematical-X, on the model of mathematical physics.

Indeed, physics appears to be one of the more suitable areas for exercising mathematics, and it is not by chance that the ongoing process of mathematisation of knowledge began with physics. Therefore, both for practical and for historical reasons, I will carry out my analysis by focusing on physics, even though the conclusions I will draw are applicable to any other field of knowledge.

In particular, I will contrast the way of doing physics typical of mathematical physics with the way of doing physics theorised, and practised, by Aristotle, which

M. Ugaglia (✉)
University of Florence, Florence, Italy
e-mail: monica.ugaglia@gmail.com

is not extraneous to mathematics—as has all too often been stated—but deals with it in a completely different manner: not as a demonstrative tool but as a reservoir of analogies.

These two different uses are the tangible expression of two different underlying metaphysics of mathematics: two incommensurable metaphysics, which give rise to two incommensurable physics. For Aristotle mathematical properties are actual properties of physical objects, but by definition they lie outside the essence of the objects, *qua* physical objects, belonging instead to their accidental features.

On the contrary, the very possibility of a mathematical physics hinges on the fact that mathematical properties are the true essence of physical objects, whereas anything that escapes mathematical formalisation is negligible.

Now, to acknowledge the existence of an ontological gap between our physics and Aristotle's one is not to deny the value of Aristotle's physics or the significant presence of mathematics within it.

Aristotle constantly resorts to mathematics, but he does so to build theories and not to formalise them. So it is not surprising that he employs mathematics in a rather unsystematic way, without exploiting its deductive power, but favouring instead its evocative capacity. And it is not surprising that, like every heuristic tool, mathematics did not leave too many traces in the final resulting theory, be it physics or philosophy in general.[1]

2   Aristotle's Physics Versus Mathematical Physics

One of the main assets of Aristotle the physicist was his being aware of the need to turn physics into scientific knowledge,[2] together with his attempt to actually do so: an attempt which led him, among other things, to develop the impressive epistemological theory of the *Prior* and *Posterior Analytics*.

One of the main limits of Aristotle the physicist was his conception of what it means to (scientifically) know. For an Aristotelian physicist, to know an object means to know what it is, grasping its essence (τὸ τι ἦν εἶναι). In order to do so, he takes the physical object and separates the essential features from the incidental ones, which can be neglected.

---

[1] In this paper I deal with Aristotle's account of scientific practice and problem-solving from a purely historical perspective. A comparison with contemporary approaches to the subject would be the much-needed follow-up to this research: I would like to thank Emiliano Ippoliti and Carlo Cellucci for their suggestions in that direction, along with all workshop participants for having discussed the subject with me. Finally, I would like to thank Ramon Masià for having made the diagrams in this article less precarious.

[2] Aristotle's final aim was to extend this principle to all other branches of knowledge, to different degrees, according to the different potentialities of their subject matters. The more a subject deals with necessity and simplicity, the more scientific it can be made (see for instance *APr* I 30; *APo* I 27).

More pragmatically—and more successfully—for a mathematical physicist to know an object means to know what it does, describing and predicting its behaviour. To do so, he too takes the physical object and separates the essential features from the incidental ones, which can be neglected (remember Galilei's *accidentari impedimenti*), but what is essential and what negligible is not the same in the two cases.

The choice does depend on metaphysical assumptions, and the metaphysics which expressly steers Aristotle's physics is completely different from the metaphysics which less openly stands behind our own mathematical physics.

For Aristotle, 'essential' means something related to change, which is the very nature of physical objects, while for the modern physicist 'essential' means mathematical: unfortunately, mathematics has nothing to do with change, as conceived by Aristotle.

2.1 Mathematical Physics

The aim of modern physics is to study phenomena, possibly predicting their behaviour. Mathematical physics consists in doing the same using mathematics.

In short, a physical phenomenon must be described in mathematical terms: any object involved must be measurable, and the range of measures must be represented by a variable. The behaviour of the objects and their interactions in their turn must be represented in terms of suitable functions of the variables. In other words, a mathematical model must be constructed and studied instead of the physical one[3]; and once the problem is solved, the result obtained must be translated back into natural language, in order to make the result accessible to a wide range of scholars, and to non-scientific readers.

Now, the possibility of studying the mathematical model instead of the physical situation hinges on the assumption that what has been put aside in building the

---

[3]The same physical situation can be described by more than one model, depending on what kind of quantities one choice to treat as independent variables. A mechanical system of N particles, for instance, admits a classical representation, where the variables are the Cartesian coordinates of the particles involved $\overline{r}_i = (x_i, y_i, z_i)$ and their time derivatives, or velocities $\overline{v}_i = \left(\frac{dx_i}{dt}, \frac{dy_i}{dt}, \frac{dz_i}{dt},\right)$ where $i = 1\ldots N$. In this case the system is described by Newton's equations of motion $\overline{F}_i = m_i \frac{d^2 r_i}{dt^2}$, namely N differential equations in 3 variables. But one can use as variables the so-called generalized coordinates q_i and $\dot{q}_i = \frac{dq_i}{dt}$, where $i = 1\ldots 3N$, so that the system is described by Lagrangian equations of motion $\frac{d}{dt}\left(\frac{\partial L}{\partial \dot{q}_i}\right) = \frac{\partial L}{\partial q_i}$. And one can use also the so-called canonical coordinates q_i and $p_i = \frac{\partial L}{\partial \dot{q}_i}$, where $i = 1\ldots 3N$, so that the system is described by Hamiltonian equations of motion $\frac{\partial H}{\partial q_i} = -\dot{p}_i; \frac{\partial H}{\partial p_i} = \dot{q}_i$.

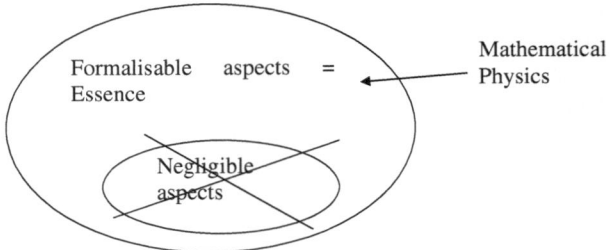

Fig. 1 Galileo's physical object

model, and neglected because it could not be forced into a formula, was actually negligible. In Galilean terms, the assumption is that it was nothing but *accidentari impedimenti.*[4]

Keeping within Galileo's framework, we must assume that our world has an underlying mathematical structure: the book of nature is written in mathematical characters. At any rate, it can be translated into mathematical characters. In Aristotle's language, this amounts to postulating that mathematics is the essence of physical objects, *qua* physical objects (Fig. 1).

But this assumption, which ultimately allows mathematical physicists to use mathematics as the demonstrative tool for physics, is a very strong metaphysical assumption. Philosophers of physics have been debating for centuries the meaning and the effective extent of such an assumption, which many people would find it hard to accept. For sure, it would not have been accepted by Aristotle, for whom essence means change, and change rules out mathematics.

2.2 Aristotle's Physics

Aristotle's physics does not consist in describing or predicting something; and definitely not in doing this by using mathematics, because a mathematical description is in any case a partial description.

---

[4]"Quando dunque si facciano simili esperienze in piccole altezze, per sfuggir più, che si può gli accidentari impedimenti de i mezzi, tuttavolta, che noi vediamo, che con l'attenuare, e alleggerire il mezzo, anco nel mezzo dell'aria, che pur è corporeo, e perciò resistente, arriviamo a vedere due mobili sommamente differenti di peso per un breve spazio moversi di velocità niente, o pochissimo differenti, le quali poi siamo certi farsi diverse, non per la gravità, che sempre son l'istesse, ma per gl'impedimenti, e ostacoli del mezzo, che sempre si augmentano, perché non dobbiamo tener per fermo che, rimosso del tutto la gravità, la crassizie, e tutti gli altri impedimenti del mezzo pieno, nel vacuo i metalli tutti, le pietre, i legni, ed insomma tutti i gravi si muovesser coll'istessa velocità?" Galilei (1718), v. III. p. 112.

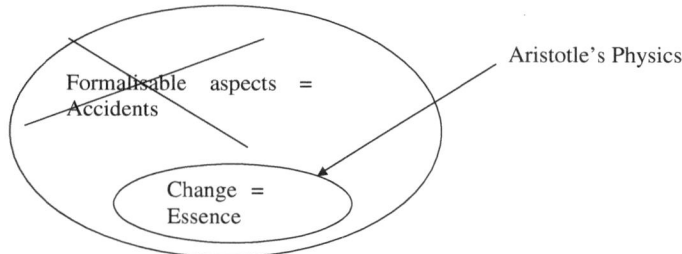

Fig. 2 Aristotle's physical object

Like any other branch of knowledge, physics consists in finding definitions, for to give a definition means to grasp the essence of something, and this is the ultimate aim of knowledge.

Now, physics means the study of nature (φύσις), and for Aristotle nature means the inner principle of change, so that the essence of a physical object, insofar as it is physical, is necessarily related to change. In other words, not only does the physicist find his object of investigation by identifying, among the objects of experience, those which contain the principle of change in themselves,[5] but it is precisely as objects of change that he studies them. In doing so, he separates what has to do with change from the other aspects, which are secondary or incidental.[6] That he cannot make use of mathematics is quite clear.

It is not because Aristotle thinks that mathematics is something extraneous to physics. Quite the opposite: mathematics is part of physics, but an insignificant one. It is part of physics because Aristotle conceives mathematical objects as being part of physical ones.[7] Take an object: insofar as it is concerned with change, it is a physical object, and its study pertains to the physicist. But subtracting change makes it a mathematical object, and hence the province of the mathematician.[8]

Now, if mathematics is what remains after subtracting change, meaning their essence, it is an accidental feature of the physical world. In particular, the mathematician is unable to help the physicist to know an object: disregarding everything that has to do with change, he disregards everything that has to do with the physical essence. It is not that the mathematician has nothing to say, but that what he says— and what it is possible for him to glean with the tools of mathematics—is incidental[9] (Fig. 2).

---

[5]The principle of changing or being changed: *Ph.* II 1, 192b13-15; III 1, 200b2-13; VIII 3, 253b5-9; VIII 4, 254b16-17; *Cael.* I 2, 268b16; *Metaph.* E 1, 1025b18-21.

[6]*Metaph.* E 1, 1025b35-1026a6; K 3, 1061b6 ff.

[7]*Metaph.* M 3, 1077b22-1078a9; *Ph.* II 2, 193b23-194a12; cf. *de An.* I 1, 403a15-16; *Metaph.* N 2, 1090a13-15.

[8]*Metaph.* E 1, 1026a14 ff.

[9]Consider for example the "mathematical" treatment of motion in terms of trajectory, velocity and time, which Aristotle develops in depth (see in particular *Ph.* VI) but which has nothing to do with

In other words, for Aristotle mathematics is not the underlying structure of the world, and in the book of nature only some minor corollary is written in mathematical terms. This ultimately prevents Aristotelian physicists from using mathematics as the demonstrative tool for physics.

2.3   The Ontological Gap

The situation is exactly the opposite of that arising in mathematical physics: what is essential for the mathematical physicist is negligible for Aristotle, and vice versa.

This inversion is something ontologically non trivial. Take Einstein's physics: by restricting it to a certain domain, Newton's physics can be obtained in the appropriate approximation.[10] Indeed, Einstein's physics and Newton's physics are different, and in some respects incompatible, but they both belong to the category of mathematical physics (Fig. 3).

On the contrary, there is no approximation leading from Newton's physics to Aristotle's physics, for no approximation can ever turn the essence into the accidents (or vice versa). Even in the right approximation—that is, by considering Newtons' laws of fallen bodies in a fluid—what one obtains is an equation of motion, which despite certain "accidental" similarities is something completely different from Aristotle's treatment of motion.[11] In other words, they are ontologically incommensurable: Aristotle's physics is not a mathematical physics, nor is it mathematisable[12] (Fig. 4).

No matter how sceptical one is about metaphysical hindrances, one has to accept empirical evidence. And the empirical evidence is that the limit of Newton' equation of falling bodies in a fluid gives a motion with constant velocity, whereas for Aristotle bodies falling in a fluid[13] increase their velocity toward the end.

---

the essence—that is, the knowledge—of motion in itself, defined in *Ph.* III in terms of power and act (ἡ τοῦ δυνάμει ὄντος ἐντελέχεια, ᾗ τοιοῦτον, κίνησίς ἐστιν, *Ph.* III 1, 201a10-11; cf. 201b-5).

[10]When the speeds involved are far from c (the speed of light in void), Einstein's equations of motion turn into Newton's laws.

[11]Similarities and differences have been studied in Ugaglia (2004), but only the similarities have been taken into account by Rovelli (2015), where any gap is denied and the coincidence between Aristotle's "equation" of motion and Newton's limit in a fluid is maintained. Even disregarding any metaphysical obstruction, this conclusion is incompatible with the fact that for Aristotle the speed of a falling body incontrovertibly increases, while in Newton's approximation it is constant.

[12]On the question see Ugaglia (2015).

[13]For Aristotle a motion necessarily occurs in a medium. But here we must be careful, because this does not happen accidentally: it is not because Aristotle's Cosmos does not contain any void that motion occurs in a medium. On the contrary, this is the way in which motion has been defined by Aristotle, namely as a *relative* notion, which requires the presence of a medium. In other words, for Aristotle the absence of void is not an imposition, but a logical consequence of his physical premises.

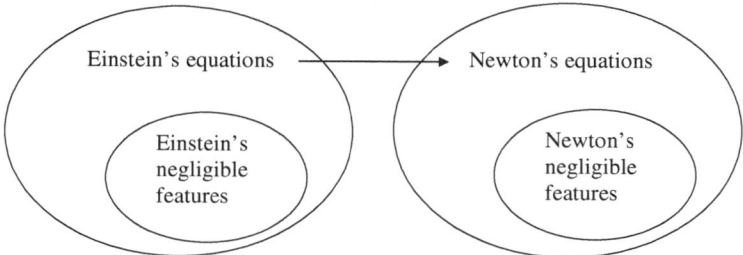

Fig. 3 Einstein's physics vs Newton's physics

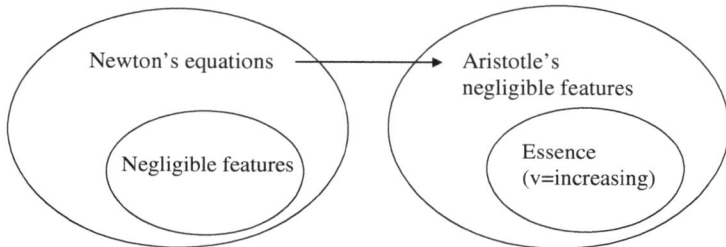

Fig. 4 Newton's physics vs Aristotle's physics

The reason for this failure is very simple: Aristotle's statement that the velocity increases is not the result of a calculation, but the logical and necessary consequence of physical, non-quantitative and necessary premises.[14]

Take the object "motion". For the modern physicist it can be completely reduced to its equation, namely to a trajectory. In particular, in this equation the relation between the moving object and the surrounding medium is represented by a numerical ratio between the values of certain parameters: the absolute weight of the movable body and the density of the medium.[15] Of course, this ratio contains all mathematical information—namely all information *tout-court*, for the modern physicist—about the relation between the body and the medium.

But for Aristotle this is not true. Even leaving aside the fact that a ratio between a weight and a density has no meaning—both for Aristotle and for any Greek mathematician—the relation between the moving object and the surrounding medium is something more than a ratio. It is an interaction at the qualitative level of

---

[14]The closer the movable object is to the end of its motion, and the more of the form it was lacking —and tends to—has been acquired, the more efficient its motion is (*Ph.* VIII 9, 265b12-16. Cf. *Ph.* V 5, 230b25-26; *Cael.* I 8, 277a27-29; II 5, 288a20-21).

[15]In a formula $v(t) = \sqrt{\frac{1}{C}}\sqrt{\frac{W_b}{\rho_m}}$ where C is a constant coefficient, depending on the shape of the moving object, W_b is the absolute weight of the moving object and ρ_m is the density of the medium.

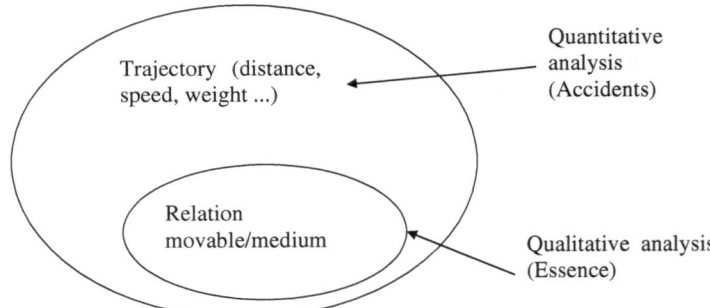

Fig. 5 Aristotle's analysis of motion

the form, which goes far beyond the level of efficient causality. Let us not forget that for Aristotle motion is *defined* as the "actuality of what potentially is, qua such", and this is something more than a trajectory (Fig. 5).

3 Aristotle's Mathematics: A Reference Structure

The conclusions reached so far do not imply that mathematics is absent from Aristotle's physics. Of course, it is present in the "corollaries" mentioned in Sect. 2.2, namely the passages where Aristotle employs mathematics exactly as we do, to prove something: for instance, in the aforementioned case of motion, the passages concerning trajectories, velocities, and spaces travelled. This is the kind of mathematics we expect to find in a physical text, and it is the only part of Aristotle's text that the modern physicist recognises as dealing with physics. For this reason it is tempting to read it as the "true" physics of Aristotle, especially for a scientist: this is "physics"—all the rest is philosophy.[16] However, as we have seen, this mathematical analysis constitutes a very small and marginal portion of Aristotle's physics[17]; and since it does not address the process of knowing an object, but at most that of describing their secondary aspects, I will leave it aside in the rest of my analysis.

---

[16]Perhaps, the fact that two synonymous terms—physics and natural philosophy—have been traditionally employed to denote the same object has contributed to create this misunderstanding among non-specialists.

[17]Take, once again, Aristotle's treatment of local motion, which is one of the most "formalised" branches of his physics: the qualitative, nonformalised theory of motion occupies the whole book IV of the *De Caelo*, together with long sections of book III and II of the *Physics*, while the quantitative features of motion are briefly mentioned in individual passages, all belonging—and designed to underpin—arguments other than those concerning motion. In order to obtain the "true" physics, in this case we must not only collect these scattered hints, but translate them into formulae, which is anything but a safe operation, as the simple example of speed in Sect. 2.3 shows.

Instead, I will discuss a completely different sense in which mathematics is present in Aristotle's philosophy—be it physics, psychology, ethics, or poetics—in a way that is pervasive and essential to knowledge. Yet despite its importance and pervasiveness, it is not so easy to correctly assess the value of such a presence, concealed as it is within a very unfamiliar and "non-mathematical" framework. I am referring here to Aristotle's use of mathematics as a paradigm, a privileged model for the construction of other sciences.

As it is well known, Aristotle considers mathematics a perfect, independent and self-contained system of knowledge. Once mathematical objects have been "separated" from their material counterpart,[18] knowing them—as mathematical objects —is a purely mathematical affair. In particular, mathematical proofs are the appropriate instruments for practising mathematics, since mathematics itself is the supporting structure of mathematics.

For this reason, even if Aristotle does not think that mathematics is the framework of nature, he is well aware of the fact that it is an interesting structure in itself: a structure that cannot be imposed (upon the natural world) but can be imitated (by the natural philosopher).

To begin with, it is imitated by Aristotle himself, who resorts to the paragon of mathematics on two levels:

– *Globally*, mathematics (and geometry in particular) is regarded as the paradigmatic example of a formal axiomatic system: as theorized in the *Posterior Analytics*, all mathematical statements—that are true, eternal, and necessary— can be obtained starting from a few, true, and basic principles (and definitions), using only necessary demonstrations—that is, scientific syllogisms.[19] For Aristotle, the resulting general system is the model of every demonstrative science: though the other branches of knowledge cannot reach the exactness and the absolute necessity of mathematics, they must tend to it as far as possible.[20]
– *Locally*, mathematics is regarded as a source of suggestions: in the whole Aristotelian corpus we find particular mathematical techniques or procedures, examples or problems which are recalled in order to clarify particularly challenging philosophical points. It is as if for Aristotle mathematics were an

---

[18]The operation is called ἀφαίρεσις, this being the standard term for *subtraction* in Greek mathematics. It must be noted that what is subtracted is not the mathematical object, but the more "physical" features of the physical object, namely the ones connected to change. Indeed, in order to subtract something, one has to know the thing to be subtracted, but to know mathematical objects before knowing physical ones leads to a Platonic position, very far from Aristotle's one. For this reason I prefer to avoid the term *abstraction*, too much connoted in this sense.

[19]See in particular *APo* I 2. In addition, Aristotle says that mathematics speaks with the greatest accuracy and simplicity (*Metaph.* M 3, 1078a9-11), and it speaks about the beautiful and the good (*Metaph.* M 3, 1078a32-b6). See *Metaph.* M and N for more general statements about mathematics *contra* Plato.

[20]τὴν δ' ἀκριβολογίαν τὴν μαθηματικὴν οὐκ ἐν ἅπασιν ἀπαιτητέον, ἀλλ' ἐν τοῖς μὴ ἔχουσιν ὕλην (*Metaph.* α 3, 995a14-15); cf. *APo* I 8, passim.

unproblematic realm, whose simplified world can serve to make philosophy just as understandable, at least locally.

It is important to note that, though adopting a mathematical language, here Aristotle is not doing mathematics: he is doing philosophy, so that in reading these passages one should not look for some stringent proof of some (internal, mathematical) truth, but at least for some rough suggestions of some (external, philosophical) way of reasoning.

The idea is simple, even though it is not so familiar to us: suppose you have to explain how sense perception works. You have a subject who senses, an object which is perceived, and a medium (or several media) connecting the two "extremes". Why not use the mathematical theory of proportions, where the function of the extremes and of the middle term is clear and universally understandable? Of course, no exact correspondence can be drawn between mathematical and physical objects, and sometime it is not even clear whether a correspondence of sort is possible; in any case, it is plain that no theorem of the theory of proportion holds in the theory of perception. It is just a matter of analogy.

Likewise, why not resort to the well-known iterative procedures of demonstration to explain the complex notion of potential infinite?[21] Or to the principles of hydrostatics when we need to explain the behaviour of the elements in their mutual relationship, and with respect to their place within the Cosmos?[22]

Finally, why not try to make the general theory of knowledge more understandable by recalling to mind the common educational experience of solving geometrical problems?

Of course, in order to use mathematical analogies properly and fruitfully one must know mathematics, and Aristotle shows that he did. In particular, the way in which he employs and exploits the analogy with the procedure of problem-solving leaves no doubt about his remarkable acquaintance with mathematicians' practice,

---

[21]The potentiality involved in the notion of potential infinite can naturally be traced back to the notion of "processuality" involved in the unending process of division of the continuum, and more generally in the iterative procedures of proof, typical of Greek mathematics. See Ugaglia (2009, 2016).

[22]Aristotle's theory of motion strikes interpreters as being amazingly naive. It certainly is, but only if it is read from a modern "kinematic" perspective, that is, against what has been the standard notional and argumentative background at least from Philoponus onwards. If one adopts the seemingly common-sense perspective that motion is essentially a translation in a space, possibly but not necessarily filled with matter, misunderstandings are unavoidable. Crucial notions such as lightness, for instance, or the non-existence of vacuum, must be introduced as ad hoc hypotheses. On the contrary, no ad hoc hypotheses are needed if one sets out from a model of Aristotle's theory of motion within the framework of hydrostatics: imagine that Aristotle devised his theory of motion in water. Less empirically, imagine that he envisaged it by analogy with what happens in water, that is, by analogy with hydrostatics, understood as a branch of mathematics. Granted, hydrostatics studies systems in a state of equilibrium, but through some extrapolation the processes of reaching equilibrium might lead to some interesting suggestions about how natural bodies move toward their proper places. The hydrostatical origin of Aristotle's theory of motion is discussed in Ugaglia (2004, 2015).

namely, with mathematics as a work in progress, and not just as a formalised system.[23] This aspect of mathematics, which advances largely by trial and error and involves perception, the grasping of things and experience,[24] cannot be formalised, and, while connected to the procedure for proving, does not coincide with it. In particular, it leaves no trace in mathematical literature.

4 Theory of Knowledge and Problem Solving

Take a geometer who has to solve a problem: he looks at the figure he has just traced wondering about what he has to do in order to solve the problem. His mental states, and the way in which they are related to his actions, are very familiar to Aristotle, who employs it in order to explain a more complex state of affairs: the one in which a generic knower passes from potentially knowing something to actually knowing it.

The basic idea is that the cognitive "gap" which separates a state in which one knows (something) only potentially from the state in which one actually knows (it) can be easily understood by considering the "gap" which separates the state of the geometer trying to solve a problem from the state of the same geometer once he has succeeded in (at least a part of) his task.

4.1 The Two-step Process of Knowing

The philosophical point to be clarified is particularly awkward: according to the process of knowing as it is formalized in *De anima*, Aristotle envisages a path from ignorance to knowledge entailing three cognitive states, and two gaps:

$$P1 \rightarrow A1 \, (=P2) \rightarrow A2$$

The first gap lies between a state P1 in which one is a potential knower, because he possesses all the prerequisites for gaining knowledge, and a state A1 in which one actually knows (=possesses knowledge).

---

[23]It is crucial to bear this distinction in mind, in order to avoid a common misunderstanding: the partial and erroneous view of ancient geometric analysis constituting the scholarly vulgata, for example, is in some measure due to a confusion between the mathematician's practice and formalisation; in turn, this confusion is due to an improper reading of Aristotle's mathematical passages (see, for example, the current interpretation of *EN* III 3).

[24]See *APr* I 41, 50a1-2.

The second gap lies between the knower in the state A1 (=P2) in which he possesses knowledge but does not (yet) employ it, and the knower in a new state A2, in which he actually exercises his knowledge.[25]

In *Metaphysics* Θ 9 Aristotle addresses the same subject. Now the context is the power/act opposition, which is to say the topic of book Θ, and Aristotle's ultimate aim is not only to illustrate the passage from power to act, but to explore all the possible modes of the opposition itself. Starting from the basic notions *(i)* of power (δύναμις) as the capacity for doing or suffering something and of act as the related operation (or activity, ἐνέργεια), he comes to his peculiar and philosophically more awkward notion *(ii)* of power (δύναμις) and act (ἐνέργεια or ἐντελέχεια) as states (of being).[26] Contextually, he proves the logical, ontological and temporal priority of the act.[27]

The philosophical challenge Aristotle has in mind is very complex and elusive, and this emerges, even at a cursory reading, from the high number of examples, similes and analogies, which he draws from different contexts, especially in the closing chapters of book Θ. In the elaborate architecture of the argument a key position is occupied by a geometrical example, where a passage from P1 to A1 (=P2) to A2 is illustrated, and the potentialities P and A,[28] read as states *(ii)*, are connected to different potentialities and actualities—let us call them p and a—to be interpreted in the framework of actions *(i)*.

4.2   The Many-step Process of Problem Solving

Take a geometer who solves a problem. As a whole, the procedure is a trivial example of P1 → A1: at the beginning the geometer does not know a result but he possesses (P1) all the geometrical instruments to obtain it. At the end of his work, having obtained it, one can say that the geometer actually knows (A1) this specific result.

---

[25]"We must also distinguish certain senses of potentiality and actuality; for so far we have been using these terms quite generally. One sense of "knower" is that in which we might call a human being knower because he is one of a class of educated persons who have knowledge; but there is another sense, in which we call knower a person who knows (say) grammar. Each of these two persons has a capacity for knowledge, but in a different sense: the former, because the class to which he belongs is of a certain kind, the latter, because he is capable of exercising his knowledge whenever he likes, provided that external causes do not prevent him. But there is a third kind of educated person: the human being who is already exercising his knowledge: he actually knows and understands this letter A in the strict sense. The first two human beings are both knowers only potentially, whereas the third one becomes so in actuality through a qualitative alteration by means of learning, and after many changes from contrary states ‹of learning›, he passes from the inactive possession of arithmetic or grammar to the exercising of it." (*de An.* II 5, 417a21-417b2, the translation is Hett's, modified).

[26]*Metaph.* Θ 1, 1046a9-13; cf. Δ 12, 1019a15-21 and 1019b35-1020a6; Θ 6, 1048a28-29; Θ 8, 1049b5-10.

[27]*Metaph.* Θ 8, passim.

[28]Henceforth, the indices are omitted when the argument holds indifferently in case 1 and 2.

Less trivially, the procedure of problem-solving involves different local instances of A1 (=P2) → A2: at any stage of the procedure the geometer knows a lot of previously obtained results, but normally they are dormant in his mind (P2). At a certain point—under some solicitation, which we will discuss later—he finds in his mind a particular result and employs it (A2).

Moreover, the states P and A are mental cognitive states (of the geometer), but the procedure of problem-solving gives a clear idea of how these mental states are connected to, and depend on, some particular practical actions. In more Aristotelian terms, it clarifies the way in which an act as a state (ἐντελέχεια) is related to an act as an action (ἐνέργεια).

Indeed, the passage P → A is not immediate: between the two extreme states of ignorance and knowledge something occurs, some practical action is done which enables the transition.

In the case of the geometer this is particularly evident, for the actions (p → a) coincide with the simple action of tracing (auxiliary) lines in a diagram.[29] Take a geometer in the state (P) of potentially knowing something. In order to move to the state of actually knowing it (A), he starts from an incomplete diagram, in which all auxiliary lines are only potentially (p) contained—that is to say, they have not been traced (yet) but could be traced.

Through the action of drawing (a) some of the lines the geometer generates the complete figure, that is, the one actually displaying the solution sought by his well-trained geometrical mind.

This allows the final passage to the state of knowledge, namely the state of possessing the solution: P—(p → a) → A. But let us see how Aristotle himself explains this situation (the comments in brackets are mine):

> Geometrical relations too are found by actualization. Indeed, one finds them by dividing (p → a): if they were already divided (a), they would be evident. Now, it is the case that they are only potentially (p) present.
>
> (1) Why does the triangle have two right angles? Because the angles round a single point are equal to two rights. If, then, a parallel to the side were raised (p → a), by simple observation, the reason would be clear (P1 → A1).
>
> (2) In general, why is the angle in a semicircle a right angle? If three are equal, the two bases and the straight line lying orthogonally upon the center (p → a), just observing would be clear (P1 → A1), if one knows this ⟨and employs it (P2 → A2)⟩
>
> So that it is manifest that what is only potentially (p) is found by being actualized (a). The reason is that thinking is an actuality. Thus potency comes from actuality, and therefore one knows (P → A) by doing (p → a); in fact, in particular cases, the actuality is later in genesis.[30]

---

[29]In this case, a diagram must be conceived not as a result—that is, as part (the right κατασκευή) of a formalized proof—but as a work in progress. Aristotle is not interested in the final diagram—such as those accompanying proofs in Euclid's *Elements*—but in the construction viewed in its process of development.

[30]*Metaph.* Θ 9, 1051a21-33.

4.2.1 To Know 2R

In Aristotle's example, the passage **(1)** describes someone moving from a state of potential knowledge of 2R to a state of actual knowledge of it: **P1 → A1**.

P1: *Potential knowledge of 2R*—The geometer does not know 2R, but he possesses all the necessary prerequisites:

- he knows the basic principles of geometry
- he knows the deductive rules
- he has a touch of imagination, insightfulness, a good eye, perseverance...

→: *Problem-solving*—In order to prove 2R, our geometer has drawn a diagram, presumably a triangle, and he is looking at his figure, but at the beginning he does not see anything but a triangle.

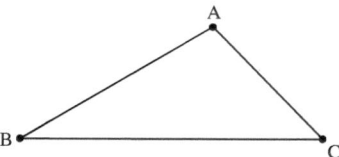

Perhaps—he thinks—it may be useful to draw (p → a) some auxiliary line, not yet traced in the diagram: for example, the line parallel to one of the sides of the triangle, passing from the opposite vertex.

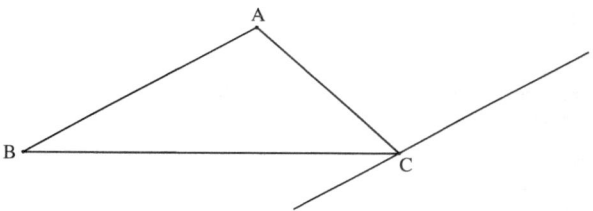

But once the line has been traced, he sees, and immediately knows, that the three angles around the vertex, whose sum is clearly two right angles, are equal to the three angles of the triangle.

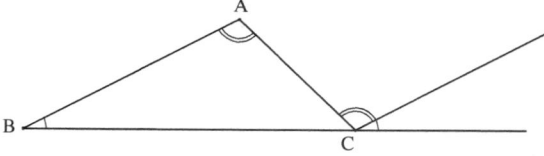

A1: ***Actual knowledge of 2R***—Hence, the geometer can say that he actually knows 2R.

4.2.2 To Use 2R (in Order to Know SC)

The same situation, namely the passage **P1** → **A1,** arises again in the same example, in passage **(2)** which describes someone moving from a state of potential knowledge of SC to a state of actual knowledge of it:

P1: ***Potential knowledge of SC***—the geometer does not know SC, but he possesses all the necessary prerequisites:

- he knows the basic principles of geometry, and 2R
- he knows the deductive rules
- he has a touch of imagination, insightfulness, a good eye, perseverance…

A1: ***Actual knowledge of SC***—Once the problem is solved, the geometer can say that he actually knows SC.

But there is more to it. Passage **(2)** also describes the geometer moving from a state of mere possession (A1) (=potential use (P2)) of the knowledge of 2R to a state of actual use (A2) of it: **P2** → **A2**:

P2: ***Potential use of 2R***—The geometer knows a lot of geometrical results, one of which is 2R. He looks at the figure he has traced for proving SC: a basic diagram, containing the data of the problem (an angle in a semicircle), and he does not see anything but an angle in a semicircle.

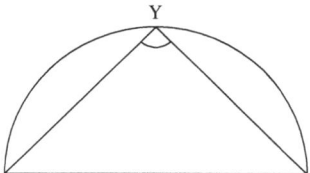

→: ***Problem Solving***—Perhaps—he thinks—it may be useful to draw (**p** → **a**) some auxiliary line: for example the radius lying orthogonally upon the center.

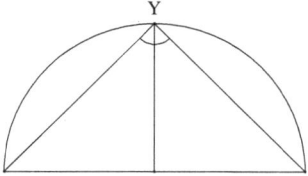

A2: Actual use of 2R—But once the line has been traced, he immediately "sees" some figures in the diagram: the three triangles XYZ, XOY and YOZ, to which he can apply the knowledge of *2R* he possesses but has not yet used.

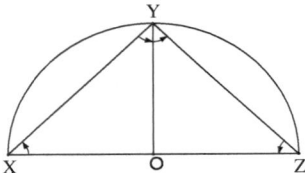

In this way, the geometer undergoes a change from a state A_1 (=P_2), of merely having knowledge, to a state A_2, of actually using his knowledge. To be more precise, he undergoes three changes of this kind: one each time he applies *2R* to a triangle.

But this is not all. The global, macroscopic change **P1 → A1,** from ignorance to knowledge of *SC,* is accomplished by means of various local changes:

(1) **p → a: drawing** OY perpendicular to XZ
(2) **deducing** that the triangles XOY and YOZ are right-angled in O (by construction)
(3) **seeing** that OX, OY and OZ are radii of the same circumference.
(4) **deducing** that XOY and YOZ are isosceles (by construction)
(5) **P2 → A2: applying** 2R to XOY and YOZ
(6) **syllogising** that in both cases the two acute angles sum up to half of two rights (using 2 and 5).
(7) **syllogising** that all the acute angles in the diagram are equal to one another (using 4 and 6)
(8) **seeing** that in particular the angles in X and Z sum up to half of two rights.
(9) **P2 → A2:** applying 2R to XYZ
(10) **concluding** that the angle at Y amounts to the remaining half of two rights, hence that it is right.

4.2.3   The Solving Idea

The steps of a procedure of problem-solving are not all on the same level: indeed, there is a step at which the geometer has *the* idea, which solves the problem.

In the specific case of proving SC, the idea is to read right as the half of two rights, as Aristotle himself explains in *Posterior Analytics*:

(3) For what reason is the angle in a semicircle a right angle? It is a right angle if what holds? Let right be A, half of two rights B, the angle in a semicircle C. B explains why A, right, holds of C, the angle in a semicircle. For B is equal to A and C to B (it is half

of two rights). Thus if B, half of two rights, holds, then A holds of C (that is, the angle in a semicircle is a right angle).[31]

In a syllogistic form:

A(right)	a	B(half of two rights)
B(half of two rights)	a	C(angle in a semicircle)
A(right)	a	C(angle in a semicircle)

Although right and the half of two rights are apparently equivalent statements, the second expression conveys more information, for it holds a clue for the way in which the result right has been obtained. Suppose we rephrase our initial question: "Why is the angle in a semicircle a right angle?" as follows: "Why is the angle in a semicircle half of two right angles?"

The reference to two rights encourages the geometer—who is in a state of A1 (=P2) knowledge of the 2R property—to make use of it ($\rightarrow$ (A2)).

But to do so, he must find in his diagram a suitable triangle, which must have the property that one of its angles is equal to the sum of the other two angles. Is this the case with the triangle drawn? Yes, and the most direct way to show it is to trace (*a*) the radius OY and take a look at the resulting angles, as in the proof sketched above.

But this is exactly what Aristotle means when he says that half of two rights is the reason why the angle in a semicircle is right. And this is nothing but another way of stating that reading "right" as "the half of two rights" is the idea, which solves the problem.

In a syllogistic form, the half of two rights is the middle term: bear in mind that for Aristotle to solve a problem means to find a suitable syllogism, which has the solution of the given problem as its conclusion and the solving idea as a middle term.

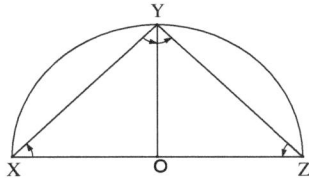

[31]*APo* II 11, 94a28-34.

$$OX = OY \rightarrow OX^\wedge Y = OY^\wedge X$$
$$OY = OZ \rightarrow OY^\wedge Z = OZ^\wedge Y$$
$$\rightarrow X\ Y^\wedge Z = O\ X^\wedge Y + O\ Z^\wedge Y$$
$$\text{But } XY^\wedge Z + OX^\wedge Y + OZ^\wedge Y = 2R(\text{th. } 2R)$$
$$\rightarrow XY^\wedge Z = 1/2 \cdot 2R\ [\text{B a C}]$$
$$\text{But } 1/2 \cdot\ 2R = R\ [\text{A a B}]$$
$$\rightarrow XY^\wedge Z = R\ [\text{A a C}]$$

4.2.4 Conclusions

The setting of the Aristotelian examples is always the same: there is a geometer, trying to solve a problem. He is equipped with a material support, on which he has drawn a diagram (p), reproducing the data of the problem, and with a mental supply (P) of known results, properties and theorems.

Progressing in his job, the geometer traces (=actualizes p $\rightarrow$ a) a line—but which line, among the many traceable (=potential p) lines?—in order to apply (=actualize P2 $\rightarrow$ A2) a result—but which result, among the many, already (=potentially P) known but not yet employed results?

He needs the right idea. Indeed, if all the required lines were already (a) traced, as in the final diagrams one finds in Euclid's *Elements*, the solution would be trivial.

But most of the time, the diagrams a geometer is accustomed to working on are obviously not of this sort, and drawing a figure, that is, progressively adding new lines in order to "see" the solution, is an essential part of the process of problem-solving. The fact is that to trace (only) those lines that are really required is not so easy.

In many cases one will add a non-useful line that just muddies the waters, as in the case of geometers who (deliberately?) draw fake diagrams (οἱ ψευδογραφοῦντες).[32]

In other cases one will add a potentially useful line, but will fail to "see" the right figure among the different combinations the diagram presents: "I know that any triangle has 2R but I do not know that *this* is a triangle", as Aristotle says in a controversial passage of *APr* II 21.[33]

---

[32]*Top.* VIII 1, 157a1-3: ἔτι τὸ μηκύνειν καὶ παρεμβάλλειν τὰ μηδὲν χρήσιμα πρὸς τὸν λόγον, καθάπερ οἱ ψευδογραφοῦντες· πολλῶν γὰρ ὄντων ἄδηλον ἐν ὁποίῳ τὸ ψεῦδος «next, stretch out your argument and throw in things of no use towards it, as those who draw fake diagrams do (for when there are many details, it is not clear in which the error lies)» The translation is Smith's.

[33]Here it is not a question of seeing something (a triangle) and not knowing that this something is a triangle, but rather of seeing nothing (interesting) at all. The geometer, sitting in front of his figure, would like to see a triangle, or some other figure, to which he could apply some (known) results. But he does not see anything interesting.

5 Mathematical Analogies and Building Science

I will conclude my analysis by placing Aristotle's analogical use of mathematics within the overall scheme of the procedure of building a science.

For Aristotle a science is both a result, i.e. a coherent formalised system of knowledge, and a work in progress, i.e. a complex of problem-solving procedures. Both these aspects are theorized in the *Analytics*, but concerning the effective exercise of the theoretical instructions he gives there, one must observe that Aristotle's treatises are for the most part devoted to problem-solving.

It is almost as if Aristotle thought that once the starting point of demonstrations and the right definitions—which are the core of any knowledge—[34] have been grasped, all the rest follows: principles and definitions contain everything, and it is a mere matter of exercise to deduce it.[35]

Now, to deduce things from principles is not what scientists actually do: it is true —in fact it is proved in the *Analytics*—that every result pertaining to a science can be deduced from its principles but this is too broad an operation, and actually a very uneconomical one. What the scientist needs, in practice, is something different: a sort of "guided" deduction.

In some sense, he has to know the problems he has to solve[36]; then he solves them in some way—maybe in a non-scientific way—and once he has obtained a result he tries to incorporate it within a deductive scheme. In other words, he tries to transform the solution he somehow has found into a scientific solution. Indeed, a scientific solution to a problem is nothing but a way to connect three terms in a scientific syllogism, i.e. to deduce the conclusion from the premises.

5.1 Theoretical Work: Induction and Deduction

Any scientific knowledge, or science, is a set of results, which can be arranged into a deductive system, grounded on first principles and definitions.

A formalized science consists of two distinct theoretical procedures: a preliminary search T_1 aimed at discovering first principles and creating definitions, and a deductive process T_2, aimed at obtaining from them all the known results.

---

[34]*APo* II passim.

[35]In this sense, a simple answer can be given to interpreters who act surprised when they find in Aristotle's *Analytics* not a *prescription* for the construction of demonstrations or for the acquisition of knowledge, but a *definition* of what a demonstration, or knowledge, is.

[36]If I do not know what I am searching for, I cannot correctly articulate my question, and my search is very difficult, as Aristotle explains in *Metaph.* Z 17 (cfr. B 1).

- T_1 is a scientific procedure leading, by means of perception, induction and memory, to the mental state Aristotle calls *nous*.[37] The exposition of the procedure is confined to the few paragraphs of *Posterior Analytics* II 19, while the procedure is employed, albeit not discussed, in all the physical treatises: *Physics*, *Metaphysics*, *De caelo* (partially)... In particular, it is not clear how T_1 applies to mathematics, but it is important to stress that in any case it has nothing to do with mathematical analysis, which is a formal, deductive procedure, and not a heuristic practice.[38]
- T_2 too is a scientific procedure, which by means of scientific syllogism or demonstration leads to the state of scientific knowledge. The exposition of the procedure occupies part of the *Posterior Analytics* but, with the exception of *De incessu animalium*, it is almost never employed in Aristotle's treatises.

The result of the composition of T_1 and T_2 is a net T of connections between ordered terms,[39] whose order is fixed and hierarchical. However, it is not in this way, by simply combining a theoretical phase T_1 with a theoretical phase T_2, that a science is built.

For Aristotle a science is the final result of a more complex procedure, which consists for the most part in practical work, which is to say: in the procedure of problem-solving, which intertwines T_1 and T_2, and which is indispensable for doing T_2. In the following sections we will see how.

5.2 Practical Work: Problem Solving

For Aristotle, to solve a problem is to give an answer to local questions of the sort "Why does B belong to A?", where establishing why B belongs to A ultimately amounts to connecting (συνάπτειν)[40] the two terms by means of a "likely" connection.

The procedure of finding such a connection—namely, the procedure of problem-solving (henceforth *PS*)—is not a scientific procedure: Aristotle himself

---

[37]I adopt here the interpretation of *nous* and *epistēmē* as intellectual states, as maintained for instance in Barnes (1993).

[38]See note 22 before.

[39]It is important to point out that for Aristotle the elements of a science are terms, and not propositions.

[40]*APr* I 23 passim.

prefers to speak of a "way" (ὁδός),[41] and its results are not necessarily scientific results. *PS* is common to all disciplines[42] and is performed by means of wisdom and sagacity (ἀγχίνοια, εὐστοχία),[43] together with some proper instruments: perception, calculus, analysis, ἀπαγωγή, βούλευσις... but also experience, training, practical arrangements, loci....[44] The exposition of *PS* occupies a whole range of treatises—*Topics, Rhetoric, Historia animalium, Posterior Analytics* and *Prior Analytics*—and is used everywhere.

The idea is to give instruction for succeeding in *PS*, and summarizing Aristotle's prescriptions, we can say that the problem-solver must: (1) have a good memory and possess good samples of data; (2) between these data he must be able of finding as many antecedents A_i of the predicate B and consequents B_i of the subject A as he can. Then he must: (3) choose among the antecedents and the consequents the more appropriate ones, and (4) act on them using similitudes, analogies, etc. in order (5) to transform the given problem into a simpler problem or a more general one.

As stated before, *PS* is not a scientific procedure, and Aristotle's prescriptions cannot be translated into unambiguous, effective logical rules.[45] They are more like a set of practical instructions, establishing a sort of hierarchy of possible solutions, ranging from extremely tenuous connections—for instance, A and B are connected by means of particular cases—to stronger ties, or syllogisms, when B and A are related by means of a common denomination, a simple sign, or a probable middle term; up to necessary connections, or scientific syllogisms, when B and A are related by means of a causal medium. Using the rules previously described, one can

---

[41]Ἡ μὲν οὖν ὁδὸς κατὰ πάντων ἡ αὐτὴ καὶ περὶ φιλοσοφίαν καὶ περὶ τέχνην ὁποιανοῦν καὶ μάθημα (*APr* I 30, 46a 3-4). On the fact that in *PS* not only formal connections, but also material connections or arguments grounded on signs and common opinions are acceptable, see for example *APo* I 6; *APr* II 23-27; *Top.* I 10 and I 14; *HA* passim. Consider in particular *Historia animalium*: with the big collection of material contained here, which consists of purely material connections, it is impossible to build a theoretical system. In order to convey this material in a scientific system it is necessary to give a "direction" to the results obtained, introducing a hypothetical necessity which steers the process of deduction.

[42]*APo* I 33; *Top.* VIII 1. As a good example of practical work without any integration in a theoretical system see Aristotle's *Problemata*, which is a collection of material which can be useful in different contexts. On Aristotle's *Problemata* see Quarantotto (2011, 2017).

[43]*APo* I 33-34; cfr. *EN* VI 5.

[44]*APr* I 27-31; *Top.* I 2; *APo* II 11-17; *SE* 16; *EN* VI 7.

[45]In fact, this may be a controversial point. At least some of the procedures Aristotle describes are the object of interesting studies in the contemporary philosophy of science; see for instance Magnani (2001) on the notion of abduction. More generally, it might be interesting to contrast Aristotle's conception of problem solving, and its relation to logic, with the recent developments in the studies of theory-building as problem-solving, where a logical toolbox for such a kind of procedure is proposed and discussed. See in particular Cellucci (2013), Ippoliti (2014) and the papers contained in this volume.

go from one problem to another, and from one solution to another, possibly transforming a weak connection into a stronger one.[46]

This is a crucial operation, for while "weak" solutions (henceforth *S*) can be used in dialectic, in rhetoric, in poetic and in whatever non-scientific discipline, or art, only the last kind of solution (henceforth S_s) involving a necessary connection, can become part of a science *T*. For example, Bryson solved the problem of the quadrature but his result cannot be inserted in any science: it is a solution but it is not a scientific explanation.[47]

In other words, the procedure of problem-solving in not enough to do science: some further condition must be added, which allows the scientist to pick up, among the obtained solutions *S*, only the scientific ones S_s.

To sum up, a scientific solution S_s differs from a generic solution *S* since the connection between the two terms A and B must be established by means of a causal medium C. Moreover, the connection between C and the other terms—namely the premises (*CaA*; *BaC*)—must be provided by either first principles, or definitions, or demonstrated theorems. In other words, the premises must be first, immediate, necessary, and proper.[48]

Under these conditions, the subset $S_s \subset S$, which is to say the scientific results of the procedure of problem-solving, can become a part of *T* (Fig. 6).

5.3   The Natural Place of Mathematics

The schematic diagram that has just been traced illustrates Aristotle's way of doing research but also applies, with few modifications, to our own way of doing research. The main modification concerns the proper instrument of the deductive step T_2: mathematical demonstration in our case, the scientific syllogism for Aristotle, where the scientific syllogism is something broader, and diversified according to the different branches of knowledge.

To deduce things using a scientific syllogism is to connect the subject and the predicate of the conclusion by means of a proper causal medium. By definition, the medium term must be a formal cause—for only a formal cause lies in the essence—

---

[46]In particular, In *Prior Analytics* Aristotle explains what a syllogism is (I 1-26 and II 1-22); how it must be constructed, i.e. how two given terms can be related (I 27-31 and II 23-27), and how a syllogism can be reduced to another, i.e. how the relation between the two terms can be usefully changed into another relation, in order to obtain a more scientific result (I 32-45).

[47]*APo* I 9.

[48]In this perspective one can say that the actual subject of *Prior Analytics* is problem-solving, while that of *Posterior Analytics* is scientific problem-solving. If all the sciences were like mathematics, there would be no need for *Prior Analytics*: since every mathematical proof may be reduced to a Barbara syllogism, it is sufficient to prove the relation in a general case and it will hold in any other (*Top.* II 3, 110a29), and will be easy to convert. On scientific problem-solving in Aristotle see Mendell (1998). The most complete account of the problem-solving approach in ancient Greek mathematics is Knorr (1986). On its relation to analysis, see Mäenpää (1993).

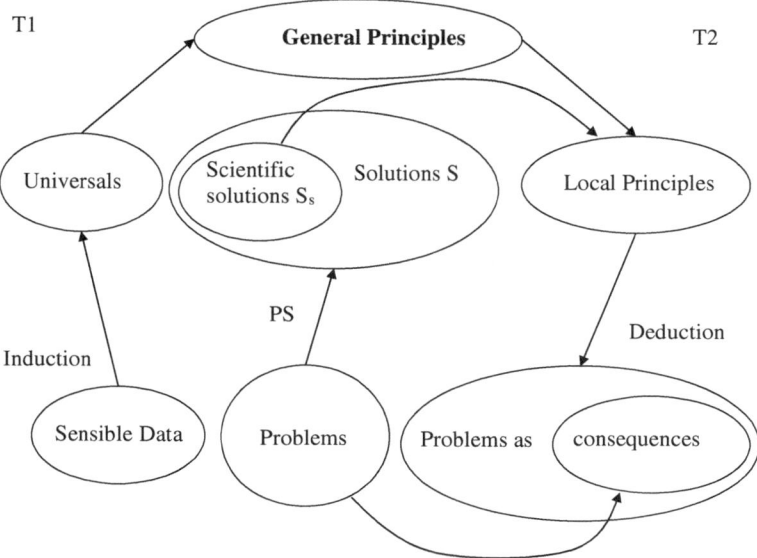

Fig. 6 Aristotle's system of knowledge

but the meaning of formal cause is not the same in philosophy and in mathematics. In physics, for example, where form is synonymous with nature, and hence with the principle of change, the formal cause coincides with the final cause and the necessity involved is a hypothetical necessity. In mathematics, instead, the formal cause is identified with the material cause, which is to say with absolute material necessity.[49]

For this reason, we find very few mathematical demonstrations in Aristotle's system at step T_2, namely the few corollaries described in Sect. 2.2.[50] On the contrary, in mathematical physics the whole demonstrative structure of T_2 is mathematical.

On the contrary, we find a lot of mathematics in Aristotle's system at step PS. As this is informal, non-deductive and heuristic, it is not surprising that mathematics is employed here in the informal, non-deductive and analogical way described in Sect. 3.

Mathematical physicists too employ mathematics in PS, namely in the heuristic phase of their work. But while Aristotle's treatises are for the most part devoted to

---

[49]Notice that, to reduce physical demonstrations to mathematical proofs would be to introduce determinism in nature, a hypothesis which Aristotle firmly rejects. On Aristotle's teleology as an alternative to determinism, see Quarantotto (2005).

[50]For Aristotle mathematical proofs can be reduced to syllogisms in Barbara (*APo* I 14). On the subject see Mendell (1998).

the description of this procedure, no traces of *PS* are left in physics textbooks, where only the deductive phase T_2 is described.

Perhaps one can find some traces of *PS* in scientific biographies of mathematical physicists, or scientists in general, and, as expected, mathematics here has the same analogical, evocative heuristic value as poetry, or music, or even LSD.[51]

It is important to note that the fact that the way of proceeding in *PS* is not scientific at all is absolutely irrelevant with respect to the final resulting science T_2: once a result, no matter how it has been obtained, can be translated in mathematical terms, and once the effective procedure of conjecturing it can be turned into a strict mathematical proof, it becomes a scientific result. Conversely, no matter how many mathematical terms the conjectural work done in *PS* displays, if the result cannot be re-obtained as the outcome of a mathematical proof, it is not a scientific result at all.[52]

Bibliography

Barnes. (1993). Aristotle, posterior analytics. Translated with a Commentary by J. Barnes. Oxford: Clarendon Press.

Cellucci. (2013). Rethinking logic. Dordrecht: Springer.

Galilei. (1718). Postille di Galileo Galilei al libro intitolato Esercitazioni Filosofiche d'Antonio Rocco, filosofo peripatetico, in *Opere*, Firenze Tartini.

Ippoliti. (2014). *Heuristic reasoning.* Berlin: Springer.

Kaiser. (2011). *How the hippies saved physics.* New York: Norton & Company.

Knorr. (1986). *The ancient tradition of geometric problems.* Boston/Basel/Berlin: Birkhäuser.

Lacan. (1977). The subversion of the subject and the dialectic of desire in the Freudian unconscious. In *Écrits: A Selection* (pp. 292–325). New York: Norton.

Mäenpää. (1993). *The art of analysis. Logic and history of problem solving.* Helsinki (dissertation).

Magrani. (2001). *Abduction, reason, and science. Processes of discovery and explanation.* New York: Kluwer Academic.

Mencell. (1998). Making sense of Aristotelian demonstration, *Oxford Studies in Ancient Philosoph, XVI,* 161–225.

Mullis. (1998). *Dancing naked in the mind field.* New York: Pantheon.

Quarantotto. (2005). *Causa finale, sostanza, essenza in Aristotele.* Napoli: Bibliopolis.

---

[51]See for instance the remarks about LSD made by the chemist and Nobel laureate Kary Mullis, who invented the polymerase chain reaction (PCR) that helps amplify specific DNA sequences, in Mullis (1998). On the role of non-conventional heuristic tools in science see also Kaiser (2011).

[52]"Thus, by calculating that signification according to the algebraic method used here, namely: $\frac{s(\text{signifier})}{s(\text{signified})} = s(\text{the statement})$ with $S = -1$ produces $s = \sqrt{-1}$. [...] Thus the erectile organ comes to symbolize the place of *jouissance*, not in itself, or even in the form of an image, but as a part lacking in the desired image: that is why it is equivalent to the $\sqrt{-1}$ of the signification produced above, of the *jouissance* that it restores by the coefficient of its statement to the function of lack of signifier (-1)" (Lacan 1977). On this and other cases of misuse of mathematics see Sokal and Bricmont (1998).

Quarantotto. (2011). Il dialogo dell'anima (di Aristotele) con se stessa. I Problemata: l'indagine e l'opera, In B. Centrone (a cura di), *Studi sui Problemata physica aristotelici* (pp. 23–58). Napoli: Bibliopolis.

Quarantotto. (2017). Aristotle's problemata-style and aural textuality. In W. Wians & R. Polanski (Eds.), *Reading Aristotle*: *Argument and exposition* (pp. 97–126). Leiden: Brill.

Rovelli. (2015). Aristotle's physics: A physicist's look. *Journal of the American Philosophical Association, 1*, 23–40.

Sokal & Bricmont. (1998). *Intellectual impostures: Postmodern philosophers' abuse of science.* London: Profile Books.

Ugaglia. (2004). *Modelli idrostatici del moto da Aristotele a Galileo.* Rome: Lateran University Press.

Ugaglia. (2009). Boundlessness and iteration: Some observation about the meaning of ἀεί in Aristotle, *Rhizai, 6*(2), 193–213.

Ugaglia. (2015). Aristotle's hydrostatical physics, *Annali della Scuola Normale Superiore di Pisa, Classe di Lettere e Filosofia, 7*(1), 169–199.

Ugaglia. (2016). Is Aristotle's Cosmos Hyperbolic? *Educação e Filosofia, 30*(60), 1–20.

Activation of the Eddy Mental Schema, Multiple Analogies and Their Heuristic Cooperation in the Historical Development of Fluid Dynamics

A. Ulazia

Abstract If we analyze the early historical evolution of the concept of eddy or vortex, it can be shown that it constitutes a mental schema and a powerful heuristic instrument in scientific thinking, with the intuitive properties of rotation and dissipation. This mental schema presents a great capacity to constitute a fruitful analogical source and to develop an analogical inference. For example, Descartes considered celestial vortexes to explain planetary motion, or Maxwell developed the electromagnetic theory via a model based on rotating vortexes. But there were more creative and detailed uses of the eddy schema which had great importance as a cooperative heuristic instrument, instead of as a mere expedient analogy. I will present two episodes to underline the activation via provocative analogy of the eddy schema, multiple roles that it can play and its heuristic adaptability: first, the *eureka* visualization of an eddy by Johann Bernoulli in the genesis of fluid dynamics; and second, Reynolds's discovery of the importance of eddies to understand the dynamic and resistance of flow.

Keywords Fluid mechanics · Heuristics · Analogy · Bernoulli Reynolds

1 Introduction

In cognitive science, a mental schema is a non-linguistic interpretation applied to an object or an image of an object that allows an investigator to gain conceptual purchase on unfamiliar ground. In this cross-domain sense, it is related to analogy. A mental schema can produce the imagery needed to adapt to the peculiarities of a new context. As a specific configuration in our brain, it is adaptable because it is not strongly tied to specific material systems. Somewhat ironically, the schema's abstract nature is what

A. Ulazia (✉)
NI and Fluid Mechanics Department, University of the Basque Country (EHU/UPV), Otaola 29, Eibar, Basque Country, Spain
e-mail: alain.ulazia@ehu.eus

© Springer International Publishing AG 2018
D. Danks and E. Ippoliti (eds.), *Building Theories*, Studies in Applied Philosophy, Epistemology and Rational Ethics 41, https://doi.org/10.1007/978-3-319-72787-5_8

allows it to offer concrete solutions. Many psychologists, cognitive scientists, and even musicians relate it to perceptual-motor schemas because our perceptions and corporeal intuitions are used to support both concrete tasks and abstract reasoning. They are embodied cognitive bases for perception and body motion.

The schematic ideas also contain the properties of the embodied mind that are the basis of Saunders Mac Lane's categorical perspective of mathematics (2012), to the extent that cultural activities and their perception and action schemas are included. Firstly, the mental schemas have related cultural expressions: in the case of the schematic consideration of the eddy, the eddy has a powerful metaphorical projection in culture, religion or literature represented by spiral or wheel forms. Secondly, these schemas used to have an axiomatic status in mathematics: the eddy schema appears in electromagnetism or fluid kinematics, but its use is analogous in both cases, and after all based on mathematically fundamental laws of differential geometry such as Stokes' theorem. This fundamental character of mental schemas is also described for arithmetic by Lakoff and Nuñez's "basic metaphor of arithmetic: numbers are groups of objects" (2000). Mac Lane elaborated the relationship between cultural activities, ideas and branches of mathematics, and in his *Activity-Idea-Formalism* table presents the activity of moving connected to the idea of change, together with mathematical formalism as rate of change and derivative. The eddy schema offers a primitive image in this sense, since it creates a spatially confined rate of change as in the case of a wheel. Herein lies one reason for its axiomatic status: even cognitively, it has a primitive nature, and is able to create powerful analogies.

Therefore, I will present mental schemas intimately related to analogical reasoning. Although it is evident that analogy is one of the most important heuristic instruments in science (and thus the cooperation of analogy and mental schemas must be frequent), there is another main reason for this intimate relation: a mental schema is cross-domain by nature, and its capacity of abstraction can establish a new comparison between supposedly distant domains. The wave concept is a good example in physics, where we can find ocean waves, light waves and acoustic waves. That is, in these cases the mental schema would be the *tertium comparationis*, the third bounding element between the analogical source (first element) and the analogical target (second element). In any case, the abstract mental schema is often considered as the source under a particular form, since it offers a concrete way (often an image) to understand something. In this way, it is possible to speak about a mental schema as a strong analogical source in the construction of a new theory.

But this habitual heuristic perception of mental schemas as powerful analogical sources that constitutes the base of an entire model or proto-model often presents too strong and too simplified an axiomatic hierarchy. This perspective is mainly post-heuristic and immersed in the context of justification. As we will see, the interest here is to study the heuristic power of the eddy schema in the context of discovery and to identify its cooperation with other heuristic instruments through the strategies developed in the construction of explanatory models. I think that this strategic view of heuristics in science must be developed more deeply than we are used to at the present time. In this sense, a mental schema could show multiple heuristic operations in different phases of theoretical construction and this variety opens the gates for different strategic decisions in creative scientific thinking. As I will describe later, I consider this perspective under the more general background of abductive reasoning.

In recent studies of the issue (Ulazia 2015, 2016a, b), the author applies these concepts in the study of the Bernoulli's creative thinking in the foundation of fluid mechanics in the XVIII century. In this time, to go beyond the vision of Newtonian particles, a new set of images was needed in order to deal with the spatial extensibility and lack of form of fluids. I point to evidence that the introduction of mental schemas via analogy was an essential abductive strategy in the creation of this imagery. But its heuristic behavior is complex: analogy can provide an initial model or proto-model that establishes the starting point of a theoretical process, but it can play other roles as well. The historical genesis analyzed by me in this previous study showed that the participation of analogy in physicists' creativity was not so restricted and that its richness opened up the field for very different roles and strategies in model-based discovery processes. Analogies, images, extreme case reasoning and thought experiments or simulations can cooperate establishing *runnable mental schemas*, and even activate these processes at origin, that is, they can play a provocative role of initialization in the form of a mental schema.

Hence, this presentation is concerned with the methodology of mental schemas, analogy and other heuristic instruments within its renewed interest for history and philosophy of science, and it is mainly focused on cognitive aspects of the creative process in science rather than on the historical and philosophical background of this creativity. There are rich conceptual issues involved in discovery, and this context needs an interdisciplinary treatment. Boundaries between logic and methodology (Aliseda 1997), and logic and psychology (Thagard 2002) are vague. According to some researchers (Hintikka 1985; Jung 1996; Paavola 2004), this leap to focus more on discovery as process needs a reconsideration of "abductive strategies". This strategic need arises from an aim to understand processes of discovery and not just finished products (Sintonen 1996).

One central point in strategy is that it cannot normally be judged only in relationship to particular moves, but the whole strategic situation must be taken into account (Hintikka 1998, p. 513). This means that "in strategies more than one step or move can and must be taken into account at the same time". This is an essential point for abductive inference. In abduction, auxiliary strategies are important because abduction on its own is weak and would generate many different hypotheses. But "a hypothesis should be searched for in relationship to various phenomena and background information" (Paavola 2004, p. 270), and we can construct an explanation on this background if we have strategies. In our opinion the cooperation at different levels of mental schemas, analogies, and other heuristic instruments may offer these strategies. Due to this generative and creative character, Clement uses the term "generative abduction" to refer to this kind of theoretical construction process and he explains "how a coalition of weak, nonformal methods" generates these constructions (Clement 2008, pp. 325–330).

If we take the original definition of abduction, Hanson distinguished three points in his logic of discovery (Hanson 1958). Abduction

1. Proceeds retroductively, from an anomaly to
2. The delineation of a kind of explanatory hypothesis which
3. Fits into an organized pattern of concepts.

We think this can be seen as a strategy, and in the third point, in which a pattern of concepts is organized, mental schema is an important cognitive element together with the analogical strategy to get a plausible fit between available concepts and facts.

2 Views of Analogy

In recent years different usages of analogy have been identified, and they have been analyzed philosophically and cognitively (Bartha 2010; Darrigol 2010; Hon and Goldstein 2012). These studies describe views of analogy as moving from a traditional proportionality scheme involving two given situations (apple is to cider, as grape is to wine), to a more generative and creative view, via historical, pioneering, and contrived analogies such as Maxwell's (1862) between electromagnetism and fluid flows. According to Hon and Goldstein, "Maxwell did not seek an analogy with some physical system in a domain different from electromagnetism; rather, he constructed an entirely artificial one to suit his needs" as "an early version of methodology of modelling" (Hon and Goldstein 2012, p. 237). In this presentation I will show that even prior to Maxwell (Johann Bernoulli) or at the very moment (Reynolds), analogy was not always proposed between established concepts and across domains; rather, it related fictional constructs to real physics via the cooperation with mental schemas and other heuristic instruments. The artificial construction of the analogical source was not as complex as in Maxwell's case. But cooperating analogies were involved with modeling the foundations of fluid dynamics around the idea of eddy.

This kind of reasoned process "involving analogies, diagrams, thought experimenting, visual imagery, etc. in scientific discovery processes, can be just called model-based" (Magnani 2009, p. 38). Similarly, Peirce stated that all thinking is in signs, and that signs can be icons, indices or symbols; in this way, all inference is a form of sign activity which includes emotions, images, and different kinds of representations. That is, thinking is often model-based, but creative thinking which implicates great conceptual changes is always model-based. Even when we are reading an array of symbols and inferring an idea from it, there are implicit icons intertwined in the mental process. This more general activity happens in what Magnani calls "the semiotic brain", in which we can consider a semiotic internal tool for organizing the brain giving predominance, according to abduction, to the most plausible ideas and pattern of concepts.

This suggests that the cooperation between symbols, analogies and other heuristic procedures integrates information from various sources for the construction of new mental models. Analogy can be treated as an operational dynamic procedure guided by what can be called creative abduction, or the capacity and the "method" of making good conjectures (Kirlik and Storkerson 2010, p. 32), in opposition to more traditional views of scientific theory construction. This dynamic view on analogy, and on the construction of new hypotheses in general, have been put forward by the so-called "friends of discovery", who move from Lakatos (1976) and Laudan (1980), and go down to Nickles (1980), Cellucci and Gillies (2005), and more recently Ippoliti (2015).

In the predominant view where the analogy is static rather than being viewed as a process, normally studies on analogical reasoning have regarded the source as being generated associatively, that is, recovered from the memory because it is familiar. This outlook is evident in Gentner's structure-mapping theory in which systems governed by higher order relations have an implicit predominance (Gentner 1983). A good example is the analogy between hydrodynamics and electricity, in which the electric potential would be equivalent to the hydraulic head, the current to the volume flow rate, the electric charge to quantity of water, etc. However, after all, these relational structures are not so static, since Gentner also specifies procedures for mapping in an attempt to introduce dynamic thinking.

Holyoak and Thagard's multi-constraint theory, in a pragmatic approach involving problem-solving theory, shows that analogy can produce new categories by abstracting the relationships between cases to a correspondence structure (Holyoak and Thagard 1989). This abstraction process is carried out under various constraints and describes a certain evolution of analogy towards greater refinement; but the theory does not take into account how analogy may cooperate with other heuristic aspects in this process in order to explain model construction. The theory does not focus on the role of analogy with regard to a developing explanatory model, that is, its role in abduction. In this way, we will consider that different roles of analogy may compose a general strategy of abductive inference in contrast to classical theories of analogy.

One problem is that the restrictive view seems to assume that the internal structure of analogy consists of three sub-processes.

1. First, a source is accessed based on a *similarity relationship* with a theoretical problem or unexplained phenomenon that constitutes the target.
2. Second, the *source-target relationship or mapping is evaluated and revised* in order to avoid focusing on some spurious aspects in the source.
3. And third, a relevant property or *relationship is transferred* from source to target with the intention of solving the problem or explaining the phenomenon.

But what is the function of analogy with regard to creating the theoretical model? These three sub-processes constitute and define analogical reasoning, but this common structure to all analogies ignores the role that analogy can perform in the evolution of the model and in the construction of the imagery that explains the target phenomena qualitatively. Furthermore, this vision ignores the possible participation of a mental schema as the third element that is previous to the source and the target. If the analogy leads to the construction of an explanatory model, and it is deve oped further, then it should be considered a third entity along with the source analog and the targeted relationship to be explained. All three entities can involve imagery, and the final model is often the result of the original analogy plus an extended evolutionary process, in which other analogies can also contribute. This view is less simplistic than the traditional one, which assumes a completed solution after a single act of transfer from a single source analog.

To investigate the relationships between analogies, mental schemas, imagery, and subsequent model evolution, Clement interviewed scientifically trained subjects thinking aloud while working on unfamiliar problems. In particular he studied how scientists use analogies and other creative processes to generate new models (Clement 1988, 1989, 2008). He found evidence of the use of analogies, physical intuition, extreme cases, thought experiments, runnable mental schemas among other processes, and attempted to explain how these processes function. He also found evidence from spontaneous gestures and other indicators that these processes can be based on imagistic mental simulation as an underlying mechanism. He also described his subjects as going through cycles of model generation, evaluation, and modification to produce progressively more sophisticated models. In that case an analogy can play the role of a 'protomodel' (starting point for a developing model). In another role, an analogy could be only expedient, in the sense that it is only useful for solving a particular problem, not developing a general model. It does not act as a starting point for an explanatory model. I will show that the heuristic intervention of the eddy mental schema is more complex and more extended than an expedient analogy process only, and that at times it involved analogies acting as protomodels in the model-based reasoning process. Thus, we will see that different versions of the eddy schema have played different roles at different times, along with multiple analogies, going beyond an expedient role and creating theoretical proto-models and models.

In this discovery as process view, Nersessian (2002) described how many of the above processes can explain the groundwork for Maxwell's development of electromagnetic theory, most recently by comparing Clement's protocols to Maxwell's historical papers (Nersessian 2008). In particular, Maxwell's use of analogies to gears, and then fluid flow and eddies, helped him to generate, criticize, and modify his initial formulations of electromagnetic theory. She also describes the way new constraints were added to the theory progressively as Maxwell went through his model modification cycles.

And in some cases, Clement's subjects actually made modifications to the problem representation they were working on in order to enhance their ability to run the mental schemas involved in mental simulations, adding to the evidence for the importance they attached to using mental simulations. He called this observed strategy "imagery enhancement". Clement's definition of mental schema included mental entities at a lower level of abstraction than the definition provided herein, because the idea of schema in this work is related to ontological classifications about the foundations of mathematical thinking. Nevertheless Clement's finding is important for this study because imagery enhancement establishes a relation between mental simulation, analogical reasoning and mental schemas in a cognitive perspective. The nexus between mental schemas and analogies is in this kind of enhancement.

Although some evidence for these points could also be found in Cement's and Nersessian's subjects, they have not emphasized the idea of analogy as heuristic that 'cooperates' with other strategies, and the contrast to the classical view of analogies that is set up by that idea. Only by comprehending the number of different heuristic reasoning processes involved can we appreciate the full complexity of the abductive model formation process and the roles that analogies can play in that process. In addition, I will emphasize that the close interplay of contexts of generation and evaluation confound the classical distinction between the context of discovery and the context of justification; that is, in our perspective both contexts are part of the process guided by the logic of discovery. Each generation or creation step needs immediately a degree of confirmation and confidence via an initial evaluation. And this evaluation does not necessarily have to be empirical.

One interesting kind of non-empirical evaluation are those that occur via mental simulation: the thinker runs an explanatory model as a way of generating an initial plausibility test for the model. For example, Nersessian describes how Maxwell's criticism of an early model of his electromagnetic field described by Nersessian, establishes a mental evaluation phase which finally repairs the initial theory. The initial Maxwellian model represented aspects of the magnetic field as disks or gears in contact and rotating; but there was a problem with the direction of rotation of adjacent gears, which was repaired by adding "idler wheels" between gears. Furthermore, this addition came to represent another useful feature of the field. An evaluation phase was developed by Maxwell using mechanistic sources in the artificial device-field analogy in order to improve the source and generate a new model in cooperation with imagistic representations and diagrams of mechanical instruments (Nersessian 2002). Although they present different explanatory model constructions, the mechanistic wheels constitute a similar mental schema as the eddy by means of a confined circular motion. As I will show in Reynolds's case, there was an innovation in the introduction of the schematic property of intrinsic energy dissipation.

The idea of weak evaluative activity (a non-formal verification of the model usually with mental experiments) during early abduction processes is intimately related to the *ignorance-preserving* character of abduction in logic (Woods 2013, p. 377). In abduction the question is not whether your information justifies a belief, "but whether it stimulates your mental strategies to produce it and the belief was well-produced". Basically, abduction is a scant-resource strategy which proceeds in absence of knowledge; it presents an ignorance-preserving character, but also ignorance-mitigating. In the application of abduction to education, Meyer (2010) writes about "discovery with a latent idea of proof", presenting beautiful examples of this ignorance mitigation. In the Fig. 1 you can see a representation that shows that the difference between the square of any entire number and the product of its contiguous numbers must be one. In every case, if we move the last column and convert it into the last row we can prove this rule. Here there is cooperation between

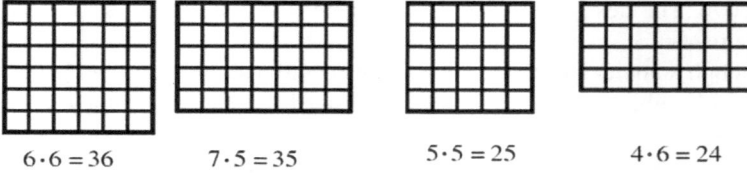

Fig. 1 Analogy between the product operation and squares or rectangles

visual representation of numbers, analogy between the product operation and squares or rectangles, spatial reorganization, and diagrams.

This cooperation invites us to discover a general rule via a strategy to believe it without an orthodox, strong mathematical proof. As Pólya says (1957, p. 83), having verified the theorem in several particular cases, we gathered strong inductive evidence for it. The inductive phase overcame our initial suspicion and gave us a strong confidence in the theorem. "Without such confidence we would have scarcely found the courage to undertake the proof which did not look at all a routine job. When you have satisfied yourself that the theorem is true, you start proving it". Thus we need confidence and belief before the final proof. This is an important aspect in the discovery context.

Taking into account these important contributions, we will attempt to show that there are significant episodes in the history of fluid mechanics with the participation of the eddy mental schema. So I want to add new perspectives on the following aspects of the theory of model-based reasoning in science:

- The examples from the history of fluid dynamics motivate recasting the role of mental schemas and analogies in science from a simplistic role to a more complex pattern in which these heuristic instruments participates with others in the abductive development of a new theoretical model.
- They can work in cooperation with other heuristic reasoning processes, such as mental simulations and diagrammatic thinking.
- The mental schemas can be also activated or provoked by analogies, so they do not have an axiomatic or initiator character in every case.
- What is more, mental schemas are not always generative, and they can play a weak but decisive evaluative function during the abduction process, to gain confidence and mitigate ignorance.
- Many of these processes appear to rely on imagery, and the representation of diagrams can be interpreted as imagery enhancement strategies to the properties of the mental schema.
- Here the symbolic representations like algebra coexist with model-based reasoning, that is, with analogy, diagrams, iconic properties of mental schemas, etc.

3   The Eureka of Bernoulli via the Contemplation of an Eddy

Johann Bernoulli's *Hydraulica* was published in 1742 as a part of his *Opera Omnia* (Fig. 2). Previously his son, Daniel's, fundamental contribution, called *Hydrodynamica*, was published, in which the famous velocity-pressure relation was properly described for the first time (Bernoulli and Bernoulli 1968). Johann also studied the forces and pressures that appear in the dynamic of fluids. But *Hydraulica* is much shorter than *Hydrodynamica*, mathematically sharper and without experiments. Daniel used the method of live forces (vis viva) based on *ascensus potentialis* and *descensus actualis* that was not acceptable from the Newtonian viewpoint. Johann wanted to explain the behavior of fluid motion from pure Newtonian principles (*vis motrix,* or Newton's second law), and as I will show now, he got the key to the solution of the problem by means of the contemplation of an eddy.

Although it can be seen as a mere continuation of his son's work, Johann's contribution must be taken into account, as Euler emphasizes: Johann Bernoulli applied Newtonian dynamics to fluid motion, and he managed to refine the concept of pressure to understand it as an internal isotropic areal force. He thought that the vis viva method was incorrect, because the main natural philosophers did not accept it. However, the idea of internal pressure was more important than this formalism, since it is the concept of pressure of current fluid mechanics. This concept allowed him to determine the dynamics of flow in any state, not only in the stationary case.

The concept of internal pressure and the division of the fluid into virtual differential elements allowed isolating a fluid element in order to follow its evolution and to determine local forces. Here was one of the important concepts for the posterior historical development of fluid mechanics. Nowadays, we call the imaginary limits of a chosen control volume of fluid "the control surface" for an identified mass of fluid. In this way, the extensibility and the perceptual lack of control of fluids acquires solid constraints for heuristic manipulation.

In the first part of *Hydraulica*, the flux of water is studied in cylindrical pipes, and in the imagery of Bernoulli the fluid is composed of many parallel cylindrical layers that are located perpendicular to the motion. He proposes a paradigmatic problem for the understanding of flow acceleration: water flowing through two connected pipes of different diameter, as you can see in the Fig. 3 [(a) is the original one and (b) is our adapted version for a more suitable explanation]. Obviously, taking into account the principle of continuity of the flow, the water in the extremes, far from the narrowing pass, flows at a velocity inversely proportional to the area of traversal sections of each pipe. So, if the area of the first pipe is twice the second, the first velocity will be half the other. That is to say, there must be an acceleration of fluid elements in the transition interval. So there was an important unsolved question to obtain a suitable Newtonian explanation: *what kind of external force was creating the needed acceleration in the transition from the wide to the narrow pipe?*

In order to keep a suitable transition velocity in the narrowing, Bernoulli imagined that an eddy was produced in the upper corner, creating a funnel shape,

DISSERTATIONIS HYDRAULICÆ

PARS PRIMA,

Agens de motu aquarum per vafa & canales cylindricos, qui ex pluribus tubis cylindricis fibi invicem adataptis funt conflati.

I.

Etur primo canalis , (Fig. 1) T A B.
A B C F D E, compofitus ex duobus LXXXIX.
tubis cylindricis, diverfæ amplitudi- CLXXXVI
nis A G D E & G B C F, quorum *Fig.* 1.
ille fundum G D apertum habeat
foramine G F, per quod commu-
nicet cum tubo anguftiori B F. Sit
vero totus canalis B E plenus liquore
homogeneo, per fe nullius gravita-
tis, fed urgeatur a parte orificii AE,
data vi motrice $=p$, quæ, æqualiter premendo, expandatur per
totam fuperficiem liquoris A E; quæritur lex accelerationis, qua
liquor per canalem profluet? Suppono autem canalem femper
manere plenum liquore, quod fit concipiendo fuppeditari ju-
giter aliunde novam materiam liquoris, eadem quovis momento
velocitate in tubum G E fubingredientis, ad refarciendum id
quod per alterum orificium G F egreditur in tubum G C, atque
ex hoc ipfo per lumen B C in auras dilabitur.

I I.

Ex Hydroftaticis affumpfi vim motricem p immaterialem ,
qua premitur fuperficies liquoris A E, propagari in inftanti, ad fuperficiem G F liquoris in tubo B F contenti ; idque five ftagnet
liquor in toto canali, five fluat; dummodo plenus maneat.

Fig. 2 First page of the *Dissertationis Hydraulica*

because no change happens suddenly, but gradually. *Natura non facit saltus.* He contemplated an eddy in the corner of an open channel and in his words experienced an *eureka* moment. He achieved his aim in the *Hydraulica* when he saw the

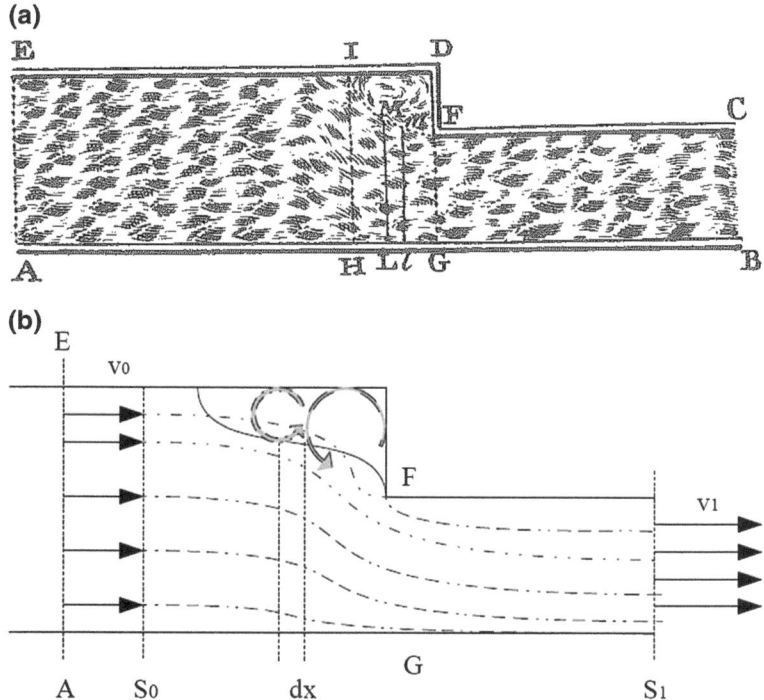

Fig. 3 a Original figure with the eddy by Benoulli (p. 397). **b** Adapted figure for our interpretation

crux of the whole matter to lie in the contemplation of the throat (whirlpool, *gurges* in Latin), "previously considered by no one".[1] This eureka insight is the initiation of a more refined thought that contains an imagery based on the eddy schema and stream lines.

We should underline the implicative function that deserves treats the stream-lines as diagrams, because they are the heuristic instruments that give sense to the direction of the eddy. The streamline representation projects virtuality, that is, gives imaginary motion to the physical concept of eddy. So, within this diagrammatic thinking, the streamlines make an allusion to the eddy suggesting a return of

---

[1]PREFATIO, p. 392: Miratus unde tanta dificultas, ut in fluidis, non aque ac in folidis, fuccedat principiorum dynamicorum applicatio; tandem rem acrius animo volvens, detexi veram dificultati originem; quan in eo consistere deprehendi, quod pars quaedam virium prementium inpemsa in formandum *gurgitem* (a me ita dictum ab aliis non anivadversum) tanquam nullius momenti fuerit neglecta, & insuper habita, non aliam ob causam quam quia gurges conflatur ex quantitate fluidi perexigua, ac veluti infinite parva, qualis formatur quotiescunque fluidum transit ex loco ampliori in angustiorem, vel vice versa ex angustiori in ampliorem. In priori casu sit gurges ante transitum, in altero post transitum.

the flow, and the explication key of the reasoning is based on the eddy: it gives to the flow the thrust needed to produce the velocity increment in the narrowing transition, and the appearance of the eddy is the Newtonian cause of the velocity change and the smooth transition within the throat form configured by it. The eddy circulation thrusts the fluid elements below producing a relevant velocity increment in a short time period.

Johann Bernoulli saw an analogy between the transition corner of his problem and the corners of open channels in which eddies were created. This analogy provokes the activation of the eddy schema and this mental schema visualizes the key of the problem in two aspects:

1. The flow's left-to-right direction needs a return when the water goes in the corner; part of the water becomes stuck circulating in an eddy. Therefore, the pressure creates another area apart from the water that goes forward, local and self-sufficient. The pressure force maintains the circulation of the eddy according to Bernoulli's imagery. So the *vis motrix* (motion force)[2] or what nowadays is known as the dynamic pressure, does not contribute directly to the transition acceleration. In this imagery, part of the pressure force is applied to the eddy in order to keep this.
2. Bernoulli says that the quantity of thrust needed to accelerate water into the narrow pipe can be calculated by integrating the *vis motrix*[3] between the eddy interval. In this way, the circulation direction is defined properly. The direction of this motion and the closed stream-lines suggest that the eddy thrusts the stream-lines of the open flow. In fact, in an infinitesimal perspective (not in the perspective of the global flow) the fluid elements must accelerate one by one in the narrowing transition. The eddy also offers an explanation at this infinitesimal level.

In this qualitative view, Bernoulli's question was about the acceleration law that the liquid obeys when it passes near the eddy. Johann establishes an imaginary domain of flow for that: the general cause of motion is the force difference between S_0 and S_1 surfaces, and the throat formed by the eddy facilitates the motion. Johann knew from hydro-statics that the transmission of pressure is instantaneous, be water static or non-static, the "immaterial guiding force" at AE surface will arrive instantaneously to GF (*Disertationis Hydraulica*, Pars Prima, p. 397–400).

In order to solve this problem mathematically, Bernoulli supposed that the flow was divided into flat surfaces perpendicular to motion. He selected a layer with infinitesimally wide dx and S area under the eddy. If we consider the mass of this

---

[2]PREFATIO, p. 373, II Def. Vis motrix est, quae quando agit in corpus quiescens, illud in motum concitat, aut quae corpus jam motum vel accelerare, vel retardare, vel ejus directionem mutare protest/"A motive force is that which, when it acts on a body at rest, excites it into motion, or which can cause a body already moving to accelerate, decelerate, or change its direction".

[3]PREFATIO, p. 374, IV Def. Vis motrix divisa per massam, dat vim acceleratricem, per hanc vero divisa, dat massam/ "The motive force divided by the mass gives the acelerative force, but divided by this gives the mass.".

layer, the force will be $\rho S dx$ times acceleration. This is the above mentioned *vis motrix* expression. Apart from that, because of kinetic elementary reasons:

$$adx = \left(\frac{dv}{dt}\right)dx = dv\left(\frac{dx}{dt}\right) = vdv;$$

And the force will be an expression of the motion force that only depends on velocity and its infinitesimal increment:

$$F = \rho S v dv.$$

Besides, the infinitesimally wide layer under analysis can have an undefined ordinate or area that reaches the eddy in its uppermost limit, but the translation of the first layer *vis motrix* implies that the ratio 'force/surface area' should be invariant. So we must multiply the force expression by the ratio S_0/S and therefore we will obtain $\rho S_0 v dv$. So the integration term is reduced to vdv, and the integration limits are established by the initial and final velocities of the narrowing transition:

$$F_0 = \int_{v_0}^{v_1} \rho S_0 v dv = \frac{1}{2}\rho S_0 \left(v_1^2 - v_0^2\right) = \frac{1}{2}\rho S_0 \left(1 - \frac{S_1^2}{S_0^2}\right)v_1^2$$

This equation describes the force needed to go from the slow velocity in the wider pipe to the rapid one, which is obligatory to flow towards the narrow pipe. Therefore, neither the eddy's form nor size is determinant to obtain the *vis motrix* that produces the eddy.

Johann Bernoulli calls this total force that guides the flow the "pressure", and, as mentioned, it is independent of the throat form generated by the eddy. According to him, this force creates the eddy and maintains its circulation, because it does not contribute to the acceleration of the total water mass (it is constant in both pipes). In this sense, the acceleration of a fluid element that is traveling through a streamline below the eddy is caused by the circulation force of the eddy. F_0 causes the eddy, and the eddy causes acceleration of elements below the eddy.

Summarizing, this is the abductive structure of Johann Bernoulli's reasoning:

1. There is a first analogy between the two-pipe problem and the corner of the rivers that associates an eddy and the subsequent throat form to narrowing transition.
2. This analogy is provocative: it activates a mental schema, the eddy schema, which is a powerful heuristic instrument in terms of visualization.
3. Then the eddy schema cooperates with diagrammatic thinking via streamlines creating a coherent imagery of the circulation sense that creates an integral imagery enhancement and serves to gain confidence as ignorance-mitigation.
4. In an infinitesimal view, this imagery can explain the acceleration of fluid elements in the transition zone, because the *vis motrix* is what produces the eddy in the corner and its rotation thrusts and accelerates the fluid element.

5. In a global view, in which the flow is constant, the eddy conforms a smooth throat in the abrupt transition, and this enables us to apply the translation of forces for the fluid layers and to obtain a more sophisticated quantitative explanation and a mathematical expression for the *vis motrix* or dynamic pressure.

4 Reynold's Eddies in Sinuous Flow

Darrgol (2005), in his reference work on the history of fluid mechanics, presents Osborne Reynolds as an "eccentric philosopher-engineer". In fact, Reynolds belonged to a new kind of engineer in the final part of the XIX century, well versed in higher mathematics and familiar with recent advances in fundamental physics. Throughout his life he maintained a double interest in practical and philosophical questions. According to Darrigol, his scientific papers were written in an unusually informal, concrete, and naive language. They rely on astute analogies with previously known phenomena rather than deductive reasoning. Even though some of these analogies later proved superficial or misleading, in most cases, like in the case of the eddy schema, Reynolds gained valuable insight from them.

Reynolds did his famous work on the transition between laminar and turbulent flow in the 1880s. Two themes of his earlier research conditioned his approach. The first was the importance of eddying motion in fluids, and the second was the dimensional properties of matter related to its molecular structure. A paper on steam boilers brought the two themes together. Reynolds puzzled over the rapidity of the transfer of heat through the surface of the boiler. Here Reynolds (1874b, p. 81) thinks on terms of *molecular philosophy:*

> Besides its practical value, it also forms a subject of very great philosophical interest, being intimately connected with, if it does not form part of, molecular philosophy.

For Reynolds, all the preliminary molecular physics of that time was not science. It was a matter of philosophy, because it only offered a metaphysical background for the explanation of macroscopic observable phenomena.

In this philosophical perspective, Reynolds had a deep interest in the kinetic molecular theory of heat and the resulting insights into transfer phenomena. The *kinetic theory* described a gas as a large number of molecules, all of which are in constant rapid random motion with many collisions with each other and with the walls of the container. Kinetic theory explained macroscopic properties of gases, such as pressure, temperature, viscosity, thermal conductivity, by means of this internal motion: the gas pressure is due to the impacts on the walls of a container, and so on.

In the boiler case, Reynolds concluded that ordinary diffusion bound to invisible molecular agitation did not suffice to explain the observed heat transfer. Then he added a new phenomenon: "the eddies caused by visible motion which mixes the fluid up and continually brings fresh particles into contact with the surface" (*ibid.*). So according to Reynolds this phenomenon of eddy formation was observable, and it happens in a different scale from microscopic and metaphysical molecular agitation. In this way, he deduced a new form of the total heat transfer rate, depending on the velocity of the water along the walls of the boiler. And he also noted the analogy of this expression with the behavior of fluid resistance in pipes.

At the same time, Reynolds encountered eddying fluid motion while investigating the racing of the engines of steamers. As seamen observed, when the rotational velocity of the propeller becomes too large, its propelling action as well as its counteracting torque on the engine's axis suddenly diminish. A damaging racing of the engine follows. Reynolds explained this behavior by a clever analogy with efflux from a vase. The velocity of the water expelled by the propeller in its rotation, he reasoned, cannot exceed the velocity of efflux through an opening of the same breath as its own. For a velocity higher than this critical velocity, a vacuum should be created around the propeller, or air should be sucked in if the propeller breaks the water surface (1874a).

According to this theory, a deeper immersion of the propeller should retard the racing (for the efflux velocity depends on the head of water), and the injection of air next to it should lower the critical velocity (for the efflux velocity into air is smaller than that into a vacuum). While verifying the second prediction, Reynolds found out that air did not rise in bubbles from the propeller, but followed it in a long horizontal tail. Suspecting some peculiarity of the motion of the water behind the propeller, he injected dye instead of air and observed a complex vortex band. Similar experiments with a vane moving through water displayed also vortex bands. So the search for eddies and its real observation was a difficult experimental endeavor.

From these preliminary observations, Reynolds inferred that hidden vortex motion played a systematic part in almost every form of fluid motion. These considerations came after Helmholtz's famous work on the theory of vortex motion, and after Maxwell's involvement in a vortex theory of matter. Reynolds convinced himself that his forerunners had only seen the tip of the iceberg. Then another amazing observation of fishermen and sailors brought invisible vortex formation to Reynolds's mind: the power that rain has to calm the sea. Letting a drop of water fall on calm water covered by a thin layer of dye, he observed the formation of a vortex ring at the surface followed by a downward vertical motion. When the drops of rain fall on agitated water, he reasoned, part of the momentum of this agitation is carried away by the induced vortices, so that the agitation gradually diminishes. Again he used dye to visualize the invisible and thinly covered world of little but powerful eddies. In this way, the eddy schema became a potential heuristic instrument that can be activated in many different new experimental and observational situations in order to obtain a first qualitative explanation. According to Reynolds's words in a popular royal conference in 1877, it was a revelation:

In this room, you are accustomed to have set before you the latest triumphs of mind over matter, the secrets last wrested from nature from the gigantic efforts of reason, imagination, and the most skillful manipulation. To-night, however, after you have seen what I shall endeavor to show you, I think you will readily admit that for once the case is reversed, and that the triumph rests with nature, in having for so long concealed what has been eagerly sought, and what is at last found to have been so thinly covered.

Reynolds thought that the failure of hydrodynamics to account for the actual motion of fluids was due to the lack of empirical knowledge of their internal 'thinly covered' motions.

Reynolds then recalled casual observations of vortex rings above chimneys, from the mouth of a smoker, and from Tait's smoke box (1876). These rings had only been studied "for their own sake, and for such light as they might throw on the constitution of matter". To Reynolds's knowledge, no one had understood their essential role in fluid motion. This he could reveal "by the simple process of colouring water", which he had first applied to elucidate the motion of water behind a propeller or oblique vane. By the same means, he studied the vortex rings formed behind a disc moved obliquely through water. The resistance to the disc's motion appeared to be caused by the continual production and release of such rings.

Reynolds emphasized that reason had failed to show such forms of fluid motion. In that time everyone knew the importance of rational hydrodynamics by famous mathematicians, but they could go further than before only with colored water. The revelation of eddies showed the way to apply mathematics most usefully. This is the key point of the introduction of his famous 1883 paper entitled "An experimental investigation of the circumstances which determine whether the motion of water shall be direct or sinuous, and of the law of resistance in parallel channels". He spoke about two appearances of water, the one like a "plate glass" and the other like that of "sheet glass". In his opinion, the two characters of surface correspond to the two characters of fluid motion. This may be shown by adding a few streaks of highly coloured water to the clear moving water. "Although the coloured streaks may at first be irregular they will, if there are no eddies, soon be drawn out into even colour bands; whereas if there are eddies, they will be curled in the manner so familiar with smoke", he said.

After that, he also commented on the other well-known distinction for laws of flow resistance in that time: tubes of more than capillary dimensions present a flow resistance proportional to the square of the flow velocity, but in capillary conducts the resistance is as velocity. The equation of hydrodynamics written by mathematicians, applicable to direct motion without eddies, showed that the resistance was as the velocity, but could explain neither the presence of eddies in water motion nor the cause of resistance varying as the square of the velocity in sensibly large pipes or around large bodies moving at high velocity. So here the eddy schema appears not as in Bernoulli's case, but as an intrinsic property of fluid that produces an internal force dissipation and subsequent increment of resistance.

We should emphasize the paradox: high viscosity fluids, by definition, present more resistance to motion, but their resistance behavior with respect to velocity is weaker than low viscosity fluids, because they form fewer eddies. Reynolds was the first scientist who recognized this paradox via the schematic association between the eddy and its intrinsic dissipation. As in Bernoulli's case the activation of the eddy schema can explain some important aspects of the problem, but in this case the mental schema was not imagistic, but associative. The empirical study of the resistance law with colored water activated this association schema, together with the analogy with respect to the heat transfer within kinetic molecular theory.

At this point he reflected philosophically about the relative nature of space and time, "to suppose that the character of motion of fluids in any way depended on absolute size or absolute velocity would be to suppose such motion outside the pale of the laws of motion". So, "what appears to be the dependence of the character of the motion on the absolute size of the tube and on the absolute velocity of the immersed body must in reality be a dependence on the size of the tube as compared with the size of some other object, and on the velocity of object as compared with some other velocity". He concluded: "What is the standard object and what the standard velocity which come into comparison with the size of the tube and the velocity of an immersed body, are questions to which the answers were not obvious. Answers, however, were found in the discovery of a circumstance on which sinuous motion depends".

After this philosophical interlude, Reynolds underlined another veiled empirical fact: "the more viscous a fluid is the less prone is it to eddying motion". The ratio between viscosity and density depends strongly on temperature, for water, at 5 °C is double what it is at 45 °C. Reynolds observed by means of his coloured streams that the tendency of water to eddy becomes much greater as the temperature rises. Hence, he connected the change in the law of resistance with the appearance of eddies and in this way delimited via dimensional analysis the search for standard distance and standard velocity. The dimension of kinematic viscosity is L^2/T, and obviously it has only kinematic units. So the idea was to combine in an adimensional way the kinematic viscosity with distance and with velocity or time. It would be an adimensional number, and Reynolds said in the introduction that similar constants are already recognized, as the flux velocity relative to the velocity of sound (Mach number nowadays).

After that, Reynolds emphasized that his previous investigation on gas transpiration was fundamental. It was an important evaluative analogy and source of confidence for his future work. Graham defines the transpiration versus the effusion in his pioneering experimental work (Graham 1846, p. 574): "The passage of gases into a vacuum through an aperture in a thin plate I shall refer to as the Effusion of gases, and to their passage through a tube as the Transpiration of gases." Reynolds tried to explain the results of Graham in capillary tubes in terms of molecular agitation and he realized that it was necessary to add something more to explain it as in the case of the heat transfer in boilers. Chaotic molecular movement was not sufficient. Graham noted that the law of transpiration became different for very small pores (small diameters), but without a theoretical understanding. In contrast, Reynolds argued that the flow of a gas depended on the 'dimensional properties of matter', that is, on the ratio between the dimensions of the flow (tube or pore diameter) and the mean free path, or the average distance traveled by a moving molecule between successive impacts.

Therefore, the essential idea was to analyze the dimensional properties of the well-known and deeply studied equation of motion. If we take the curl of the equation and eliminate the pressure from it, the time derivative of the vorticity (the curl of the velocity) has two terms:

$$\frac{\partial w}{\partial t} = \nabla \times (v \times w) + \frac{\mu}{\rho}\Delta w$$

There are two terms, and we can obtain an expression of both according to velocity (U) and length (L) dimensions. The first term's dimension is

$$\frac{1}{L} \cdot U \cdot \frac{U}{L} = U^2/L^2,$$

and the second one's

$$\left(\frac{\mu}{\rho}\right)\left(\frac{U}{L^3}\right).$$

The relative value of this term, $\rho LU/\mu$, must be adimensional because of the obvious dimensional homogeneity of any equation in physics. Reynolds wrote that "this is the definite relation of the exact kind for which I was in search". This was only the relation, without showing in what way the motion depends on the mathematical integration of the equation. However, without knowing the particular cause, this integration would show the birth of eddies to depend upon what we currently call the Reynolds number, one of the most important adimensional numbers in physics.

After this strong hypothesis, Reynolds tested the relations between terms of his number, U, L and kinematic viscosity. The crucial test was done in a straight smooth transparent tube in order to identify the change of the law of resistance from variation as velocity to variation as the square of the velocity (Fig. 4). This transition would present the birth of eddies by means of ink streaks in water. The variation of his number could be done by be the variation of the flow velocity, the temperature of water (that is, kinematic viscosity), or of the diameter of the tube. So there were different manners to obtain the critical point. And the result was surprising: approximately, for a value of 2000 the transition in the law of resistance and the resultant birth of eddies happened. That is, if you increase the velocity but you increase proportionally the viscosity, the regime of flow remains because Reynolds's number is the final indicator. This is a very strong scientific conclusion in terms of generalization because Reynolds's number is able to establish quantitatively a fundamental regime division that nowadays is extended for all type of fluids.

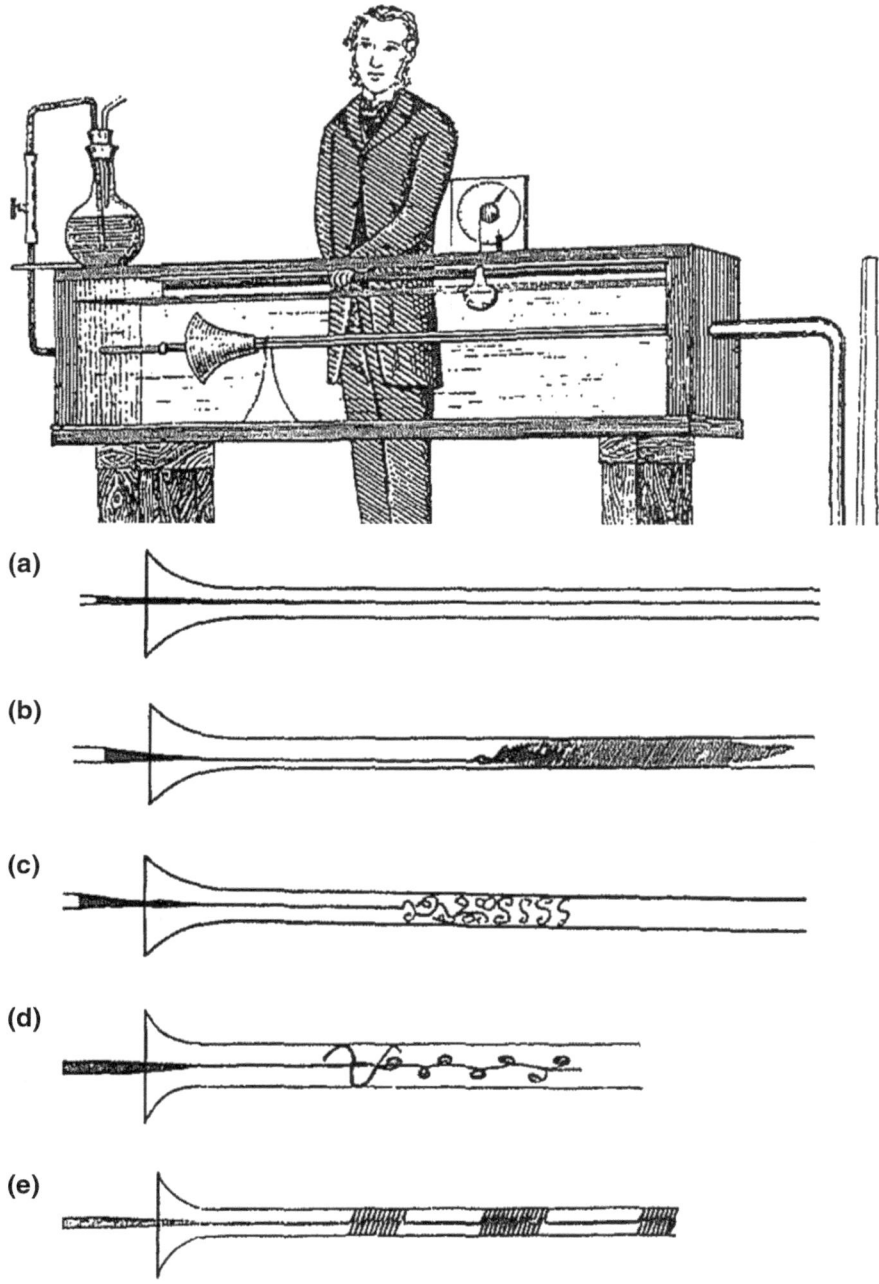

Fig. 4 Reynolds's experiment with colored water showing the transition from direct to sinuous flow. Original picture in Reynolds (1883)

Summarizing, Reynolds's abductive reasoning structure is the following one:

1. Reynolds knows various unexplained observations in ship propellers or calmed sea by rain which suggests the importance of eddies as an intrinsic dissipation source in fluids.
2. This establishes a mental schema that relates the concept of eddy with sinuous motion and the increment of resistance.
3. By means of his ink streak experimental technique, he observes that the more viscous a fluid was the less prone was it to eddy motion.
4. The analogy between gas transpiration and liquid resistance law in pipes has, first, an evaluative function that increases confidence as ignorance-mitigation and then, second, works in generative terms:

   a. The analogy cooperates via the idea of dissipation with the eddy schema;
   b. Generalizes the relevance of the kinematic viscosity because of its dimensional properties with a direct analogy to the ratio between pore diameter and mean free path,
   c. And therefore, activates the dimensional calculus in the equation of motion.

5. The kinematic dimension of the ratio between the viscosity and density suggests the search for a standard velocity and a standard length in flow motion and eddy formation.
6. Reynolds finally obtains his number and identifies quantitatively the transition point from direct to sinuous flow with the birth of eddies that he can visualize.

5 Conclusions

Several strategies for developing theoretical models and explanations have been identified around the eddy schema with its main schematic properties of kinetic rotation (in the case of Johann Bernoulli) and energy dissipation (in Reynolds's case). Instead of considering only a passive role as a projection over the finished theoretical product, we can study its several roles and its cooperation with other abductive strategies such as analogy in the creative process. Multiple analogies have collaborated in both historical episodes with other heuristic instruments as diagrams, inventing problems analogous to the observation of streams and whirlpools on the river, or relating the internal eddies with energy dissipation by means of the analogy between the pipe resistance law and gas transpiration. Both can be considered as imagery enhancement strategies from analogies to schemas involving mental simulations.

This variety transcends the simplistic, restrictive view of a mental schema as analogical source of the entire theoretical model with a mere expedient character. Instead, it emphasizes the heuristic power in each strategic decision in both generative and evaluative terms. That is, it mixes the discovery and justification contexts. The heuristic process does not derive on a full knowledge acquisition but on a modulated ignorance-mitigation, and it establishes a gradual process of model evolution and confidence that is indispensable in order to continue with the discovery process until the final qualitative model.

References

Aliseda, A. (1997). *Seeking explanations: Abduction in logic, philosophy of science and artificial intelligence*. Amsterdam: Universiteit van Amsterdam.

Bartha, P. (2010). *By parallel reasoning*. Oxford: Oxford University Press.

Bernoulli, D., & Bernoulli, J. (1968). *Hydrodynamics, by Daniel Bernoulli & Hydraulics, by Johann Bernoulli*. New York: Dover.

Cellucci, C., & Gillies, D. (2005). *Mathematical reasoning and heuristics*. London: King's College Publications.

Clement, J. (1988). Observed methods for generating analogies in scientific problem solving. *Cognitive Science, 12*(4), 563–586.

Clement, J. (1989). Learning via model construction and criticism: Protocol evidence on sources of creativity in science. In J. Glover, R. Ronning & C. Reynolds (Eds.), *Handbook of creativity: Assessment, theory and research* (pp. 341–381). New York: Plenum.

Clement, J. J. (2008). *Creative model construction in scientists and students: The role of imagery, analogy, and mental simulation* (Softcover edition, 2009). Springer: New York. ISBN: 9048130239978904813023.

Darrigol, O. (2005). *Worlds of flow: A history of hydrodynamics from the Bernoullis to Prandtl*. Oxford: Oxford University Press.

Darrigol, O. (2010). The analogy between light and sound in the history of optics from the ancient Greeks to Isaac Newton. *Centaurus, 52*(2), 117–155.

Gentner, D. (1983). Structure-mapping: A theoretical framework for analogy. *Cognitive Science, 7*(2), 155–170.

Graham, T. (1846). On the Motion of Gases. *Philosophical Transactions of the Royal Society of London, 136*(1846), 573–631.

Hanson, N. R. (1958). The logic of discovery. *The Journal of Philosophy, 55*(25), 1073–1089.

Hintikka, J. (1985). True and false logic of scientific discovery. *Communication and Cognition, 18*(1/2), 3–14.

Hintikka, J. (1998). What is abduction? The fundamental problem of contemporary epistemology. *Transactions of the Charles S Peirce Society, 34*(3), 503–533.

Holyoak, K. J., & Thagard, P. (1989). Analogical mapping by constraint satisfaction. *Cognitive Science, 13*(3), 295–355.

Hon, G., & Goldstein, B. R. (2012). Maxwell's contrived analogy: An early version of the methodology of modeling. *Studies in History and Philosophy of Science Part B: Studies in History and Philosophy of Modern Physics, 43*(4), 236–257.

Ippoliti, E. (Ed.). (2015). *Heuristic reasoning*. Berlin: Springer.

Jung, S. (1996). *The logic of discovery*. New York: Peter Lang.

Kirlik, A., & Storkerson, P. (2010). Naturalizing Peirces semiotics: Ecological psychology's solution to the problem of creative abduction. In *Model-based reasoning in science and technology* (pp. 31–50). Berlin: Springer.

Lakatos, I. (1976). In J. Worrall & E. Zahar (Eds.), *Proofs and refutations: The logic of mathematical discovery*. Reediting: Cambridge University Press (2015).

Lakoff, G., & Núñez, R. E. (2000). *Where mathematics comes from: How the embodied mind brings mathematics into being*. New York: Basic Books.

Laudan, L. (1980). Why was the logic of discovery abandoned? In *Scientific discovery, logic, and rationality* (pp. 173–183). Netherlands: Springer.

MacLane, S. (2012). *Mathematics, form and function*. Springer: Science & Business Media.

Magnani, L. (2009). *Abductive cognition: The eco-cognitive dimension of hypothetical reasoning*. New York: Springer.

Maxwell, J. C. (1862). XIV. On physical lines of force. *The London, Edinburgh, and Dublin Philosophical Magazine and Journal of Science, 23*(152), 85–95.

Meyer, M. (2010). Abduction—A logical view for investigating and initiating processes of discovering mathematical coherences. *Educational Studies in Mathematics, 74*(2), 185–205.

Nersessian, N. J. (2002). Maxwell and the method of physical analogy: Model-based reasoning, generic abstraction, and conceptual change. *Essays in the History and Philosophy of Science and Mathematics*, 129–166.

Nersessian, N. J. (2008). *Creating scientific concepts*. Cambridge: MIT press.

Nickles, T. (1980). *Scientific discovery: Case studies* (Vol. 1, 2). United Kingdom: Taylor & Francis.

Paavola, S. (2004). Abduction as a logic and methodology of discovery: The importance of strategies. *Foundations of Science, 9*(3), 267–283.

Pólya, G. (1957). *How to solve it. A new aspect of mathematical method*. Princeton: Princ. UP.

Reynolds, O. (1874a). The cause of the racing of the engines of screw steamers investigated theoretically and by experiment. *PLPSM, also in ReP, 1,* 51–58.

Reynolds, O. (1874b). On the extent and action of the heating surface of steam boilers. *PLPSM, also in ReP, 1,* 81–85.

Reynolds, O. (1883). An experimental investigation of the circumstances which determine whether the motion of water shall be direct or sinuous, and of the law of resistance in parallel channels. *Proceedings of the Royal Society of London, 35*(224–226), 84–99.

Sintonen, M. (1996). Structuralism and the Interrogative Model of Inquiry. In *Structuralist theory of science. Focal issues, new results* (pp. 45–75). Berlin/New York: Walter de Gruyter.

Tait, P. G. (1876). *Lectures on some recent advances in physical sciences* (2nd edn.). London.

Thagard, P. (2002). *Coherence in thought and action*. Massachusetts: MIT press.

Ulazia, A. (2016a). Multiple roles for analogies in the genesis of fluid mechanics: How analogies can cooperate with other heuristic strategies. *Foundations of Science, 21*(4), 543–565.

Ulazia, A. (2016b). The cognitive nexus between Bohr's analogy for the atom and Pauli's exclusion schema. *Endeavor, 40*(1), 56–64.

Woods, J. (2013). *Errors of reasoning: Naturalizing the logic of inference*. London: College Publications.

Part III
New Models of Theory Building

TTT: A Fast Heuristic to New Theories?

Thomas Nickles

Abstract Gigerenzer and coauthors have described a remarkably fast and direct way of generating new theories that they term *the tools-to-theories heuristic*. Call it *the TTT heuristic* or simply *TTT*. TTT links established methods to new theories in an intimate way that challenges the traditional distinction of context of discovery and context of justification. It makes heavy use of rhetorical tropes such as metaphor. This chapter places the TTT heuristic in additional historical, philosophical, and scientific contexts, especially informational biology and digital physics, and further explores its strengths and weaknesses in relation to human limitations and scientific realism.

Keywords Heuristics · Tools-to-theories · Gigerenzer · Discovery
Innovation · Scientific realism · Digital physics · Informational biology

Thanks to Thomas Sturm and other members of the Department of Philosophy, Autonomous University of Barcelona; to Jordi Cat, Jutta Schickore, and the Indiana University Department of History and Philosophy of Science and Medicine, where I presented an early version of some of these ideas; and to Emiliano Ippoliti for the invitation to develop them at the Rome-Sapienza workshop, which travel problems prevented me from attending. I am indebted to Marco Buzzoni and to the Herbert Simon Society for the opportunity to present at the University of Macerata and in Turin, Italy, where Gerd Gigerenzer and Riccardo Viale made helpful comments. Thanks to David Danks for final improvements.

T. Nickles (✉)
Department of Philosophy Emeritus, University of Nevada, Reno, USA
e-mail: nickles@unr.edu

© Springer International Publishing AG 2018
D. Danks and E. Ippoliti (eds.), *Building Theories*, Studies in Applied Philosophy, Epistemology and Rational Ethics 41, https://doi.org/10.1007/978-3-319-72787-5_9

1   What Is TTT?

In fascinating case studies from cognitive psychology, Gerd Gigerenzer and sometime coauthors, David Murray, Daniel Goldstein, and Thomas Sturm, have identified what they term "the tools-to-theories heuristic."[1] For convenience, I shall call it *the TTT heuristic* or simply *TTT*. This remarkable heuristic claims to generate (or to provide the crucial seed-ideas for generating) entire, deep theories from methodological tools already entrenched in scientific practice.[2] Gigerenzer's purpose is to call attention to a discovery route that differs from standard accounts of scientific discovery as either data-driven (bottom-up) or else theory-driven (top-down).[3] TTT also interestingly couples discovery and justification.

As far as I know, Gigerenzer was first to articulate TTT and to point out its philosophical and scientific interest. But is TTT too good to be true? Is it reliable enough to be defensible? Where can TTT be most usefully applied?

What sort of heuristic would generate a significant theory, sometimes in a single step? According to Gigerenzer and Sturm (2007, 309, their emphasis), the tools-to-theories heuristic has two components:

1. *Generation of new theories*: The tools a scientist uses can suggest new metaphors, leading to new theoretical concepts and principles.
2. *Acceptance of new theories within scientific communities*: The new theoretical concepts and assumptions are more likely to be accepted by the scientific community if the members of the community are also users of the new tools.

In another publication, Gigerenzer's thesis is that "*scientists' tools for justification provide new metaphors and concepts for their theories*" (2003, 54, his emphasis). He stresses that it is new tools and not new data that are the drivers here. Metaphor is the primary vehicle for projecting the scientists' tools onto the target system. However, it turns out that these metaphors are more than that, since the scientists in question often take them literally.

The basic idea here is that the tools that scientists develop to handle a body of phenomena of a target system then get projected on the target system itself, as the correct explanation of why the system behaves in the manner that it does. The

---

[1]See Gigerenzer (1991a, b, 2003); Gigerenzer and Murray (1987); Gigerenzer and Goldstein (1996); Gigerenzer and Sturm (2007).

[2]Wider cultural orientations can also have a role. Gigerenzer and Goldstein (1996) note how different today's concepts of computers and computation are from those of the era of the Jacquard loom in which Babbage worked. The attempt to understand current mysteries (such as "mind") in terms of the latest technology, as culturally understood, may be a near universal human cultural tendency.

[3]Examples of data-driven discovery are Baconian induction and Pat Langley's early BACON programs (Langley et al. 1987). Examples of theory-driven discovery are Einstein's theories of relativity. By theory-driven, Gigerenzer also has in mind Popper's "logic of discovery" via nonrational conjectures (Popper 1963).

system itself is supposed, at some level, to be doing just the kind of science that the researchers themselves are doing. The reader will find several examples below.

Gigerenzer and Sturm note that community acceptance does not mean justification in the full, epistemological sense. They are making descriptive statements about the behavior of the communities they have studied, not (until later) making normative judgments. Since the issue of scientific realism is important in the normative evaluation of TTT, I shall, for simplicity, distinguish two versions of TTT. *Strong realist TTT* interprets expert community acceptance as sufficient epistemological justification of the projected theory, as literally true of the target system, or very nearly. *Weak TTT* retains the acceptance/epistemic justification distinction, where 'accepted' means only 'judged a hypothesis or research model seriously worth pursuing'. For weak TTTers, TTT functions as a model or theory *candidate* generator, the products of which are plausible or at least pursuitworthy, subject to further evaluation (Nickles 2006). Gigerenzer and Sturm subsequently point out the dangers of what I am calling the strong realist version. They are correct to do so in my view (see Sect. 8 and 9). Of course, acceptance can be a matter of degree, so intermediate positions between strong and weak TTT are possible. At the end of the chapter, I shall return briefly to the idea of a continuum of realist and hence TTT positions.

2 Examples from Cognitive Psychology

The TTT thesis remains rather vague, given the centrality of metaphor and the variety of scientific tools. The general TTT strategy is pluralistic in encompassing a cluster of paths to theory construction. Gigerenzer and colleagues sometimes have in mind specific techniques (e.g., specific statistical methods applied to certain types of databases). Sometimes the guiding idea is a metaphorical concept, such as the mind as a digital computer. In some cases, specific forms of instrumentation or computational tools are involved. It is not evident, in the abstract, how we get from tools to theories. Concrete examples will help to clarify the TTT strategy. One purpose of my chapter is to further broaden the range of examples that Gigerenzer and associates have provided. It will not be my purpose to provide a crisp definition of TTT.

Gigerenzer and Murray (1987) have developed a set of detailed examples from 20th-century cognitive psychology. Gigerenzer terms these developments "the inference revolution" (2003, 58). The principal examples are versions of the mind as an intuitive inferential statistician: Neyman-Pearson statistics postulated as the basis for intuitive decision-making by ordinary people, and Fisher's analysis of variance (ANOVA) as the basis of ordinary causal reasoning by laypersons. Since the mind is supposedly producing these judgments subconsciously, they are intuitive. They pop into our consciousness without our being aware of the underlying statistical reasoning steps.

Another series of examples brings out still more clearly what TTT involves, namely the vision research in the unconscious inference tradition, from Helmholtz to Richard Gregory (and, I would add, Tomaso Poggio). Here the basic idea is that the visual centers of the brain are, on a scale of milliseconds, generating and testing zillions of hypotheses against the incoming data stream, with the winning hypothesis of the moment popping into consciousness as the represented scene.

In each of these cases, the mind is treated as if it were either a single scientist, a research team, or an entire community of scientists in the given specialty area.[4]

  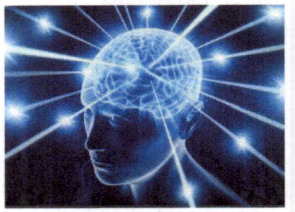

The methodological strategy in each case is to develop computational routines for analyzing the data gathered in experimentation and then to postulate that the target system produces the observed behavior by performing the very same kinds of computational processes.[5]

Another sort of example is the cognitive revolution of the 1960s that treated the mind as a digital computer. Interestingly, the cognitive scientists who used computers or computational ideas in their research were those who embraced the computer model most readily (Gigerenzer and Goldstein 1996; Gigerenzer and Sturm 2007).[6]

Although not one of Gigerenzer's examples, it seems to me that prominent work on the early cognitive development of children also partially fits the TTT pattern, at a kind of metalevel. Susan Carey, *Conceptual Change in Childhood* (1985) and Alison Gopnik et al., *The Scientist in the Crib* (1999), are prominent here in adapting *philosophical* models of scientific change to models of the cognitive development of young children. It is not surprising that an implicit hypothetico-deductive model of generate-and-test has been attributed to children, but so has Kuhn's model of the allegedly discontinuous cognitive changes that scientists undergo in a scientific revolution (Kuhn 1962/1970)—now projected as a

---

[4]Note that this model incorporates a Popperian conjectures and refutation idea without explaining where the conjectures ultimately come from. In this respect, the model is regressive, for what are the established tools of the *target* system (the mind itself) for generating theory candidates, and where did they come from? Stated more generally, the regress point is that TTT reduces the problem of theory generation to the problem of method or tool generation, without resolving the latter. Sometimes theory suggests new methods and new instruments (see Sect. 8), a widely recognized sequence compared to the one Gigerenzer calls to our attention.

[5]Thanks to the U.S. National Institutes of Health for the graphics.

[6]Here the tool had a hardware dimension as well as a computational (software) dimension, immediately raising the question of whether (mostly unknown) biological brain organization could be construed as a digital computer. Analysts such as Fodor (1968 and after) were happy to assume that the brain was sufficiently complex to implement one or more digital computers, and hence that biological embodiment issues could simply be ignored. See Shapiro (2004) for criticism of the multiple realizability thesis.

theory of major cognitive steps in child development. Kuhn's work especially caught the attention of developmental psychologists because of the prominence of Piaget's previous postulation of discontinuous stages.[7] Interestingly, Piaget played a large role in Kuhn's own thinking leading up to *The Structure of Scientific Revolutions*.[8] The psychologists who made this TTT move were clearly familiar with, and influenced by, Kuhn's model (as well as Piaget's) in their own practice. To date, however, there is no consensus in the child development field.

I suspect that any number of other approaches in cognitive psychology employ the TTT move, or something close to it. A typical example may be face recognition work using "face space" software models (Meytlis 2011). Once scientists have a model that works well to capture test subject behavior, it is tempting to suppose that the brain itself has implemented just such a model. On the other hand, we should not forget Leslie Orgel's caution that "Evolution is cleverer than you are" (Dennett 2013, 225).

3 Why Is TTT Important?

Before moving on to a wider range of examples, let us pause to review why Gigerenzer and colleagues find TTT important and interesting (my exposition).

1. TTT articulates a previously unnoticed, heuristic mode of theory formation prevalent in cognitive psychology and elsewhere, one different from both Baconian, data-driven science and neo-Romantic, Popperian, theory-driven science. TTT is driven by methodological tools (including material technologies), and familiar practices,
2. hence *explains where* (some) new theories or principles come from,
3. and *why* the new theories get accepted by members of the expert community.
4. Thus the TTT heuristic embraces *both* "discovery" (theory-generation) and "justification" (at least in the acceptance sense) and thereby helps to overcome the old, invidious distinction between context of discovery and context of justification, in six ways:

   a. by making context of discovery a legitimate subject for methodology.
   b. For example, the historical path dependence of TTT is incompatible with the positivist and Popperian view that the origins of a theory are irrelevant, that only testing matters (Gigerenzer and Sturm 2007, 68).
   c. by intimately coupling generation and justification processes rather than by treating them as separate stages of the old sort.
   d. Where TTT is applied, "justification" (accepted practice) *precedes* and *explains* generation of the theory proper and inspires it. "[T]he context of justification explains the context of discovery" (Gigerenzer 2003, 72).

---

[7]See also the special issue of *Philosophy of Science* 4 (1996) and Bishop and Downes (2002).
[8]See Galison (2016) and Kaiser (2016).

    e. By coupling methodological requirements with empirical content, TTT accelerates research. The new tools and hence newly generated theories can feature new kinds of data (and exclude competing kinds) in a way that is mutually supporting, in a kind of bootstrap process. But overly strong TTT coupling invites the charge of circularity.

    f. For in strong cases, the mutual support (coupling) of method and (allegedly true) theory can lead (and historically has led) to dogmatic entrenchment by promoting the specific research tools to exalted normative status as the very definition of "the scientific method" for that specialty. In such cases, the projected theory can now be interpreted as rationally generated by *the* method of science, which, in turn, is supported by the truth of the new theory.[9] Alternative approaches may be depreciated as unscientific, or no longer up-to-date.

5. In one stroke, TTT moves from epistemology (order of knowing), in the form of methodology-based scientific practice, to ontology (order of being), in the form of a postulatory theory of the target system and its behavior.
6. Given the role of metaphor, TTT challenges another invidious distinction in analyzing scientific work—that between logic and rhetoric.
7. By highlighting the importance of historical path, critical attention to TTT improves theory evaluation and presents an opportunity to develop alternatives. Metaphors and heuristics both have limits. Historical perspective in the use of TTT can help anticipate weaknesses in theories and help us to critically evaluate processes by which metaphors become dead metaphors, if and when they do.
8. Gigerenzer's cases are "[c]onsistent with the marketing-methods thesis, which implies that new tools spread from applied fields into basic research" (2003, 60).

    I shall elaborate on some of these points below.

4 Other Examples of TTT: Informational Biology?

What is the scope of the TTT heuristic? Can the TTT move be plausibly made in fields other than psychology? Has it already been made? Obviously not, it may seem, because the TTT move involves a kind of transfer or projection of scientists' forms of cognition onto other *cognitive* systems. It is a transfer from one cognitive system to another—from mind to mind, so to speak. It might work in some areas of artificial intelligence, perhaps, but surely not in areas where there is no question of cognitive activity.[10]

---

[9]See my comment on Peirce in Sect. 8.

[10]Computability theory, molecular neuroscience, and the extended cognition movement make it more difficult than ever to say what counts as cognition.

Or does it? Let's first have a quick look at information-theoretic biology, where ordinary cognition is not in play, then at the extreme case of digital physics, where we are dealing with basic physical processes. In both these cases there is, as yet, no community acceptance as opposed to prominent individuals making TTT-like moves; so, again, one component of the full TTT heuristic is missing. My purpose is to expand on Gigerenzer's suggestion that TTT-like moves have a wider play in the history of science. That TTT is no longer a human-mind-to-human-mind relation may limit the intuitive attractiveness of the projection, but it probably shouldn't, now that mind-body dualism is dead.

Some biologists and philosophers consider information processing to *constitute* various basic biological processes in nature. The three strongest cases are:

1. genetics from a gene-centric, "central dogma," information-theoretic stance. Here we find Richard Dawkins (1976), George Williams (1992), and John Maynard Smith (2000) on genes as essentially information. The physical gene is merely the carrier.
2. biological development as governed by a "developmental program" encoded in the genes. For examples, see Maynard Smith and Szathmáry (1995) and Sansom (2011).
3. evolution itself as an algorithmic process, (e.g., Dawkins 1995; Maynard Smith and Szathmáry 1995; Dennett 1995; Chaitin 2012).

As before, the basic idea here is that scientific methods of analysis are now projected onto the target biological entities themselves. It is information transmission, reception, and control processes that are what are said "really" to lie behind genetic reproductive and developmental processes, as well as evolution whole hog.

Informational biology has its critics. It remains an open question how fertile information-level accounts will be in biology. With information-theoretic approaches to any subject, there is always a question of how to define or characterize information. In this case, some of the writers have in mind the technical concept of information introduced by Claude Shannon in 1948. (If so, then evolution was indeed cleverer than we are, at least until 1948!) Critics such as Peter Godfrey Smith (2007, 107) contend, to the contrary, that Shannon information is not sufficiently rich for informational biology, on three grounds: (1) shannon information is purely syntactic, where most biological authors intend a semantic conception; (2) shannon information is correlational and hence symmetric, whereas the information needs to be unidirectional in order to be causal; and (3) scientists employ information in a descriptive sense, whereas the ontic projections interpret it prescriptively or imperatively, as giving instructions—a program.[11] Sahotra Sarkar (1996, 187) and Paul Griffiths (2001, 395) loudly complain, as Sarkar puts it, that

---

[11]For other criticisms and clarifications see Artmann (2008), Collier (2008), Griffiths (2001), Griesemer (2005), Godfrey-Smith (2007), Godfrey-Smith and Sterelny (2016), Oyama (2000), Ridley (2000). Some informational biologists are attempting to develop richer conceptions of information, e.g., naturalized teleosemantic conceptions.

there is no clear, technical notion of "information" in molecular biology. It is little more than a metaphor that masquerades as a theoretical concept and ... leads to a misleading picture of possible explanations in molecular biology.

Further, since most authors would argue that the informational processes must be embodied in real, physical systems, a basic question arises as to where to locate the causal efficacy. To critics, information is too abstract to play this role. Still unresolved "reduction" questions arise about the relation of the informational level of description to that of the material substrate (e.g., Shapiro 2004, Chaps. 3, 4). Some critics regard injecting information theory in the TTT manner as using a cannon to kill a sparrow: it is not necessary and introduces needless complexities.

In sum, there is no consensus about how to extend Shannon information to semantic information in either the declarative (descriptive) or procedural sense (Floridi 2015). The present, fluid situation summarized here could of course change with future progress. For example, Jantzen and Danks (2008) offers a promising account of biological information in terms of a graph-theoretical representations of causal relations.

In the biological case, we see a prospective TTT that lacks one or more components of the full TTT heuristic. For we have a mere scattering of individuals rather than community consensus, and some proponents seem to have borrowed a tool for which they are/were not already expert users (in itself a disciplinary crossover phenomenon fairly common in the history of science). Below we shall meet another sort of partial case of TTT, in which it is individual principles that are imposed on a system (sometimes the universe as a whole) that serve to constrain detailed theorizing. Partial *modeling* of a target system is a related, more local sort of case.

What about the realism issue? Some of the biological authors are clearly strong realists in holding that we *know* that the biological processes concerned are "essentially" or "constitutively" informational. But surely (I claim), it is premature to say that strong realism is justified in these cases.

5   Other Examples of TTT: Digital Physics?

Now I turn to the extreme case of digital physics, a field as far removed from cognitive psychology as one can get. Here it may seem that there is no question of a TTT move involving a projection of one cognitive system onto another.[12] I look at this extreme case in order to stretch the TTT idea to the limit, a move that will perhaps illuminate its weaknesses as well as strengths. Galileo may have anticipated

---

[12]But wait! Newell and Simon's "physical symbol system hypothesis" states that a physical system, as they define it, "has the necessary and sufficient means for general intelligent action" (1976, 116). And Deutsch's position (see below) is based on his claimed extension of the Church-Turing thesis to the physical universe (Deutsch 1985).

a TTT approach to the physical world when he said, in *Il Saggiatore*, that the book of the universe is written in mathematical language.

Digital physics treats the universe as *digital* and *informational* at bottom, and also as *computable* in some sense, usually by a quantum computer (Zenil 2013; Aguirre et al. 2015). There is no question that digital, informational approaches have been fruitful. I am not criticizing the tools themselves. The question is to what extent these humanly-developed research tools can be projected onto nature itself as a description of what is really going on there.

Several digital physicists claim that, in principle, a universal quantum computer *could simulate* the entire universe, computing each successive state. Others go so far as to suggest that the universe really *is* a simulation. (Pop writers have suggested that we may be part of a simulation or video game invented by brilliant extraterrestrials!) But some digital physics advocates go further still, to claim that the universe itself *is* a quantum computer. Here are some examples.

In the late 1980s, one of the most important physicists of the 20th century, John Archibald Wheeler, raised the question of whether we can get "it" from "bit." A few years later, in a 1990 article, Wheeler said that information may not be just what we *learn* about the world. It may be what *makes* the world.

> **It from bit.** Otherwise put, every 'it'—every particle, every field of force, even the space-time continuum itself—derives its function, its meaning, its very existence entirely… from the apparatus-elicited answers to yes-or-no questions, binary choices, bits. 'It from bit' symbolizes the idea that every item of the physical world has at bottom … an immaterial source and explanation; that which we call reality arises in the last analysis from the posing of yes-no questions and the registering of equipment-evoked responses. [1990, 310; Wheeler's emphasis]

> [W]hat we call existence is an information-theoretic entity. [Ibid. 313]

Here is the late Jacob Bekenstein of the Hebrew University, Jerusalem, and a former student of Wheeler:

> Ask anybody what the physical world is made of, and you are likely to be told "matter and energy." Yet if we have learned anything from engineering, biology and physics, information is just as crucial an ingredient.
>
> …
>
> Indeed, a current trend, initiated by John A. Wheeler of Princeton University, is to regard the physical world as made of information, with energy and matter as incidentals. [2003, 59][13]

---

[13]It was Edward Fredkin of MIT and Carnegie Mellon University who originally reversed the abstraction hierarchy and said that information is more basic than matter and energy. Interestingly, biological theorist George Williams (1992) held that the informational gene is as basic as matter and energy.

Oxford quantum computationist David Deutsch reports:

Many computer scientists have been so impressed with recently discovered connections between physics and computation that they have concluded that the universe *is* a computer, and the laws of physics are programs that run on it. [1997, 346]

Writes G. Mauro D'Ariano of the University of Pavia, in a prize-winning paper:

David Deutsch in his seminal paper "Quantum Theory, the Church-Turing principle and the universal quantum computer" rephrased the Church-Turing hypothesis as a truly physical principle. In short: every piece of physical reality can be perfectly simulated by a quantum computer. But now: what is the difference between Reality and its simulation? It's a matter for metaphysics: if Reality isindistinguishable from its simulation, then *it is* its simulation. The Universe is really a huge quantum computer: the computational universe of Seth Lloyd. [2012, 130; D'Ariano's emphasis]

What does Seth Lloyd of MIT himself say? In *Programming the Universe*, he writes:

The universe is made of bits. Every molecule, atom, and elementary particle registers bits of information. Every interaction between those pieces of the universe processes that information by alternating those bits. That is, the universe computes, and because the universe is governed by the laws of quantum mechanics, it computes in an intrinsically quantum-mechanical fashion: its bits are quantum bits. The history of the universe is, in effect, a huge and ongoing quantum computation.The universe is a quantum computer.

...

As the computation proceeds, reality unfolds. [2006, 3]

Computer engineer Tommaso Toffoli of Boston University:

In a sense, nature has been continually computing the "next state" of the universe for billions of years; all we have to do—and, actually, all we can do—is "hitch a ride" on this huge ongoing computation. [1982, 165]

In *Our Mathematical Universe*, MIT physicist Max Tegmark writes:

Whereas most of my physics colleagues would say that our external physical reality is (at least approximately) *described by* mathematics, I'm arguing that it *is* mathematics (more specifically, a mathematical structure).

...

I'm writing *is* rather than *corresponds to* here, because if two structures are equivalent, then there's no meaningful sense in which they're not one and the same .... [2014, Chap. 11; Tegmark's emphasis]

However, according to quantum physicist Anton Zeilinger, University of Vienna, quantum mechanics is *only* about information, not directly about reality. His view is based on his interpretation of Bohr's Copenhagen interpretation.

[I]nformation is the basic concept of quantum physics itself. That is, quantum physics is only indirectly a science of reality but more immediately a science of knowledge. [Brukner and Zeilinger 2005, 47]

...

> If, as we have suggested above, quantum physics is about information, then we have to ask ourselves what we mean by a quantum system. It is then imperative to avoid assigning any variant of naïve classical objectivity to quantum states. Rather it is then natural to assume that the quantum system is just the notion to which the probabilities in (3.4) and (3.5) refer and no more. The notion of an independently existing reality thus becomes void. [Ibid., 56]

Polymath Stephen Wolfram (1984, 2002), argues that the world is a cellular automaton, or at least can be fully simulated by one. Wolfram regards computation as the fundamental concept that should drive research.

To anticipate some critical doubts: Rodney Brooks, former head of the MIT Robotics Lab, counters with "Computation as the Ultimate Metaphor":

> Those of us who are computer scientists by training, and … scientists of other stripes, tend to use computation as the mechanistic level of explanation for how living systems behave and 'think'. I originally gleefully embraced the computational metaphor.

> If we look back over recent centuries we will see the brain described as a hydrodynamic machine, clockwork, and as a steam engine. When I was a child in the 1950's I read that the human brain was a telephone switching network. Later it became a digital computer, and then a massively parallel digital computer…. The brain always seems to be one of the most advanced technologies that we humans currently have.

> The metaphors we have used in the past for the brain have not stood the test of time. I doubt that our current metaphor of the brain as a network of computers doing computations is going to stand for all eternity either. [Brooks 2008] [14]

6 Some Difficulties for TTT Digital Physics

What is going on in the quoted statements and many others like them? Are these digital physicists and computer scientists just being sloppy in their semi-popular writings,[15] each trying to out-radicalize the others? Do they mean only to be advancing hypotheses rather than making strong realist claims? Sometimes we see

---

[14]*Objection!* We can say that even an old-fashioned balance scale or spring scale can compute weights, so why not say that neurons in the brain compute visual outputs and that quantum physical processes compute the solutions to physical problems involving causal forces and laws of nature? *Reply.* The objection uses 'compute' in a very broad sense. For starters, we can make a distinction between those entities that employ a symbol system or code with symbol-transforming rules and those that do not.

[15]Deutsch (2011) presents himself as a strong realist who believes that his account is the eternal truth about the universe. In these passages, he seems to believe in cumulative truth against future revolution or even long-term, transformative evolution of scientific results. Yet, in other passages, he insists that scientific progress will continue for centuries. Moreover, he agrees with Popper (his favorite philosopher of science) that we cannot know now what we shall only know later. As an illustration, he even cites Michelson's strong realist commitment to late 19th-century physics, a strong realism soon undermined by the relativity and quantum revolutions (Chap. 9). It is hard to see how transformative progress can continue if we already have the fundamental truth (Nickles 2017 and forthcoming a and b).

Bohr-inspired philosophical idealism and sometimes pragmatic "don't cares," meaning, roughly, "Metaphysical questions are not relevant to my scientific practice, and I am talking only about my practice."[16]

I hope the above quotations are enough to indicate that some leading experts are genuinely committed, as strong realists, to digital physics as a true ontological account of the world—and hence to a strong TTT sort of postulatory move. In other words, they believe that they have established the fundamental truth about reality, not merely constructed a valuable simulation or generated a bold hypothesis for consideration.

The sources of my quotations are admittedly informal writings, but it is precisely in this venue where physicists tend to disclose their overall world views. And when they do, we find a great diversity of disagreement among even the best physicists, who can disagree fundamentally about what is really going on at the micro-, nano-, or Planck-scale levels (Brockman 2015). This fact alone suggests that there is far from a consensus about the nature of reality among top physicists, and that making strong realist claims about any such ontic view is extremely premature. Thus a key component of the Gigerenzer-Sturm account of TTT is lacking, unless we restrict it to a subset of the digital physics community. Nonetheless, we can appreciate that influential individuals are making strong TTT moves and encouraging other researchers and science analysts to follow them.

Some of the challenges faced by informational biology also apply here, although physicists tend to remain closer to a syntactical conception of information, which they relate to thermodynamic entropy. (Shannon, too, spoke of entropy in a more abstract sense.) Whether abstract quantum information is sufficient to provide *causal* explanations is a key question. A related difficulty is that some forms of digital physics seem to leave the universe-computer without a physical platform, as if the universe could be software "all the way down." Another challenge is to explain how the universe-computer could have developed from the big bang, i.e., a prior state in which the computational model fails.

Finally, I'll mention *the simulation illusion*. This is the unwarranted slide from simulation to reality, precisely the slide from research tools to ontic assertion. Underlying this slide is the ambiguity of 'model' and 'to model X as Y' and, more basically, of 'representation' and 'to represent X as Y'. As one critic put it, we must not confuse a model of nature with the nature of our model. Simulations can be valuable research tools that enable us to learn much about reality, even using them to generate "empirical" data. (There is much discussion about how this is possible.) But the simulation is still a tool, not reality itself. If we cannot tell the difference between how we handle our simulation and how we handle real phenomena,

---

[16]The ontic interpretation of digital physics resurrects old questions of the sort raised by the Copenhagen interpretation of quantum mechanics and old questions involving hypothetical constructs, intervening variables, and postulatory theories. However, physicists have succeeded in making direct connections of therodynamics to information theory, for example. More controversially, the Landauer principle or limit establishes a lower bound energy cost for nonreversible computational processes such as erasure, Deutsch has provided a physicalist version of the Church-Turing principle, and cosmologists speak of a principle of conservation of information.

doesn't that just show that we are using our usual research tools in both cases, that the tools at best help to generate a promising hypothesis about reality, and that the strong realist conclusion is scientifically otiose? If so, then strong realism adds nothing to the scientific toolkit. To state the worry in a provocative form: identifying reality with what is basically our own (group) cognitive function is a form of *idealism*—reality as a product of our own "minds" (forms of scientific behavior). How strange that TTT strong realism begins to smack of a sort of socio-technical idealism. Reality is now our own cognitive construction, driven by a metaphor strong enough to make the physical universe practically disappear!

7 More General TTT Difficulties

I shall postpone Gigerenzer and Sturm's (2007) major worry about circularity until Sect. 8 on dynamical coupling.

As Gigerenzer and coauthors characterize it—and name it—the TTT heuristic projects, transfers, or converts research tools into theories, methods-*cum*-instruments into ontic postulations. Since the tools or methods in question are often correlational, in such cases TTT pretends, instantly, to transform *correlational* "theory" (in the disciplinary sense of theory) into a specific *causal-postulatory* theory of the target system.[17] That's an easy way to turn a merely correlational science into a causal-explanatory one. Surely too easy!

Second, and relatedly, research tools being instruments in a broad sense, TTT involves a transition from a sort of instrumentalist stage of research to full-blown realism. For scientists' instrumental methods of data processing are now postulated as the way in which the target system itself really works.[18] Unless and until further probing is done, the strong TTT projection is really just neo-instrumentalism masquerading as strong realism.

Again, relatedly, the TTT move can be verificationist in confusing a (postulated) thing or process with the evidence for it, or with the way of analyzing the evidence. This makes the postulation seem far more justified than it in fact is. For example, there is a slide from saying that an electron is nothing more than a node in the network of experimental tools and expert practices to saying that an electron is an independent, objective, theoretically postulated, unobserved entity, to which those

---

[17]This move is especially familiar in the social and behavioral sciences, where so much methodological ink has been spilled over the status of unobserved, hypothetical constructs versus intervening variables.

[18]A reverse process occurs when a once-established postulatory theory is demoted, deflated, to a mere tool, as has happened with Newtonian mechanics and several other historical successes. Here the move is from an intended, literal description of underlying reality to a mere "as if"; e.g., as if there were Newtonian forces in the world that acted instantaneously at a distance. The now-deflated theory is reduced to an instrument for calculation. Of course, well-established theories also serve as useful tools.

tools and practices offer fleeting access. It is a slide from a quasi-definition of 'electron' to a bold empirical claim.

As already hinted, there are also potential fallacies of ambiguity and vagueness over the meaning of 'model,' 'representation', 'information', and 'computation', but here I shall highlight the rhetorical fallacy of reification or hypostatization. The basic mistake is to start with diverse, concrete materials, to provide an encompassing description or label at a more abstract level of description, then to treat the abstraction as the real reality, with causal powers. It treats an abstraction as a distinct, concrete, denizen of the universe. It is what Whitehead called *the fallacy of misplaced concreteness*. Information, for example, is not an entity over and above the detailed physical processes.[19]

I suspect that this sort of thinking pervades much scientific thinking and writing about science, via a sort of *partial* TTT projection, the projection of a methodological principle, a metatheoretic guide to theorizing in a specific domain, onto the universe.[20] Einstein famously inspired the process of doing physics in terms of basic, high-level principles, such as the principle of relativity and the equivalence principle. Soon there was much fruitful work relating invariance and symmetry and conservation principles to each other. Much valuable theorizing in high energy physics has been driven by this approach. However, when a physicist writes that "gauge symmetry implies that there must be an electromagnetic force," does this genuinely *explain* what brought such a force into existence? Is gauge symmetry a concrete causal process? Does it really *causally explain why* anything at all exists or occurs?[21] It seems to me that here there is sometimes a confusion of *explanation-why* with *explanation-that*. Again, please note that I am not denying the great fertility of such principles in theory development and the fact that they can be used as premises in arguments, only that they should not be given distinct causal-referential status without further ado.

General nonrealist considerations also challenge TTT. When we are dealing with bold, highly theoretical claims, a few good confirmations, no matter how precise and subjectively impressive, fall short of establishing that the theory in question is ontologically true, or nearly so. When used alone, the strongest use of the TTT heuristic is worse than realist in this traditional sense, insofar as it is supposed to obviate the need for competitive testing of the theory produced and insofar as it places road blocks in the way of alternative approaches.

---

[19]Chapter 5 of Deutsch (2011) is titled "The Reality of Abstraction".

[20]The positivists sometimes made the reverse move, insisting that universal causation and simplicity (for instance) were methodological principles rather than basic claims about the universe.

[21]No more than we can explain why (causally) a flagpole is 15 m high in terms of the length of its shadow and the angle to the sun.

8 Dynamical Coupling and Bootstrapping, Good and Bad

Coupled dynamical variables can produce interesting system dynamics, because they introduce interaction terms and hence nonlinearity. The Lotka-Volterra predator-prey equations are a simple, stock example. The same points can be made (perhaps a bit metaphorically) at the general methodological level. Here, coupling two or more methodological factors can be a research accelerator, but tight coupling can be dangerous. In my view, the coupling involved in the strong realist version of the TTT heuristic is too tight, too intimate, too direct. Gigerenzer and company agree.

Traditional empiricism and rationalism were linear theories of knowledge in the sense that new claims had to be justified in terms previous results. If A justified B, then B could in no way help to justify A. However, in nature and artifact we find mutual support of many kinds, so the idea of mutual support is not intrinsically circular. Think of an arch, in which the two sides and the keystone are mutually supportive. Ditto for the steel beams providing the support for a building. The account of such support is not circular in a vicious sense until we get to extreme cases, as a literal bootstrap operation would be: the hand lifts the boot off the ground by pulling up on the strap, and the strap simultaneously pushes the hand upward![22] But there is a milder, legitimate form of epistemological bootstrapping, in which each of two or more claims or practices can help to support the others. Philosophers of science have realized this, for example, in the complex relations between theory and observation. The observations support the theory, but the theory both indicates which observations are significant and how to find and interpret them.

Coupling at this epistemological or methodological metalevel used to be forbidden for fear of vicious circularity. Recall the big debate of the 1960s and '70s over the theory-ladenness of observation reports (e.g., Scheffler 1967). According to traditional methodological accounts, the observational data had to be completely theory neutral. In fact, observation, theory, and method statements as well as those about goals, standards, and instrument design all had to be (more or less) logically independent of one another. We now know that science does not work like that. For example, efficient instrumentation often has some theory built right into its logic. Think of particle detectors in high-energy physics, or the specific methods used by Bahcall and Davis to detect and count solar neutrinos and by Cowan and Reines to detect antineutrinos. What scientists count as an observation or detection often depends on several theory-laden inference steps (Shapere 1982; Kosso 1989). Without a delicate interweaving of theory and experiment, modern science could not exist.

As Charles Peirce wrote in his early essay, "The Fixation of Belief," "questions of fact and questions of logic [method] are curiously interlaced" (CP 5.364). Nor is

---

[22]I am making a general point about mutual support and our account of it. I am not referring specifically to the bootstrap method of Glymour (1980).

it surprising that a successful theory and new instrumentation can alter methods, standards, and goals. As Peirce also wrote, in reference to Darwin's theory as introducing an implicit statistical methodology: "each chief step in science has been a lesson in logic" (ibid.). Here Peirce could regard the development of a new method, not merely a new law or theory, as the ultimate goal of the discovery process. In this case we get new methods from successful theories rather than new theories from successful methods. It is TTT in reverse.[23]

A degree of mutual support in such cases is reasonable. The requirement that method be completely universal (in the sense of being theory- and data-neutral, and applicable to all sciences at all times) condemns it to being extremely weak.[24] As Herbert Simon often emphasized (e.g., Simon 1990), efficient problem solving, efficient research, must take into account the structure of the task environment, and this will require substantive premises.

Nonetheless, if the coupling is too close, difficulties arise. Kuhn's *Structure of Scientific Revolutions* (1962/1970) was criticized precisely for coupling the dominant theory not only to observational methods but also to goals and standards and conceptual language in such a way that the complex becomes *too* self-reinforcing, all items being constituents of the same paradigm, thereby producing an exclusionary normal science that can only be escaped via radical revolutions (Scheffler 1967; Laudan 1984). Although they did allow theory-ladenness of observation, Popperians and Lakatosians complain that many coupling moves are blatantly ad hoc, since, for example, the proposed theory is directly motivated by extant empirical knowledge rather than backed by rigorous tests via novel predictions. For them, this is a form of circularity in which the same information is used twice—to generate the theory and then to "test" it.[25,26] At the limit, circularity represents a *logical* coupling of the tightest sort.

---

[23]There is a distant connection here to what I have called "discoverability" as opposed to original discovery (Nickles 1985). Once a sufficiently rich body of domain information and investigative tools have been developed, by whatever means, it may then be possible to go back and rationally reconstruct a "discovery" process leading to the current theory. In this way, the theory is supported generatively (by rationally reasoning *to* it from prior premises, in part results made possible by having in hand that very theory), not merely consequentially (by testing predictions logically derived *from* it).

[24]One may sense a conflict here with Gigerenzer's program of fast and frugal heuristics, which, as explicitly stated, include little or no domain knowledge (Gigerenzer et al. 1999). However, Gigerenzer's ABC Group do not regard such heuristics as universal in scope. On the contrary, a major part of their research is to explore their limits of application, in terms of the structure of the task environment or domain involved (see Nickles 2016). The more that is known about the (statistical) structure of the environment, the better the chance of finding fast and frugal heuristics that work well in that context. In this sense, fast and frugal heuristics are very domain specific.

[25]For criticism of the Popper-Lakatos position on coupling, see Nickles (1987).

[26]This sort of coupling does provide resistance to the fast and frugal approach, which aims to apply minimal information, at least overtly. But against a rich background of information about the statistical structure of the environment, simple algorithms (as in heart attack triage) can work. On the other hand, this means that such fast and frugal methods are not as fast and frugal as they appear, except in post-discovery application.

Gigerenzer and Sturm (2007) share the circularity worry in a more direct form, since, with TTT, the postulatory theory is not merely constructed so as to agree with known data. Instead, the resulting theory is a mirror image of the already established tools and procedures that claim to support it. Thus, we have something close to *self*-support rather than mutual support. Worse, the detail they provide in their rich case studies shows how some scientific subcommunities have used TTT in a power play to take over a specialty, by *excluding* data from competing approaches while championing its own, thereby presupposing its own, unique methodological correctness. We might say that the cognitive scientific practices that concern Gigerenzer and Sturm are "Kuhnian" in the sense that the successful methodological tools and related practices bring with them new scientific standards that can be dogmatically applied by the normal scientists in power to exclude alternative approaches as unscientific or at least substandard. The pre-justification of the projected theory has a cost of possible vicious circularity that closes off research options rather than opening them up. In the strongest instances, TTT rigs the game in its own favor.

9 Weak Versus Strong TTT: Is There a Cost?

There is one place where I disagree with Gigerenzer and Sturm. Worrying about how seriously to take the realist invitation of TTT, they write:

> One might try to avoid such difficult problems by biting the antirealistic bullet. Is it not better to view the theoretical concepts and claims of cognitive psychology as "mere" constructs or as "as-if" models? One may do so, but there is a price to be paid here. [2007, 313]

The price, they say, is retreat from the fertility of the cognitive revolution back to the precognitive days of behaviorism and instrumentalist positivism.

I think they are being too binary here, for there exists a whole spectrum of positions, going from strong realism to rabid antirealism. I am a *non*realist in deep theoretical matters, an agnostic rather than a true *dis*believer. My anti-strong realism is not anti-realism simpliciter. I have no problem with researchers being *intentional* realists, i.e., regarding their theories as attempts to provide true accounts of the world and their theoretical terms as meaningful (possible) references to hidden entities and processes. Intentional realism is a relaxation from the strictures of positivism, one that has stimulated research in many quarters.

But intentional realism is a far cry from strong realism—the view that we now *know* that our mature scientific theories are true (or approximately true) and that our theoretical terms *do* refer to real entities and processes (Psillos 1999)—and that we *understand* the processes that the theories describe. For the strong realist, our mature theories provide an objective (completely human-independent) account of reality. Recall that Popper was an intentional realist but definitely not a strong realist. Popper rejected the positivist view that theoretical terms were either meaningless or required to be quasi-definable in terms of laboratory practices and results. The nonrealist or agnostic or anti-strong realist is not committed to a positivist theory of meaning, positivist instrumentalism, or the old positivist distinction of observational from theoretical terms.

By contrast, the weak (realist and nonrealist) versions of TTT claim only to generate *candidate* models or theories in the "proof of concept" or "how-possibly?" sense—asserting that the projected tools are sufficient in principle to produce the observed behavior. This claim is justified, assuming that the scientists' tools do provide adequate regimentation of the known phenomena. While the strong realist TTTer claims already to have the truth about the target system, or something close, the weak TTTer claims only to have provided a rationally-generated hypothesis that is worth serious consideration. The weak TTTer will point out that the justification of the projection itself, while endowing the hypothesis with a degree of antecedent credibility, does not amount to rigorous testing against competitors, or even the attempt to find such competitors. I conclude that the projection must be subject to further heuristic and epistemic appraisal, in competition with alternatives, like any other plausible hypothesis (Nickles 2006).

That said, how strong is the presumption in favor of the projected theory is a matter of degree. It is legitimately stronger insofar as scientists' best attempts to develop alternative methods and theories fails. To minimize the old problem of underdetermination of theory by evidence (or, in the TTT case, also by method), stronger TTTers must mount something like a convincing inference-to-the-best-explanation (IBE) argument at the methodological level. (Of course, IBE arguments for realism remain controversial.) TTT moves can be fertile starting points to further methodological and theoretical inquiry, but they should not bring inquiry to a close. And, from a longer historical perspective, we should keep in mind Orgel's and Shakespeare's caution that there may well be more things in heaven and earth than are dreamt of in our current science.

The main point of this section is that there is no question of having to retreat to pre-cognitive days. The cognitivists can postulate all they want, even in the absence of testing against competitors, as long as they stick to weak TTT and are not methodologically dogmatic. Weak TTTers have exactly the same set of research tools as strong TTTers. Adopting strong TTT does not provide strong realists with an additional tool. The toolboxes are the same.

In those cases in which we get good enough tools to give us very good experimental access, control, and theoretical understanding, I am willing to shift in the direction of realism; but then, thanks to those research advances, it has become a rather shallow theoretical realism. So, as a pragmatic nonrealist, I think that we can have our cake and eat it, too. We can be realists (or nonrealists who allow intentional realist theorizing) without being strong realists.

The spectrum of TTT positions, from weak to strong, all differ from inductivist and from Popperian conceptions of theory production, and both have the virtue of immediately integrating a new hypothesis with existing practice. The question, in a given case, is whether the coupling is *too* close.

10 Conclusion: Human-Centered Science and the Simon Tradition

Gigerenzer and colleagues have provided a rich account of the TTT heuristic and some of its applications. I have attempted to extend the discussion to current research that has not, so far, achieved community acceptance, or even practical expertise, and in fields far from cognitive psychology. Since our mind-to-mind folk intuitions are not here in play, these cases perhaps serve to bring out sharply the weaknesses as well as the strengths of TTT. I have also considered coupling issues relevant to the dynamics of scientific change.

In my judgment, the weak TTT moves are legitimate, but stronger realist ones, by themselves, become increasingly problematic. In the language of Herbert Simon (e.g., Simon 1990), I see strong theoretical realism as a kind of *optimizing* position, one that is rarely warranted, whereas our historicity forces us to be pragmatic *satisficers* if we wish to progress rapidly. We must make do with what is available at our stage of historical development. We cannot search through a haystack that today conceals items that will only become available to us in the, perhaps distant, future, let alone those "possible" items that we'll never hit upon.

References

Aguirre, A., Foster, B., & Merali, Z. (Eds.), (2015). *It from bit or bit from it?* Springer, Dordrecht: On Physics and Information.

Artmann, S. (2008). Biological information. In S. Sarkar & A. Plutynski (Eds.), *A companion to the philosophy of biology* (pp. 22–39). Oxford: Blackwell.

Bekenstein, J. (2003, August). Information in the holographic universe. *Scientific American,* 58–65.

Bishop, M., & Downes, S. (2002). The theory theory thrice over: The child as scientist, superscientist or social institution? *Studies in History and Philosophy of Science, 33*(1), 117–132.

Brockman, J. (Ed.). (2015). *This idea must die: Scientific theories that are blocking progress.* New York: Harper Perennial.

Brooks, R. (2008). *Computation as the ultimate metaphor.* https://www.edge.org/response-detail/ 11249. Accessed 31 July 2017.

Brukner, Č., & Zeilinger, A. (2005). Quantum physics as a science of Information. In A. Elitzur et al. (Eds.), *Quo vadis quantum mechanics?* (pp. 47–62). Berlin: Springer.

Carey, S. (1985). *Conceptual change in childhood.* Cambridge, MA: MIT Press.

Chaitin, G. (2012). *Proving Darwin: Making biology mathematical.* New York: Pantheon Books.

Collier, J. (2008). Information in biological systems. In P. Adriaans & J. van Benthem (Eds.), *Handbook of the philosophy of science, vol. 8: Philosophy of information.* Elsevier: Amsterdam.

D'Ariano, G. M. (2012). A quantum-digital universe. *Advanced Science Letters, 17,* 130–135. Updated version in Aguirre et al. (2015), pp. 25–35.

Dawkins, R. (1976). *The selfish gene.* Oxford: Oxford University Press.

Dawkins, R. (1995). *River out of Eden: A Darwinian view of life.* New York: Basic Books.

Dennett, D. (1995). *Darwin's dangerous idea.* New York: Simon & Schuster.

Dennett, D. (2013). *Intuition pumps and other tools for thinking*. New York: Norton.

Deutsch, D. (1985). Quantum theory, the Church-Turing principle and the universal quantum computer. *Proc. Royal Society of London A, 400,* 97–117.

Deutsch, D. (2011). *The beginning of infinity*. London: Allen Lane.

Floridi, L. (2015). Semantic conceptions of information. In E. N. Zalta (Ed.), *The Stanford encyclopedia of philosophy* (Fall 2016 Ed.). http://plato.stanford.edu/archives/fall2016/entries/information-semantic/.

Fodor, J. (1968). *Psychological explanation*. New York: Random House.

Galison, P. (2016). Practice all the way down. In Richards & Daston (Eds.), (pp. 42–69).

Gigerenzer, G. (1991a). From tools to theories: A heuristic of discovery in cognitive psychology. *Psychological Review, 98,* 254–267.

Gigerenzer, G. (1991b). Discovery in cognitive psychology: New tools inspire new theories. *Science in Context, 5,* 319–350.

Gigerenzer, G. (2003). Where do new ideas come from? A heuristic of discovery in cognitive science. In M. Boden (Ed.), *Dimensions of creativity* (pp. 53–74). Cambridge, MA: MIT Press.

Gigerenzer, G., & Goldstein, D. (1996). Mind as computer: Birth of a metaphor. *Creativity Research Journal, 9,* 131–144.

Gigerenzer, G., & Murray, D. (1987). *Cognition as Intuitive Statistics*. Hillsdale, NJ: Erlbaum.

Gigerenzer, G., & Sturm, T. (2007). Tools = Theories = Data? On some circular dynamics in cognitive science. In M. Ash & T. Sturm (Eds.), *Psychology's territories* (pp. 305–342). Hillsdale, NJ: Erlbaum.

Gigerenzer, G, Todd, P, ABC Research Group. (1999). *Simple heuristics that make us smart*. New York: Oxford University Press.

Glymour, C. (1980). *Theory and evidence*. Princeton: Princeton University Press.

Godfrey-Smith, P. (2007). Information in biology. In D. Hull & M. Ruse (Eds.), *The Cambridge companion to the philosophy of biology* (pp. 103–113). Cambridge: Cambridge University Press.

Godfrey-Smith, P., Sterelny, K. (2016). Biological information. In E. N. Zalta (Ed.), *The Stanford encyclopedia of philosophy* (Summer 2016 Ed.). http://plato.stanford/edu/archives/sum2016/entires/information-biological/. Accessed 15 August 2016.

Gopnik, A., Maltzoff, A., & Kuhl, P. (1999). *The scientist in the crib*. New York: William Morrow.

Griesemer, J. (2005). The informational gene and the substantial body: On the generalization of evolutionary theory by abstraction. In M. Jones & N. Cartwright (Eds.), *Idealization XII: Correcting the model: Idealization and abstraction in the sciences* (pp. 59–115). Amsterdam: Rodopi.

Griffiths, P. (2001). Genetic information: A metaphor in search of a theory. *Philosophy of Science, 68,* 394–412.

Jantzen, B., & Danks, D. (2008). Biological codes and topological causation. *Philosophy of Science, 75,* 259–277.

Kaiser, D. (2016). Thomas Kuhn and the psychology of scientific revolutions. In Richards & Daston (Eds.), (pp. 71–95).

Kosso, P. (1989). *Observability and observation in physical science*. Dordrecht: Kluwer.

Kuhn, T. S. (1962). *The structure of scientific revolutions*. Chicago: University of Chicago Press. 2nd ed with postscript (1970).

Langley, P., Simon, H. A., Bradshaw, G., & Zytkow, J. (1987). *Scientific discovery*. Cambridge, MA: MIT Press.

Laudan, L. (1984). *Science and values*. Berkeley: University of California Press.

Lloyd, S. (2006). *Programming the universe*. New York: Knopf.

Maynard-Smith, J. (2000). The concept of information in biology. *Philosophy of Science, 67,* 177–194.

Maynard-Smith, J., & Szathmáry, E. (1995). *The major transitions in evolution*. New York: Freeman.

Meytlis, M. (2011). A model of face space. *Visual Cognition, 19*(1), 13–26.

Newell, A., & Simon, H. A. (1976). Computer science as empirical inquiry. *Communications of the ACM, 19*(3), 113–126.

Nickles, T. (1985). Beyond divorce: Current status of the discovery debate. *Philosophy of Science, 52,* 177–206.

Nickles, T. (1987). Lakatosian heuristics and epistemic support. *British Journal for the Philosophy of Science, 38,* 181–205.

Nickles, T. (2006). Heuristic appraisal: Context of discovery or justification? In J. Schickore & F. Steinle (Eds.), *Revisiting discovery and justification: Historical and philosophical perspectives on the context distinction* (pp. 159–182). Dordrecht: Springer.

Nickles, T. (2016). Fast and frugal heuristics at research frontiers. In E. Ippoliti, F. Sterpetti, & T. Nickles (Eds.), *Models and inferences in science* (pp. 31–54). Springer International: Switzerland.

Nickles, T. (2017). Is scientific realism a form of scientism? In M. Pigliucci & M. Boudry (Eds.), *Science unlimited? The challenges of scientism.* Chicago: University of Chicago Press.

Nickles, T. (forthcoming a). Cognitive illusions and nonrealism: objections and replies. In E. Agazzi, M. Alai (Eds.), *Varieties of scientific realism,* forthcoming, Springer.

Nickles, T. (forthcoming b). Do cognitive illusions tempt strong scientific realists? In W. J. González (Ed.), *New approaches to scientific realism.*

Oyama, S. (2000). *The ontogeny of information.* Durham, NC: Duke University Press.

Peirce, C. S. (1931–1935). The collected papers of Charles Sanders Peirce. In: C. Hartshorne & P. Weiss (Eds.). Cambridge, MA: Harvard University Press.

Popper, K. R. (1963). *Conjectures and refutations.* New York: Basic Books.

Psillos, S. (1999). *Scientific realism: How science tracks truth.* London: Routledge.

Richards, R., & Daston, L. (Eds.). (2016). *Kuhn's structure of scientific revolutions at fifty.* Chicago: University of Chicago Press.

Ridley, M. (2000). *Mendel's demon: Gene justice and the complexity of life.* London: Weidenfeld & Nicholson.

Sansom, R. (2011). *Ingenious genes: How gene regulation networks evolve to control development.* Cambridge, MA: MIT Press.

Sarkar, S. (1996). Decoding 'Coding'—Information and DNA. *Biosciences, 46,* 857–864.

Scheffler, I. (1967). *Science and subjectivity.* Indianapolis: Bobbs-Merrill.

Shapere, D. (1982). The concept of observation in science and philosophy. *Philosophy of Science, 49,* 485–525.

Shapiro, L. (2004). *The mind incarnate.* Cambridge, MA: MIT Press.

Simon, H. A. (1990). Invariants of human behavior. *Annual Review of Psychology, 41,* 1–19.

Tegmark, M. (2014). *Our mathematical universe.* New York: Knopf.

Toffoli, T. (1982). Physics and computation. *International Journal of Theoretical Physics, 21,* 165–175.

Wheeler, J. A. (1990). Information, physics, quantum: The search for links. In W. Zurek (Ed.), *Complexity, entropy, and the physics of information* (pp. 309–336). Redwood City, CA: Addison-Wesley.

Williams, G. (1992). *Natural selection: Domains, levels, and challenges.* New York: Oxford University Press.

Wolfram, S. (1984, September). Computer software in science and mathematics. *Scientific American,* 188–203.

Wolfram, S. (2002). *A new kind of science.* Champaign, IL: Wolfram Media.

Zenil, H. (2013). *A computable universe: Understanding and exploring nature as computation.* Singapore: World Scientific.

Heuristic Logic. A Kernel

Emiliano Ippoliti

Abstract In this paper I lay out a non-formal kernel for a heuristic logic—a set of rational procedures for scientific discovery and ampliative reasoning—specifically, the rules that govern how we generate hypotheses to solve problems. To this end, first I outline the reasons for a heuristic logic (Sect. 1) and then I discuss the theoretical framework needed to back it (Sect. 2). I examine the methodological machinery of a heuristic logic (Sect. 3), and the meaning of notions like 'logic', 'rule', and 'method'. Then I offer a characterization of a heuristic logic (Sect. 4) by arguing that heuristics are ways of building problem-spaces (Sect. 4.1). I examine (Sect. 4.2) the role of background knowledge for the solution to problems, and how a heuristic logic builds upon a unity of problem-solving and problem-finding (Sect. 4.3). I offer a first classification of heuristic rules (Sect. 5): primitive and derived. Primitive heuristic procedures are basically analogy and induction of various kinds (Sect. 5.1). Examples of derived heuristic procedures (Sect. 6) are inversion heuristics (Sect. 6.1) and heuristics of switching (Sect. 6.2), as well as other kinds of derived heuristics (Sect. 6.3). I then show how derived heuristics can be reduced to primitive ones (Sect. 7). I examine another classification of heuristics, the generative and selective (Sect. 8), and I discuss the (lack of) ampliativity and the derivative nature of selective heuristics (Sect. 9). Lastly I show the power of combining heuristics for solving problems (Sect. 10).

Keywords Heuristics · Logic · Discovery · Reasoning · Problem-solving

1 Reasons for a Heuristic Logic

Heuristic logic stems from the attempt to connect logic back to one of its primary and original purposes, as defined by Aristotle among others (see Ippoliti and Cellucci 2016), namely the development and refinement of a method to obtain *new* knowledge. In order to do that, we have to answer two basic questions:

E. Ippoliti (✉)
Sapienza University of Rome, Rome, Italy
e-mail: emiliano.ippoliti@uniroma1.it

© Springer International Publishing AG 2018   191
D. Danks and E. Ippoliti (eds.), *Building Theories*, Studies in Applied Philosophy,
Epistemology and Rational Ethics 41, https://doi.org/10.1007/978-3-319-72787-5_10

1. Is it possible to account for *how* knowledge is extended?
2. Is it possible to build *rational* tools to obtain *really* new knowledge?

Heuristic logic answers those questions positively and draws on a conception of logic different from the one put forward by mathematical logic, and builds upon a problem-oriented approach[1] to the advancement of knowledge.

More specifically, heuristic logic argues that knowledge advances by generating *local* hypotheses to solve bottom-up *specific* problems, and that there is an inferential way to accomplish this task, i.e. rules and rational procedures. These rational procedures are the heuristics, that is, an open set of rules for discovery (of course the notion of rule here is not the same adopted by classical logic, as we will see). A heuristics is a way of constructing (and not simply a way of *reducing*) the *problem-space* and the several paths through it that lead to the formulation of good hypotheses to solve the problem at hand.

So, this paper sets out to argue that:

- There is such a thing as a logic of discovery, a heuristic logic (in a specific sense of 'logic'), which provides us a rational guide to extend our knowledge by employing the non-deductive rules that enables us to generate hypotheses.
- There is such a thing as a method of discovery, a version of the analytic method (see Cellucci 2013), which proceeds bottom-up, from the problem, along with other data and available knowledge, to the hypotheses.
- A hypothesis is a sufficient condition for the solution to a problem. Accordingly it is local, provisional and plausible.

Bottom line: it is possible to produce a systematic treatment of heuristic reasoning and to build a heuristic logic (in a sense of 'logic' that I will make explicit later) that is an open set of rules capable of providing rational procedures to guide us in the advancement of knowledge.

2 Heuristic Logic: A Theoretical Framework

By heuristic logic, here I mean the study of the (mental) operations employed in solving problems, where problems are intended as the essential engine for advancing knowledge. A robust framework for a rational account of the advancement of knowledge is the one provided by a version of the analytic method examined by Cellucci (see Cellucci 2013 for a full account of it). This framework employs three building blocks:

---

[1]See in particular Lakatos (1976, 1978), Laudan (1977, 1981), Nickles (1978, 1981), Popper (1999), Jaccard and Jacoby (2010), Cellucci (2013), Ippoliti and Cellucci (2016).

1. A *problem-solving approach*, that is the idea that knowledge advances by solving problems (see Newell and Simon 1972; Lakatos 1976, 1978; Laudan 1977, 1981; Nickles 1978, 1981; Popper 1999; Weisberg 2006; Cellucci 2013).
2. A version of the *analytic method for problem-solving*. In effect, it states that we have a method, a rational procedure, to solve problems: we formulate a hypothesis (a proposition that implies the statement of the problem at hand) by means of non-deductive inferences and then we test this hypothesis by comparing it with the existing knowledge (see in particular Cellucci 2013).
3. An *'infinitist' solution for the problem of regress*. The hypothesis introduced to solve a problem in that way, in turn, represents a new problem (for it must be justified), which is solved in the same way, that is by introducing a new hypothesis and so on *ad infinitum*. Now, a typical counter-argument to such an 'infinitist' solution to the regress problem states that we do not produce knowledge at all by proceeding in that way, since the regress, that is, the passage from a proposition to another one that justifies it and so on, is unstoppable.

This counter-argument can be blocked by noting that:

(a) We do not produce knowledge only if the regress is *arbitrary*.
(b) We can go through deeper and deeper lines of the regression, and therefore we do extend our knowledge.

In order to justify (a) and (b) the notion of *plausibility* is introduced (see e.g. Ippoliti 2006; Cellucci 2013), which builds upon Aristotle's notion of *endoxa*. Plausibility offers us a guide (not an algorithm) for the selection of the hypotheses in the form of an evaluation (not a calculus) of the reasons *pro* and *contra* a given hypothesis, and hence of the reasons for the passage from one specific proposition to another one during the search for a solution to a problem. So this passage is not arbitrary.

This framework implies that the discovery is a process characterized by two moves at each step. The first, bottom-up, goes from the problem and the data up to the hypothesis, and it is put forward in a non-deductive way. The second, top-down, goes from the hypothesis to the data. This second move is important for determining the plausibility of a hypothesis and employs deductive as well as non-deductive reasoning. More in detail, at this stage we can perform a sort of 'plausibility test', which employs both deductive and non-deductive inferences. The first step of this test is a deductive one: we draw conclusions from the hypothesis and then we confront them with each other in order to check if they produce a contradiction or not. The second step involves deductive as well as non-deductive reasoning, since we compare the conclusions with other hypotheses regarded as plausible, other findings or experimental observations. Such a comparison can require heuristic reasoning too, for instance the non-deductive patterns of plausible inference identified by Polya. One example is the following (see Polya 1954, vol. II, p. 9), where H is a hypothesis:

$H \rightarrow B_{n+1}$
B_{n+1} is (very) different from the previous verified consequences of H, B_1, B_2, ..., B_n
B_{n+1} is true
_____

H is more credible (plausible)

Another example is the following one, based on analogy:

H is similar to B
B turns out to be true
_____

H is more credible (plausible)

Such a framework, of course, needs a complement, namely rational means to generate hypotheses. This is provided by the heuristic logic, but before examining the kernel of a heuristic logic, we need to specify the notion of 'method', or better in what sense a heuristic logic provides a method for scientific discovery.

3   A Matter of Method: When the Method Matters

The statement that there is such a thing as a method for scientific discovery requires a clarification. The first thing to be clarified is the meaning of the term 'method.' As we draw on a problem-solving approach, here the term 'method' simply denotes a means to solve problems. Moreover, when we say that there is a rational way to discovery, we are stating that we can build procedures that are not purely psychological, subjective or intuitive—thus contrasting the hypothesis of 'romantic genius' or the 'black box' explanation (see for example Popper 1934; Campbell 1960; Wertheimer 1982; Einstein 2002). On the other side, we do not mean that these procedures, and the resulting method, are algorithmic: no mechanical rules allow us to find the solution to a non-trivial problem, nor do they allow us to *find* such a problem. Even if the formal tradition argues that the purpose of the heuristic is to formulate mechanical rules to be implemented on a machine, this idea is hard to defend and, in a sense, misleading, since the purpose of heuristics is to find non-mechanical rules that guide the solution to problems, even if it takes some skill to apply them.[2]

_____

[2]The most famous defender of this view is Herbert Simon (see Simon et al. 1987), who argued for an 'algorithmic discovery', a computational model for discovery, implemented in his software BACON. In the end we can consider this view untenable. At most it can model the result of the construction of a hypothesis—the only one that can be treated effectively by a computer, namely the one after the conceptualization, selection of data and choice of variables, has already been made by humans (for a critical appraisal of Simon's approach see in particular Nickles 1980; Kantorovich 1993, 1994; Gillies 1996; Weisberg 2006).

In addition, the notion of method that shapes heuristic logic takes into account the *no free lunch* issue. The idea of a *general* (domain-independent) problem-solver seems untenable also because it is contrary to some formal results, known as the no free lunch theorems (NFLT), which seem to support a 'local' view on methods. As a matter of fact, it is necessary to incorporate a lot of domain-specific knowledge in a method in order to solve a problem. This point is stressed by Tom Nickles (see in particular Nickles 2014), based on an interesting and legitimate interpretation of the no free lunch theorems.

These theorems establish some interesting properties of algorithms that solve problems in machine learning, and search and optimization. In effect "much of modern supervised learning theory gives the impression that one can deduce something about the efficacy of a particular learning algorithm (generalizer) without the need for any assumptions about the target input-output relationship one is trying to learn with that algorithm" (Wolpert 1996, p. 1341).

The NFLTs state that if one algorithm (e.g. hill climbing) performs better than another on some problem, there will be other problems where this relationship is reversed. Tellingly, it means that if one algorithm performs better than another on a class of problems, it will perform worse on all the remaining classes of problems. It is just in this sense that there is no 'free lunch': there is always a cost to be paid for the effectiveness of a method (algorithm) on certain classes of problems, and the cost is the loss of effectiveness over other classes of problems. Therefore, the theorems tell us "what can(not) be formally inferred about the utility of various learning algorithms if one makes no assumptions concerning targets" (Wolpert 1996, p. 1344).

As a consequence, the NFLTs support the idea that a "blind faith in one algorithm to search effectively in a wide class of problems is rarely justified" (Wolpert and Macready 1996, p. 4) and refute a seemingly reasonable expectation for algorithm and methods in solving problems, namely "that there are pairs of search algorithms A and B such that A performs better than B on average, even if B sometimes outperforms A" (Ibid., p. 2) and "if we do not take into account any particular biases or properties of our function cost, then the expected performance of all the algorithms on that function are *exactly* the same (regardless of the performance measure used)" (Ibid.).

Of course, once a bit of information about the domain, even a very generic one, is specified in the algorithm, it will enable some algorithms to outperform others on average. But there are still other algorithms, and classes of problems, that will remain undistinguished on the average, and the way of telling them apart is to specify in the algorithm additional information about the domain, and so on. That is, the issue is an iterative one and the way to gain better performances is to specify in the algorithm more and more information about the domain. This means the way of improving problem-solving is to incorporate more and more domain-specific knowledge.

The NFLTs support also another interesting conclusion: since an algorithm must incorporate domain-specific knowledge in order to be effective in solving specific problems, it will always be 'partial' and selective. It follows that it will be effective only to the extent that it matches the characteristics of the domain to which it is applied. So the NFLTs therefore support the interesting conclusion that for a given problem, a method (i.e. an algorithm) can get more or different information than other methods from the very *same* dataset. In other words, the *same* set of data generates *different* information when explored by *different* approaches.

Bottom line: there is no such a thing as a well performing universal meta-heuristics, but a *specific* heuristics has to be selected and adjusted to the problem at hand by using available knowledge and data about it, that is, by using a theory. This point is iterative. Once you have specified domain-specific knowledge, and distinguished some algorithms from others, the problem represents itself. The better knowledge about the domain is specified in the algorithm, the better its performance.

3.1 Analysis and Deduction

It is necessary to clarify the role of deduction in order to be able to assert that the analytic method is how knowledge advances. In effect, deduction does not provide a way of extending our knowledge, despite some attempt of arguing contrariwise (see Musgrave 1988; Zahar 1989). Since the rules of deductive reasoning are non-ampliative, their conclusion cannot go beyond the premises and, as a consequence, they cannot generate really *new* knowledge. But this fact does not imply that deductive reasoning has no role at all in the advancement of knowledge. As a matter of fact, deductive reasoning plays an important role in it: it facilitates the comparison of our hypotheses with experience and existing knowledge, and thus helps us in assessing the plausibility of the hypothesis formulated in an analytic way.

Moreover, deduction is a way of compacting a domain, that is, what we *already* know. In fact deduction can be used, as indeed it is, to move more quickly from one piece of knowledge already known to another one (already known), that is, to set up new relationships between known entities or properties. Representation theorems[3] are a stock example in this sense.

Just like discovery and justification are not separated—the latter is included in the former—even the axiomatic method plays a role in the formation of new hypotheses, by helping to establish their plausibility. The two methods have to work together.

---

[3]A representation theorem is a theorem that states that an abstract structure with certain properties can be reduced to, or is isomorphic to, another structure.

4 Heuristic Logic: A Characterization

If we accept the thesis that the logic stems from the method, namely that it is a means for gaining new knowledge, then a 'heuristic logic' not only is not an oxymoron, but it has a legitimate and crucial place in the study of scientific method.

Heuristic logic aims to provide rules for solving problems, and interesting problems, or at least those at the frontier of knowledge, are difficult, nay almost impossible, to solve by means of algorithmic methods. Thus, heuristic logic is different from classical logic. The latter is a closed set of mechanical rules (an algorithm) to solve a given problem. These rules do not require special skills to be applied, are syntactic and formal, and guarantee the solution to a problem when they can be applied. On the other hand, heuristic logic is an open set of non-mechanical procedures (rules) that guide us in the search for a solution to a given problem: it requires some skill to apply these rules, which are semantic, content-related, and do not guarantee the solution to a problem when applied.

4.1 Heuristics: Ways of Creating Problem-Spaces

A heuristics is a tool for *modelling* the research space of hypotheses, i.e. the problem-space. Since it is possible to combine ideas and concepts in many ways, and in virtue of the data-hypothesis plurality (under-determination), a combinatorial explosion of the problem space will occur. A heuristics is a means to deal with such a combinatorial space. The formal literature on the subject suggests that a heuristics is a tool to *reduce* the space, by limiting the search to a certain region, making it more manageable (see e.g. Simon et al. 1987). As a matter of fact, a heuristics is not simply a way of reducing the problem space, but a tool to *build* such a space. This space essentially depends on the existing knowledge and the available data. It is determined by the way the data is processed using heuristics. It is therefore a provisional object, which changes over time (it is *positional*) as data and existing knowledge change. Thus, a heuristics can reduce as well as expand the space. Indeed, in a sense it must also expand (mostly combining knowledge from different fields of knowledge); otherwise some hypotheses that could solve the problem could not be formulated. Accordingly, the problem space of a problem is essentially *dynamic*, i.e. it does not *precede* the conclusion.

4.2 The Role of Background Knowledge

A crucial role in the construction of a problem-space is played by the background knowledge (consisting of taxonomies and classifications), which is essential to build the paths of the problem-space. This means that background knowledge shapes the whole problem-solving process, namely both sides of it: how a problem is *found* and how is *solved*. In effect I argue for the *unity* of problem-solving and problem-finding (see Sect. 4.3). This hypothesis has been specifically formulated by Simon (see Simon et al. 1987), who argued that the procedures that enable us to discover new 'laws' can also be a means to discover new problems, new observational tools and new representations.

To illustrate this point, Simon stresses that a phenomenon can often be represented by a relationship between *n* terms of a set of objects, and examines the case of the discovery of penicillin by Fleming. The fact that a mold causes the lysis of some bacteria can be expressed formally with $L(m, b)$, where *m* is a mold, *b* a bacterial culture and the relationship L stands for '*m* causes the lysis of *b*'. This representation generates several problems that guide our research. For example:

a. find a set of molds that can produce these effects;
b. find a set of bacteria or other organisms that may be affected;
c. investigate how much the *intensity* depends on the pair (m, b).

In effect, all these problems were parts of the research program that Fleming himself developed after the accidental discovery of the phenomenon. The central point here is that "to apply this problem-generation strategy one must be able to designate candidates for the sets to which *m* and *b* belong" (Ibid., 306): and it is here that the background knowledge comes into play. Simon observes that "generally prior knowledge, in the form of classifications, will provide the candidates. For example, the fact that the Penicillium, the originally observed body, was a mold suggests that the experiment can be extended to other molds. The fact that the target body is a kind of bacteria suggests the generalization to other bacteria" (Ibid., 306–7). This is an example of how the background knowledge, as well as heuristic tools, shapes the construction of the problem space, from which Simon draws the conclusion that "every field of human knowledge sufficiently developed can be used as a basis for the generalizations and for the subsequent generation of candidates" (Ibid., 307).

The thesis of *unity* of problem-solving and problem-finding, of course, holds also for the several heuristics that we can produce: they can be employed to both solve and generate new problems. Some of these heuristic procedures are well known and have been used and explicitly theorized since at least Plato and Aristotle (See Quarantotto 2017). The two main kinds of heuristics are analogy and induction, or better, various kinds of analogy and induction (induction by a single case, induction by multiple cases),[4] which are the foundations of any heuristics. The other basic heuristics rules are generalization, specialization, metonymy, and metaphor.

---

[4]See Cellucci (2013), Chap. 20.

4.3   The Unity of Problem-Solving and Problem-Finding: A Few Examples

These heuristic rules have been used to reach a lot of results and acquire new knowledge in various fields of human knowledge. As already mentioned, they can also be used to find *new* problems. Let us examine first the case of analogy as a way of finding new problems.

A remarkable example of generation of a problem by analogy is the Poincaré conjecture, formulated by the famous French scientist in 1904 and then solved by the Russian mathematician Grigori Perelman (see Perelman 2002, 2003a, b). Poincaré developed a mathematical tool to distinguish, and accordingly classify in a topological sense, all the varieties of dimension 2. He then conjectured that the same thing was possible also for dimension 3, namely that such a tool can also distinguish a three-dimensional sphere from other varieties of dimension 3. In this way he had built a new problem by means of an analogy with the solved problem.

Even induction, of course, can be used to generate new problems. In this case the notion of hypothesis and problem collapse. A nice example of generation of a problem by induction is the Goldbach's conjecture, which states that any even integer equal to or greater than the number 2 can be written as the sum of two primes (not necessarily distinct). Goldbach examined various ways of representing integers as sums, looking at properties that could emerge. To this end he examined many examples of integers. For example, the number 2 can be expressed as the sum $1 + 1$. The number 4 can be written as the sum $2 + 2$. Again, the number 6 can be represented as $3 + 3$. The number 10 can be represented as $3 + 7$ or $5 + 5$. 14 can be expressed as $7 + 7$ or $11 + 3$. Since all these sums are sums of prime numbers, except for the number 2, by induction we can ask whether that property holds for all the natural numbers—getting Goldbach's conjecture.

5   Heuristics: Primitive and Derived

Heuristics can be classified in several ways. The very first one is the distinction in *positive* and *negative* heuristics (see Lakatos 1978). Roughly, positive heuristics guide us in the construction of the *admissible* paths in the search for a solution to a problem, while negative heuristics *prevent* us from building certain paths (by preventing the *modus tollens* on a part of the theory). Of course what is a positive or a negative heuristics is provisional, and can change over time.

Another useful classification is the one that tells apart primitive and derived heuristics (see Ippoliti and Cellucci 2016).

5.1 Primitive Heuristics: The Building Blocks

Primitive heuristics provide the building blocks of the construction and ampliation of knowledge. Basically they are analogies,[5] disanalogies, inductions and combinations thereof—also internally, for example analogies between analogies.

These inferences are a primitive heuristics in the sense that in order to investigate and know something, first of all we have to spot similarities and dissimilarities between what we are investigating and the objects and conceptual systems that we already know (i.e. analogies and disanalogies). Secondly we look for regularity within the collected data in order to interpret a given phenomenon (i.e. inductions). All the heuristics that we can develop are variants, combinations or juxtapositions of these primitive heuristics, and so they presuppose them. A kind of relevant primitive heuristics is multiple analogy (see Shelley 2003), i.e. the one that uses more than one source to reason about a certain target. One of the main advantages of multiple analogy is the fact that it provides a way to build knowledge on a certain subject that may be stronger and more cogent than the one provided by a single source analogy: an analogical transfer based on a single source may be misleading in a way that can be corrected or improved by the use of multiple sources.

6 Derived Heuristics: The Scaffoldings

From the primitive heuristics we can build a more articulated and complex heuristics—the derived one. The set of derived heuristics is the scaffolding of the construction of knowledge. Of course this set is open, and new rules can be added as new problems are solved (*ars inveniendi adolescit cum inventis*); nonetheless we can provide a first classification of it by using labels such as inversion heuristics (Sect. 6.1), heuristics of switching (Sect. 6.2), the use of figures and diagrams, scenario building, thought experiments, the analysis of extreme cases, and the analysis of a deviant case (see for example Jaccard and Jacobi 2010, Chap. 4 for a list of derived heuristics).

I will examine some examples of derived heuristics and then I will move on to consider how it can be derived, and in what sense, from primitive heuristics.

---

[5]An analogy is a type of inference that concludes, from the fact that two objects are *similar* in certain respects and that one of them has a certain property, that the other has the same property. It therefore enables the transfer of certain properties from a *source* (object or set of objects) to a *target* (object or set of objects). According to the meaning that we attach to the expression "to be similar", we can define different kinds of analogies (e.g. analogy for quasi-equality, separate indistinguishability, inductive analogy, proportional analogy).

6.1   Inversion Heuristics

Inversion heuristics is a class of heuristics based on a specific change of viewpoint of a problem at hand, that is an inversion of it: it examines a 'negative' (or 'positive') in order to acquire knowledge and formulate hypotheses about a 'positive' (or 'negative') target. Thus it is based on explicit disanalogies. More in detail, it requires the investigator to pose as dissimilar several aspects of the phenomenon under investigation. Which aspect to pose as dissimilar can vary according to the target of our investigation. The main kinds of inversion heuristics are: *reframing the problem in terms of its opposite* (or *invert the focus of a problem*), and *making the contrary assumption*.

6.1.1   Reframing the Problem in Terms of Its Opposite

This heuristic simply inverts the focus of the problem or the goal of our investigation. Its main advantage is that it provides a tool to remove, in the exploration of a problem, potential blocks posed by the particular definition that we may have adopted. Thus, just to give a simple example, if we are trying to figure out how to *attract* more customers, such a heuristic will suggest to focus on the ways that push customers away. A nice practical example of this heuristic is the police operation known as 'the sting'. In 'a sting', criminals who are well known to the police are told that they have won a prize and therefore they are personally invited to an event to get it. When they show up, they are arrested. Instead of approaching the problem in terms of how to go out to catch a criminal, here the focus is inverted: the focus is how to push criminals to come to the police.

6.1.2   Make the Opposite Assumption

This heuristic process tells us to take an explicit assumption of the problem under investigation and reformulate it in terms of its opposite. For example, if a phenomenon is considered to be stable, think of it as unstable. If we believe that the two variables are related, make the assumption that they are independent. If two phenomena coexist, assume that they do not coexist and see what would happen in this case. If X is believed to cause Y, consider the possibility that Y causes X.

6.2   The Heuristics of Switching

The heuristics of switching is based on a specific change of viewpoint, which we get by switching from one order of analysis to another one. Thus it is based on explicit analogies and disanalogies. More in detail, it requires the investigator to keep constant, or similar, several aspects of the phenomenon under investigation and to explicitly conjecture others as dissimilar. Which aspect to pose as similar and which one as dissimilar can vary according to the target of our investigation. The main kinds of inversion heuristics are: changing the *unit* of analysis, changing the *level* of analysis, and focusing on *processes* versus focusing on *variables*.

6.2.1 Changing the Unit of Analysis

This heuristic is based on a change of the *unit* of analysis of a problem, or better, of what was posed as such. In effect a powerful heuristic move, especially in social science, is to pass from the investigation of individuals to the study of couples or aggregates of individuals. For example, during the investigation of the performance of a sports team, we could try to understand the factors that influence the efficiency of pairs or triads of players within the team rather than individual players.

6.2.2 Changing the Level of Analysis

This heuristic is based on a change of the *level* of analysis of a problem, or better, of what was posed as such. A way of characterizing a *level* of analysis is, e.g., the switch from a *micro* to a *meso* or a *macro* level and *vice versa*. Another way is to switch between several determinants of the phenomenon under investigation. Both of these are common in social sciences. For instance, a common way of examining the voting pattern of people is by political orientations of voters. A change of level would require the investigator to examine how certain contexts (e.g. media, job, education, neighbourhood, etc.) impact the political orientations of individuals. This heuristic move could bring out new aspects about political orientation that we might have not considered, or new ways of thinking about it that may help us to better explain their relationship with the vote.

6.2.3 Processes Versus Variables

This heuristic is based on a specific change of representation of a problem, the one we get by passing from a variable-oriented representation to a process-oriented one. A *process* is characterized as a set of activities that unfold over time to produce change or maintain equilibrium in a system, or to get from event x to event y. For instance, instead of thinking about stock market dynamics in terms of variables (e.g. change in price value), a process-based analysis will view the formation of a price as consisting of a sequence of actions and procedures. An approach in terms of variables is like taking a picture of the system at a single point in time, while an approach based on process is looking at how a system flows in time.

6.3   Other Examples of Derived Heuristics

6.3.1   Extreme Cases

This heuristic procedure is to keep some of the variables of the problem constant, and to let other variables vary to extreme values: this move offers us useful information on the way to solve a problem, since it can provide a first tentative solution that can be extended or adjusted to other values or cases. Basically it is a form of *specialization*, i.e. the passage from the examination of a given set to the examination of a smaller one, which is contained as a subset of it.

Some problems can often be solved much more easily by considering extreme cases of a situation. When we examine extreme cases, we have to be careful (1) to consider only the extremes that do not change the nature of the critical variables of the problem, and (2) to vary a variable that does not affect other variables.

This heuristic is also a very powerful tool for solving mathematical problems. Pólya (1954) illustrates the power of this derived heuristic with a toy but instructive example:

> Two men are sitting at a table with the usual rectangular shape. One puts a cent on the table, and then the other does the same, and so on in turn. Every cent must lie flat on the table (not straight), and cannot be placed on another cent previously put on the table. The player who puts the last coin on the table will take all the money. Which player should win, provided that each one plays the best strategy?

This problem can be solved by examining an extreme case in the following way. Let us consider a table small enough to be covered by a single cent. In this scenario (case), of course, the first player wins, and it goes without saying that he would win only his money (and the game would be uninteresting). But such an extreme case provides a clue to construct the solution to the problem: it provides a first answer and so we can try to generalize it. To this end, let us now imagine to be allowed to vary the size of the table by gradually increasing it. Now, if the first player puts the first penny right in the centre, and if the table is large enough to hold a penny next to the first penny on one side, then it will be big enough to contain another cent on the opposite side. Generalizing this line of thought, we can see that, regardless of the size of the table, if the first player places his first cent exactly in the middle and then mirrors the move of the other player (on the opposite side of the table), he will win. It is worth noting that the problem can also be solved by using another strategy, that is symmetry. This point can be raised for many similar cases.

The heuristic of the extreme case analysis resembles another strategy that is often employed to solve problems: the analysis of the 'worst case scenario', namely, to imagine what is the worst that can happen if our solution does not work. In effect, the use of the worst-case scenario is a kind of analysis of extreme cases.

6.3.2 Thought Experiment

Thought experiment is one of the most famous tools for generating hypotheses and solving problems, and it has been extensively employed both in science and in philosophy since antiquity—a nice example is the experiment of 'lead fish' in Aristotle's *Physics* (see e.g. Ugaglia 2004). This procedure was also extensively employed by Albert Einstein (e.g. the twin paradox, the experiment of the train, or the light ray experiment). A thought experiment puts forward the construction of hypothetical experiments that are conducted 'in the mind', by imagining collecting data and then processing the consequent outcomes, or better by imagining certain results and then examining what might be their implications for the explanation of the phenomenon under investigation. Basically, it requires thinking of variables or scenarios, and then considering the possible effects of different variations or manipulation of them, just as if we were performing an experiment.

The main benefits of this heuristic stem from its 'mental' nature, and are twofold:

1. It enable us to overcome technical or technological limitations, and hence to consider situations that cannot be set up practically at the time of their conceptualization (e.g. a macroscopic object moving to the speed of light).
2. It is particularly effective to highlight surprising or paradoxical consequences of a theory or hypothesis.

A well-known instance of this heuristic is the so-called 'counterfactual': a conditional statement whose antecedent states a hypothesis contrary to what really happened, and whose consequent sets out the implications that would be derived from it.

6.3.3 Scenario-Building

As the 'thought experiment' heuristic suggests, the search for a solution involves building scenarios (scenario-planning). A *scenario* is normally defined as a scheme of a natural or expected course of events. Scenarios are therefore stories, or narratives, which describe plausible projections of a specific part of the future. They answer the question *what if*, and usually a few (three or four) scenarios are developed in order to investigate what consequences a decision taken now will generate. The construction of a scenario is therefore based on a storyline: a combination of events whose unfolding can be foreseen and of other events for which it cannot be foreseen. Scenarios are therefore projections of potential future: they are not predictions (a projection should not be confused with a prediction). The scenarios do not in fact intend to prove that what is projected will take place: they simply highlight possible trajectories for critical events.

From a logical viewpoint, a scenario has to be simply *plausible,* that is, based on assumptions and pieces of evidence that are *possible* (they show that the narrative projected *can* take place), *credible* (to show *how* the plot can take place), and *relevant*. This implies that, of course, every part of a scenario can be questioned.

Scenario-building can be put forward essentially in two ways: *future backward* and *future forward*. In the former, we select interesting futures and then we try to shape paths leading to them starting from the actual state of the world. In the latter, we start from the actual world and project plausible futures on the basis of an analysis of the actual 'forces' and a plausible evolution of them. In both cases, the starting point for a construction of scenarios is the current state of the world and knowledge of the past. In order to build plausible futures, of course, we must use knowledge about the past, or better about how, in the past, events have unfolded in circumstances similar to those that characterize the current state of the world, and then we can project a part of the past onto the future. This process can be applied to any bifurcation of possible alternatives.

A way to build scenarios is to assume the explicit violation of one of the forces, circumstances or conditions that holds in the current state of the world.

A stock example of the usefulness and effectiveness of the scenario-building heuristics is the one provided by Shell in 1983, where scenario-building was employed to take a decision about a large investment in a gas field in northern Norway (Troll).

7 Reducibility of the Derived Heuristics

This open set of heuristics is defined as 'derivative' since it can be accounted for as a variation, combination, or juxtapositions of the set of primitive heuristics. In this sense, primitive heuristics is a *prerequisite*. It is worth noting that both primitive and derived heuristics can be applied also to entities and inferential outcomes that are already the result of an application of primitive as well as derived heuristics (e.g. respectively analogies between analogies and analogy between scenarios).

In order to show how derived heuristics are reducible to primitive ones, I will examine the cases of *scenario-building* and *analysis of extreme cases*.

The latter case is the result of the application of an analogy (step 1) followed by an inductive process (step 2). By initially focusing on a subset of the original set of the problem (which, as such, has certain similarities with the initial state of the problem), the analysis of extreme cases first builds upon an internal analogy. The examination of the properties of this subset provides us with a tentative, plausible solution to the problem, which we can try to apply to other cases, up to all possible cases (induction). In Polya's example, the first step is to focus on an extreme case (a table as small as a cent, which is internally similar to the initial one), from which we get a tentative answer to the problem, which we then try to extend to all possible cases of the game.

Scenario-building is a heuristic that builds upon explicit analogies, disanalogies, and inductions. In order to construct a plausible future (a plot) we must use the knowledge about the past—how events evolved in past circumstances *similar* to those of the current state—and then project a part of this past onto a 'plausible' future. This process is applied to each bifurcation in the plot. A main device to build a plot, namely the explicit violation of one of the forces, circumstances or conditions holding in the current state of the world, is simply an explicit disanalogy with the past or present state of the world.

In a similar way, it is possible to show how primitive heuristics shape each derived heuristic.

8 Heuristics: Generative and Selective

Another useful classification of heuristics is the one that tells apart *generative* and *selective* heuristics (see Ippoliti and Cellucci 2016). The former enables us to *generate* new options or hypotheses; the latter enables us to *choose* between options already given. Generative heuristics is essentially a primitive or derived heuristics, an inference employed to *build* hypotheses to solve certain problems. A generative heuristics is ampliative. A selective heuristics is a way to draw conclusions or take decisions about options that are *already* available: it essentially aims at making a choice under conditions of lack of resources (e.g. time, information, computational power). It is not ampliative, in a logical sense, since the option to choose has to be known in advance in order to be selected—at most, it could be argued that it is ampliative only in a psychological sense.

Selective heuristics have been investigated and refined by two well-known and different traditions in heuristic reasoning, the *Heuristics and Biases* approach (H&B, see Kahneman and Tversky 1979, 1986, 1992) and the *Fast & Frugal* approach (F&F, see Gigerenzer and Goldstein 1996; Gigerenzer and Todd 1999; Gigerenzer 2000, 2008; Gigerenzer and Selten 2001).

The H&B approach focuses on inference and decision-making that depart from 'optimizing' rationality. A heuristics in this sense is a form of reasoning that does not fit the rational choice theory (RCT), and accordingly produces mistakes because of several distractors (such as *risk aversion*). When we apply an H&B heuristics (e.g. availability, or anchoring & adjustment) we assess and calculate in the wrong way the chances that our choices will produce certain consequences and we evaluate poorly each of the final states that might arise.

The F&F approaches revises the H&B one, and shows that a heuristics is not simply a mistake or a departure from 'optimizing' rationality, but it is a kind of reasoning that produces better outcomes than RCT *under certain circumstances*. The F&F, by building upon the findings of evolutionary psychology (e.g. the massive modularity of our brain, and the adaptive, domain-specific nature of these modules), characterizes rationality in ecological terms: even though heuristics at first sight seem to contradict the RCT, as a matter of fact they are appropriate to those contexts (environments) because they were developed precisely to solve problems in that context. In other words, they embed knowledge about that domain, and this fact makes them effective in such a context and in contexts *similar* to it.

The F&F tradition has identified many kinds of selective heuristics, for example follow-the-crowd, tallying, satisfaction, tit-for-tat, fluency, heuristics-of-equality (or 1/N), and default-heuristics.

9 Ampliativity and the Derivative Nature of Selective Heuristics

A selective heuristics is not a primitive one; it is derived: it is an induction or analogy of some kind, or a combination of them.

It is worth noting that the distinction between generative and selective heuristics is not rigid: that is, it is not always possible to separate the two classes. As a matter of fact, not only can a given generative heuristics be used to select a hypothesis, but a selective heuristics can point at a *new* hypothesis. Nonetheless the meaning of *new* here has to be clarified: a selective heuristics can only search between *already* existing hypotheses within a set of alternatives. *New* has to be meant in a weak sense. The point at stake here is the ampliativity of a selective heuristics: since it requires that the several alternatives are already known or knowable, it does not introduce a new one. It cannot. Take for example a well-known F&F heuristics in social sciences: *follow-the-crowd*. It tells us to act like the crowd (most-option) when we are not sure what to do. In order to work, this kind of reasoning requires that an option already exists, and that it is accessible somehow. It is not a way to introduce a new option. Of course at the very beginning of problem-solving, the particular problem-solver may not know it, but it does not imply that such an option is *new* and that the reasoning leading to it is ampliative. As a matter of fact it is at most a physiological ampliativity, not a logical one. Bottom line: an F&F heuristics is not logically ampliative.

Selective heuristics raise a crucial question in the study of heuristic, namely the importance of time. In effect, whenever we have to make a choice, ignoring the negative side effects of the time needed to choose a hypothesis may be fatal. Sometimes, when an urgent decision is required, the costs generated by the delay that the development of a detailed theory requires are unjustifiably ignored. The search for a hypothesis by means of a process guided by optimising rationality could delay a decision so much as to compromise the entire process and its outcome. The situation worsens even more when the use of an F&F heuristics produces better results: in this case, not only does it time, but it enables us to achieve even better results than RCT. This problem raises a further issue, namely the problem of a 'meta-heuristics': what procedure to employ in order to choose which heuristics to use to solve a problem (under scarcity of time). Of course there are plenty of cases where, although the cost of the detailed development of a hypothesis or theory is heavy, it would make sense to pursue this development.

10   The Power of Heuristic Reasoning: Combining Heuristics

Most of the times, solving a problem requires a combination of several kinds of heuristics: it is rare that a problem, especially a complex one, can be solved by a *single* application of a heuristic. As a matter of fact, in order to find a solution, usually it is necessary to integrate in an appropriate and cogent way different kinds of heuristics. This combination of heuristics can be iterative, in the sense that it might require the combination of the very same heuristics, for example an analogy between analogies or a concatenation of analogies. In order to better illustrate this point, I will discuss the case of multiple analogies.

10.1   Multiple Analogies

The use of multiple analogies is a very effective way of combining the very same heuristic. One of the most useful kinds of multiple analogies is *concatenation*, since every new analogy can correct or modify flaws or misleading aspects of the previous analogy. Every analogy in the chain often uses aspects of previous analogies (even though it may be totally new in some cases), gradually improving the conclusions about the target.

It is possible to distinguish several ways of combining analogies. The most important are the following (see Spiro et al. 1989 for a detailed examination)[6]: *integration* with a new analogy, *correction* with a new analogy, *alteration* of an earlier analogy, *refinement* of a previous analogy, *magnification* ('zoom in'/'zoom out'), *viewpoint shift*, *juxtaposition*, and *competition*.

This taxonomy illustrates also how a primitive heuristics, i.e. analogy, can be internally combined in order to generate derived heuristics. Science and mathematics, of course, have plenty of cases of discovery obtained by such a combination of *different* kinds of heuristics. Here I will recall two remarkable examples. The first comes from the social sciences: the combination of two kinds of heuristics, *metaphor* and *reframing the problem in terms of its opposite*, used by William McGuire (see McGuire 1968) in order to account for the problem of persuasion. The second comes from physics: the combination of *analogy, thought experiment, extreme cases* and *diagrams* employed by Bernoulli to produce the first model of 'internal pressure' in a liquid (see Ulazia 2016).

In the first example, McGuire approaches the problem of accounting for persuasion by examining the way people can be made resistant to attitude-change and *not* by exploring the ways in which someone can be persuaded (thus by *reframing*

---

[6]In their paper Spiro et al. discuss these ways of combining analogies only by means of didactical examples: they do not offer examples of real cases of scientific discovery, or hypotheses, obtained by these combinations. Nonetheless their taxonomy is effective and can be put in use.

the problem in terms of its opposite). In addition, he employs a *metaphor* (from biology) to understand how to make people resistant to persuasive communications. Such a metaphor uses information from a source domain (immunology) in order to understand the target domain (persuasion). The target tells us that people are made resistant to many diseases by introducing a small quantity of a contaminating virus into their body: in this way the body produces antibodies to resist the virus. Similarly, in order to understand how people resist an attitude-change, McGuire designed a kind of 'immunization' experiment, whereby people are exposed to "short, challenging, counter-attitudinal messages that would make them think through their attitude in more detail and organize their defences to counterarguments" (Jaccard and Jacobi 2010, p. 61). These short messages do not produce a change of attitude, but are strong enough to trigger counterarguments and defence mechanisms to combat future persuasive attempts.

In the second example, Johann Bernoulli (see Ulazia 2016 for a detailed examination of it) approaches the problem of accounting for fluid's pressure and advances the hypothesis that eventually leads to the notion of internal pressure, which is the conceptual starting point for Euler to prove the isotropy of pressure in fluids.

The focus of the problem was the behaviour of a fluid's pressure within a pipe, and Johann Bernoulli advanced the hypothesis of an 'immaterial force'. This *qualitative* hypothesis stated that consecutive portions of the fluid are pushed against each other and their contact is shaped by the action and the reaction of an 'intermediate force', defined by Beroulli as 'immaterial' because it is not specific to any portion of the fluid. This 'force' pushes forward the preceding fluid's portion and pushes backward the next portion. The inferential path leading to this conjecture is shaped by analogies. In effect the very first step is an analogy between fluids and solids: the force acting in fluids is similar to the one involved in the contact between two solid bodies, since an intermediate force regulates "the accelerative force of each, by diminishing the one and increasing the other" (Ulazia 2016, p. 554). Such a first inferential step is refined by a further analogy, or better, disanalogy. In effect, solids and fluids differ from each other, since in the former the force works in a single direction (extends itself like a straight elastic), whilst in the latter the immaterial force works in every direction (like elastic air). It leads to the conclusion that "although pressure is an isotropic force it acts just in the axial direction of the pipe due to the confinement imposed by the sidewalls" (Ibid., p. 556). This inferential path, shaped by a combination of analogies, produces the "dynamic imagery of the first model of internal pressure in liquids" (Ibid., p. 556).

These two stock examples show how the combination of heuristics is the *machinery,* and the engine, of problem-solving, and hence display how scientific discovery and ampliation of knowledge can be rationally and inferentially pursued, and reached.

Acknowledgements I would like to thank David Danks, Carlo Cellucci, the two anonymous referees, and the speakers at the conference 'Building Theories' (Rome, 16–18 June 2016) for their valuable comments on an early version of this paper.

References

Campbell, D. T. (1960). Blind variation and selective retention in creative thought as in other knowledge processes. *Psychological Review, 67,* 380–400.

Cellucci, C. (2013). *Rethinking Logic*. Dordrecht: Springer.

Einstein, A. (2002). Induction and deduction in physics. In Albert Einstein (Ed.), *Collected papers* (Vol. 7, pp. 108–109). Princeton: Princeton University Press.

Gigerenzer, G., & Goldstein, D. G. (1996). Reasoning the fast and frugal way: Models of bounded rationality. *Psychological Review, 103,* 650–669.

Gigerenzer, G., & Todd, P. M. (1999). *Simple heuristics that make us smart*. New York: Oxford University Press.

Gigerenzer, G. (2000). *Adaptive thinking: Rationality in the real world*. New York: Oxford University Press.

Gigerenzer, G., & Selten, R. (Eds.). (2001). *Bounded rationality: The adaptive toolbox*. Cambridge, MA: MIT Press.

Gigerenzer, G. (2008). Why heuristics work. *Perspective on Psychological Science, 3*(1), 1–29.

Gillies, D. (1996). *Artificial intelligence and scientific method*. Oxford: Oxford University Press.

Ippoliti, E. (2006). *Il vero e il plausibile*. Morrisville (USA): Lulu.

Ippoliti, E., & Celluci, C. (2016). *Logica*. Milano: Egea.

Jaccard, J., & Jacoby, J. (2010). *Theory construction and model-building*. New York: Guilford Press.

Kantorovich, A. (1993). *Scientific discovery: Logic and tinkering*. New York: State University of New York Press.

Kantorovich, A. (1994). Scientific discovery: A philosophical survey. *Philosophia, 23*(1), 3–23.

Kahneman, D., & Tversky, A. (1979). Prospect theory: An analysis of decision under risk. *Econometrica, 47*(2), 263.

Kahneman, D., & Tversky, A. (1986). Rational choice and the framing of decisions. *The Journal of Business, 59*(S4), S251.

Kahneman, D., & Tversky, A. (1992). Advances in prospect theory: Cumulative representation of uncertainty. *Journal of Risk and Uncertainty, 5*(4), 297–323.

Lakatos, I. (1976). *Proofs and refutations: The logic of mathematical discovery*. Cambridge: Cambridge University Press.

Lakatos, I. (1978). *The methodology of scientific research programmes*. Cambridge: Cambridge University Press.

Laudan, L. (1977). *Progress and its problems*. Berkeley and LA: University of California Press.

Laudan, L. (1981). A problem-solving approach to scientific progress. In I. Hacking (Ed.), *Scientific revolutions* (pp. 144–155). Oxford: Oxford University Press.

McGuire, W. J. (1968). Personality and susceptibility to social influence. In E. F. Borgatta & W. W. Mabert (Eds.), *Handbook of personality: theory and research* (pp. 1130–1187). Chicago: Rand McNally.

Musgrave, A. (1988). Is there a logic of scientific discovery? *LSE Quarterly, 2–3,* 205–227.

Newell, A., & Simon, H. A. (1972). *Human problem solving*. Englewood Cliffs, NJ: Prentice Hall.

Nickles, T. (1978). Scientific problems and constraints. *PSA Proceedings of the Biennial Meeting of the Philosophy of Science Association, I,* 134–148.

Nickles, T. (Ed.). (1980). *Scientific discovery: Logic and rationality*. Boston: Springer.

Nickles, T. (1981). What is a problem that we may solve it? *Synthese*, Vol. 47, No. 1, *Scientific Method as a Problem-Solving and Question-Answering Technique*, pp. 85–118.

Nickles, T. (2014). Heuristic appraisal at the frontier of research. In E. Ippoliti (Ed.), *Heuristic reasoning* (pp. 57–88). Berlin: Springer.

Perelman, G (2002). The entropy formula for the Ricci flow and its geometric applications. arXiv: math. DG/0211159.

Perelman, G. (2003a). Ricci flow with surgery on three-manifolds. arXiv:math.DG/0303109.

Perelman, G. (2003b). Finite extinction time for the solutions to the Ricci flow on certain three-manifolds. arXiv:math.DG/0307245.

Polya, G. (1954). *Mathematics and plausible reasoning*. Vol. I—*Induction and analogy in mathematics*), Vol. II—*Patterns of plausible inferences*. Princeton: Princeton University Press.

Popper, K. (1934). *The logic of scientific discovery* (1959). London: Hutchinson & Co.

Popper, K. (1999). *All life is problem solving*. London: Routledge.

Quarantotto, D. (2017). Aristotle's problemata style and aural textuality. In R. Polansky & W. Wians (Eds.), *Reading Aristotle* (pp. 75–122). Leiden: Brill.

Shelley, C. (2003). *Multiple analogies in science and philosophy*. Amsterdam: John Benjamins B.V.

Spiro, R. J., Feltovich, P. J., Coulson, R. L., & Anderson, D. K. (1989). Multiple analogies for complex concepts: Antidotes for analogyinduced misconception in advanced knowledge acquisition. In S. Vosniadou & A. Ortony (Eds.), *Similarity and analogical reasoning* (pp. 498–529). New York: Cambridge University Press.

Simon, H., Langley, P., Bradshaw, G. L., & Zytkow, M. (1987). *Scientific discovery: Computational explorations of the creative processes*. Boston: MIT Press.

Ugaglia, M. (2004). *Modelli idrostatici del moto da Aristotele a Galileo*. Roma: Lateran University Press.

Ulazia, A. (2016). Multiple roles for analogies in the genesis of fluid mechanics: How analogies can cooperate with other heuristic strategies. *Foundation of Science, 21*(4), 543–565. https://doi.org/10.1007/s10699-015-9423-1.

Weisberg, R. (2006). *Creativity: Understanding innovation in problem solving, science, invention, and the arts*. Hoboken: John Wiley & Sons Inc.

Wertheimer, M. (1982). *Productive thinking* (Enlarged Ed.), Chicago: University of Chicago Press.

Wolpert, D. H. (1996). The lack of a priori distinctions between learning algorithms. *Neural Computation, 8,* 1341–1390.

Wolpert, D., & Macready, W. (1996). No Free Lunch Theorems for Search. *Technical Report SFI-TR-95-02-010*. Sante Fe, NM, USA: Santa Fe Institute.

Zahar, E. (1989). *Einstein's Revolution: A Study In Heuristic*. La Salle (Ilinois): Open Court.

The Noetic Account of Scientific Progress and the Factivity of Understanding

Fabio Sterpetti

Abstract There are three main accounts of scientific progress: (1) the epistemic account, according to which an episode in science constitutes progress when there is an increase in knowledge; (2) the semantic account, according to which progress is made when the number of truths increases; (3) the problem-solving account, according to which progress is made when the number of problems that we are able to solve increases. Each of these accounts has received several criticisms in the last decades. Nevertheless, some authors think that the epistemic account is to be preferred if one takes a realist stance. Recently, Dellsén proposed the noetic account, according to which an episode in science constitutes progress when scientists achieve increased understanding of a phenomenon. Dellsén claims that the noetic account is a more adequate realist account of scientific progress than the epistemic account. This paper aims precisely at assessing whether the noetic account is a more adequate realist account of progress than the epistemic account.

Keywords Epistemology of modality · Factivity · Knowledge
Noetic account · Scientific progress · Understanding

1 Introduction

Scientific progress is still one of the most significant issues in the philosophy of science today, since, as Chang states, neither "philosophers nor scientists themselves have been able to" settle this issue "to general satisfaction" (Chang 2007, p. 1). And this state of facts can be explained considering the "immense difficulty" of the topic.

F. Sterpetti (✉)
Campus Bio-Medico University of Rome, Rome, Italy
e-mail: fabio.sterpetti@uniroma1.it

F. Sterpetti
Sapienza University of Rome, Rome, Italy

© Springer International Publishing AG 2018
D. Danks and E. Ippoliti (eds.), *Building Theories*, Studies in Applied Philosophy, Epistemology and Rational Ethics 41, https://doi.org/10.1007/978-3-319-72787-5_11

Certainly, one of the main difficulties that arises in developing an account of scientific progress is to find the way to coherently take into account both of the two main characteristic features of science development: (1) *theory change*, which seems to unequivocally emerge every time we carefully analyse one of the main transitions in the historical development of science, and (2) the striking *empirical success* of our best scientific theories, which, according to scientific realism at least, i.e. the nowadays mainstream view in philosophy of science, cannot be adequately explained without referring to the (approximate) truth of our best theories (Niiniluoto 2015b; Chakravartty 2015).

Those two features of science development pull in different directions, hence the difficulty of giving a satisfying account of progress. Indeed, if a radical theory change occurs, the past theory should be regarded as false. But realists consider empirical success a good indicator of truth, and claim that successful theories are (approximately) true. So, if a theory enjoys empirical success, it should not undergo a *really* radical change and finally be dismissed as false. But history of science seems to provide examples of once-successful theories that have nevertheless been successively dismissed, and for which it is not easy to demonstrate that there is some continuity between the new theories and the replaced ones.[1] Did those past false theories constitute instances of genuine scientific progress? It depends on how we conceive of progress, namely on which requirements we think have to be fulfilled in order for a theory to be regarded as *progressive*. For example, if you take a realist stance on progress, you probably will require a theory to be (approximately) true in order to consider it an instance of progress.

In other words, the debate about scientific progress intersects the debate about scientific realism, i.e. the central topic in philosophy of science (Chakravartty 2015). For example, if we take the debate over scientific realism to be about what the aim of science is,[2] then the relation between these two debates may be described as follows:

> X is the aim of science just in case science makes progress when X increases or accumulates. (Dellsén, 2016, p. 73)[3]

---

[1]Cf. e.g. Niiniluoto (2015b, Sect. 3.5): "many past theories were not approximately true or truthlike. Ptolemy's geocentric theory was rejected in the Copernican revolution, not retained in the form 'approximately Ptolemy'. Indeed, the progressive steps from Ptolemy to Copernicus or from Newton to Einstein are not only matters of improved precision but involve changes in theoretical postulates and laws."

[2]Claiming that the debate over scientific realism is about what is the aim of science is just one of the many possible ways to define such debate that have been proposed so far (Chakravartty 2015), and it is used here just for illustrative purpose. Which characterization of scientific realism is the most adequate one does not impinge on the present article, since for any possible characterization of the debate over scientific realism, it is possible to define how this debate and the debate over scientific progress intersect each other in a way similar to the one presented here. For similar reasons, it is not relevant here to survey the different proposals that have been advanced on what is the aim of science.

[3]See also Niiniluoto (2015b), Bird (2007).

Thus, determining the aim of science is relevant for the investigations on scientific progress, since it may give us a sort of criterion to determine whether progress occurred or not. For example, if in analyzing a historical case C, we find that there has been an increase in X during C, and X is the aim of science, we can conclude that C did constitute a case of progress. But even determining whether progress occurred or not is relevant in order to support or attack a specific view on what is the aim of science. For example, if we are able to show that in case D, although it is uncontroversial that D constitutes progress, X did not increase, we can affirm that X cannot be taken to be the aim of science.

Recently, Dellsén (2016) proposed a new conception of scientific progress, the noetic account,[4] according to which an episode in science is progressive when there is an increase in scientists' *understanding* of a phenomenon. Dellsén's noetic account is mainly devoted to overcoming some of the inadequacies that afflict the epistemic account of progress developed by Bird (2007). Bird's proposal, notwithstanding the criticisms it received (see e.g. Rowbottom 2010; Saatsi 2016), has had wide resonance, since it revived a debate about scientific progress in which the two main accounts were the semantic account and the problem-solving account (see Sect. 2), and it still remains one of the accounts of progress more congenial to scientific realists.

Dellsén claims that the noetic account is a more adequate realist account of scientific progress than the epistemic one. This article aims precisely at assessing whether the noetic account is a more adequate realist account of progress than the epistemic one. The article is organized as follows: the three main accounts of scientific progress are presented (Sect. 2); then Dellsén's proposal is illustrated (Sect. 3) and the concept of understanding is analysed in some detail (Sect. 4); an argument is then proposed, which aims to assess whether the noetic account is an adequate realist view of progress by testing it against an uncontroversial yet problematic case of progress (Sect. 5); it is pointed out that Dellsén's view of understanding is wanting, and it is examined whether the modality-based view of understanding recently proposed by Rice (2016) may represent a valid option to 'fix' the noetic account (Sect. 6); since this option is shown to be available only at a very high cost, i.e. the commitment to some form of possible-worlds modal realism, it is examined whether another more promising realist account of modality, i.e. modalism, may be of use to support Dellsén's account (Sect. 7); finally, some conclusions are drawn (Sect. 8).

---

[4]On the reason why Dellsén named his view 'noetic', cf. Dellsén (2016, p. 72, fn. 2): "'Noetic' as in the Greek 'nous', which is often translated into English as 'understanding'."

2   The Main Accounts of Scientific Progress

Three main accounts of scientific progress may be found in the extant literature:
(1) the epistemic account, according to which an episode in science constitutes
progress when there is an increase in knowledge (Bird 2007); (2) the semantic
account, according to which progress is made when either the number of truths or
the verisimilitude of a theory increases, depending on which variant of this account
we are dealing with (Popper 1963; Niiniluoto 2015a); (3) and the problem-solving
account, according to which progress is made when the number of problems that we
are able to solve increases (Kuhn 1970; Laudan 1977).

Those conceptions mainly differ for the concept they take to be central in order
to account for scientific progress: (1) the epistemic account is based on the concept
of "knowledge"; (2) the semantic account is based on the concept of "truth" (or
"verisimilitude", depending on which formulation of this approach we adopt);
(3) the problem-solving account is based on the concept of "problem-solving".
Being based on different concepts, those accounts lead to different outcomes in
determining whether an episode has to be regarded as a genuine instance of scientific progress or not.

Each of the above described accounts of progress has received several criticisms
in the last decades (see Niiniluoto 2015b for a survey). Each of these accounts,
indeed, seems unable, relying on its proper criteria of progressiveness, to account
for the progressiveness of some episodes that are instead usually taken to represent
genuine cases of scientific progress. Finding out this sort of counterexample to the
definition of progress given by rival accounts has been (and is still) the main
business in the dispute over scientific progress.

To better illustrate this way of attacking rival accounts, and to present more in
detail Bird's conception of progress, i.e. the conception that is mainly discussed by
Dellsén, and from which Dellsén moves to develop his own proposal, we will
describe some of the arguments given by Bird to support the inadequacy of the
semantic conception (Sect. 2.1) and the inadequacy of the problem-solving conception (Sect. 2.2).

2.1   Bird's Criticism of the Semantic Account

Bird's defense of the epistemic view of progress, a view that can be traced back to
Bacon (Bird 2007, p. 87, n. 1), begins with the consideration that this view is the
one that better accounts for the intuitive meaning of scientific progress. Indeed,
according to Bird, if we ask ourselves what scientific progress is, the intuitive
answer is simple: science "makes progress precisely when it shows the accumulation of scientific knowledge; an episode in science is progressive when at the end
of the episode there is more knowledge than at the beginning" (Bird 2007, p. 64).

In other words, in Bird's view, the epistemic conception is better than other accounts of progress, because those episodes in the history of science that according to our intuition represent genuine cases of progress are all cases in which an increase in knowledge occurs. At the same time, it is not always possible to observe in such cases a corresponding increase in the number of truths or in the number of problems that we are able to solve. So, the epistemic account fares better than other accounts of progress, and has to be preferred.

It is worth specifying that Bird takes knowledge to be not merely "justified true belief"; rather he takes it to be "true belief justified in a non-accidental way". In this way, he tries to avoid the possibility that an accidentally justified true belief may be regarded as an instance of genuine scientific progress:

> we know that knowledge is not justified true belief, thanks to Gettier's counter-examples. Are then truth and justification jointly sufficient for a new scientific belief adding to progress? No, for precisely the same reasons that they do not add to knowledge. We may construct a Gettier style case of a scientific belief that is accidentally true and also justified [...]. Such a case will not be a contribution to progress. (Bird 2007, p. 72)

So, in Bird's view, while truth and justification are necessarily required for a new scientific belief to be regarded as an instance of scientific progress, that belief merely being true and justified is not sufficient to make it a genuine instance of scientific progress.

It is exactly following this line of reasoning that Bird develops his attack on the semantic view of progress. Indeed, in order to point out the implausibility of the semantic view, Bird asks us to consider the following scenario:

> Imagine a scientific community that has formed its beliefs using some very weak or even irrational method M, such as astrology. But by fluke this sequence of beliefs is a sequence of true beliefs. [...]. Now imagine that at time t an Archimedes-like scientist in this society realises and comes to know that M is weak. This scientist persuades (using different, reliable methods) her colleagues that M is unreliable. [...]. The scientific community now rejects its earlier beliefs as unsound, realising that they were formed solely on the basis of a poor method. (Bird 2007, p. 66)

The problem for Bird is that if we adopt the semantic view, we should describe this case as follows: this community was experiencing progress until time t, because the number of truths held by the community was increasing, while after time t the community started experiencing a regress, because it gave up the true beliefs previously accumulated, and so their number decreased. According to Bird, this way of representing this scenario is unacceptable, since it contradicts our intuition, according to which things go exactly the other way around: giving up beliefs produced by an unreliable method is a progressive episode, and should not be judged as regressive. On the other hand, merely accumulating beliefs through an unreliable method cannot represent a real instance of progress, despite the accidental truth of those beliefs.

Thus, according to Bird, the trend of progress growth matches the trend of knowledge growth, and not the trend of growth of the number of truths.

2.2 Bird's Criticism of the Problem-Solving Account

Bird (2007) elaborates a different scenario in order to criticize the problem-solving account of progress. He constructs his argument starting from an example made by Laudan (1977) in order to support the problem-solving account.[5]

Consider the following historical episode: Nicole d'Oresme and his contemporaries believed that hot goat's blood would split diamonds. Now, for Kuhn (1970) a puzzle is solved when a proposed solution is sufficiently similar to a relevant paradigmatic puzzle-solution. According to Laudan, a problem P is solved when the phenomenon represented by P can be deduced from a theory T. But Laudan does not require that either P or T be true: "A problem need not accurately describe a real state of affairs to be a problem: all that is required is that it be *thought to be* an actual state of affairs" (Laudan 1977, p. 16). In this perspective, if Oresme's solution to the problem of splitting diamonds is sufficiently similar to a relevant paradigmatic solution in his historical context, according to Kuhn's standards, Oresme's solution provides a genuine solution to that problem, and thus, since progress amounts to problem-solving, it represents scientific progress. Moreover, if Oresme and his contemporaries were able to give a theory from which the splitting of diamonds by hot goat's blood is deducible, Oresme's solution represents scientific progress according to Laudan's standards as well.

The main problem with the problem-solving account of progress, according to Bird, is that both Kuhn and Laudan, i.e. the main supporters of this account, do not think of solving a problem as involving "knowledge", if knowledge is understood in the classical way as requiring (at least) truth. And this fact, in Bird's view, leads those who adopt this account to judge certain historical episodes or hypothetical scenarios in a way that contradicts our intuition about what scientific progress is. Consider again Oresme's scenario:

> imagine that some second scholar later comes along and proves at time *t* by impeccable means that Oresme's solution cannot work. Whereas we had a solution before, we now have no solution. [...]. By Laudan and Kuhn's standards that would mark a *regress*. (Bird 2007, p. 69)

This way of representing the dynamic of scientific progress in this scenario is, according to Bird, unacceptable, since it contradicts our intuition, according to which things go exactly the other way round: since Oresme's solution was not a real solution, because it is demonstrably ineffective, the community was not

---

[5]Bird's presentation of the problem-solving account does not do justice to the theoretical richness of this approach. For reasons of space, we follow Bird. For some recent works that innovate the problem-solving view, see Cellucci (2013), Ippoliti (2014), Ippoliti and Cellucci (2016), who advocate the heuristic view, according to which knowledge increases when, to solve a problem, a hypothesis "is produced that is a sufficient condition for solving it. The hypothesis is obtained from the problem, and possibly other data already available, by some non-deductive rule, and must be plausible [...]. But the hypothesis is in its turn a problem that must be solved, and is solved in the same way" (Cellucci 2013, p. 55).

experiencing any progress until time t, while when at time t such alleged solution was demonstrated to be ineffective by reliable means, that very fact constituted progress for the community.

In other words, according to Bird, while Oresme's solution "might reasonably have seemed to Oresme and his contemporaries to be a contribution to progress, it is surely mistaken to think that this *is* therefore a contribution to progress (Bird 2007, p. 69). In order to decide whether a solution is a real solution, and so whether it represents a genuine progressive episode, we have to refer to our own current standards on what is true in that domain, i.e. to our knowledge.

It is worth underlining here that Bird explicitly refers to our *current* knowledge in order to rebut Oresme's solution: indeed, he admits that it is reasonable to claim that such solution might have *seemed* to Oresme and his contemporaries, according to their system of beliefs, an instance of genuine knowledge, even if it is *in fact* not really an instance of genuine knowledge, according to our current knowledge.

Thus, according to Bird, the trend of progress growth matches the trend of knowledge growth, and not the trend of growth of the number of solved problems.

3   The Noetic Account of Scientific Progress

Recently, Dellsén (2016a) maintained that Bird's account of scientific progress is inadequate, and proposed the noetic account, according to which an episode in science constitutes progress when scientists achieve increased *understanding* of a phenomenon.

The peculiarity of this account with respect to its rivals is that it is based on the concept of "understanding", rather than on the concept of knowledge, truth, or problem-solving.

Generally speaking, "understanding" has to be understood here as it is usually understood in the debate over "understanding and the value of knowledge" that has spread in the last decade both in epistemology and philosophy of science.[6] More precisely, according to Dellsén, "understanding" has to be construed as the ability of a subject to explain or predict some aspect of a phenomenon. In this perspective, an agent has some "scientific understanding of a given target just in case she grasps how to correctly explain and/or predict some aspects of the target in the right sort of circumstances" (Dellsén 2016a, p. 75). Thus, in Dellsén's view, "an episode in science is progressive precisely when scientists grasp how to correctly explain or

---

[6]For a survey on the issue of understanding, see Baumberger et al. (2017) and de Regt et al. (2009). On the related issue of the value of knowledge, see Pritchard and Turri (2014) for a survey. With regard to the debate over understanding in epistemology, see Elgin (2007, 2009), Kvanvig (2003), Zagzebski (2001); with regard to the debate over understanding in philosophy of science, see de Regt (2009, 2015), de Regt and Gijsbers (2017), Mizrahi (2012), Khalifa (2011) and Grimm (2006).

predict more aspects of the world at the end of the episode than at the beginning" (Dellsén 2016a, p. 72).

Dellsén claims that, since the noetic account rests on the concept of understanding instead of the concept of knowledge, the noetic account fares better than the epistemic one, because it is able to account for two classes of events that the epistemic account is not able to account for, namely: (1) cases in which progress occurs, while no increase in knowledge occurs (Sects. 3.1 and 2) cases in which an increase in knowledge occurs, while no progress is made (Sect. 3.2).

3.1   Progress Without Knowledge Increase

In order to point out the inadequacy of the epistemic account, Dellsén (2016a) considers Einstein's explanation of Brownian motion in terms of the kinetic theory of heat, presented in one of his famous *annus mirabilis* papers, "On the Movement of Small Particles Suspended in a Stationary Liquid Demanded by the Molecular-kinetic Theory of Heat" (Einstein 1905/1956). Einstein's paper's first paragraph reads:

> In this paper it will be shown that according to the molecular-kinetic theory of heat, bodies of microscopically-visible size suspended in a liquid will perform movements of such magnitude that they can be easily observed in a microscope, on account of the molecular motions of heat. It is possible that the movements to be discussed here are identical with the so-called 'Brownian molecular motion'; however, the information available to me regarding the latter is so lacking in precision, that I can form no judgment in the matter. (Einstein 1905/1956, p. 1)

Dellsén maintains that if we adopt the epistemic account of scientific progress, we should conclude that Einstein's contribution does not represent a case of genuine progress. Indeed, the epistemic account rests on the concept of knowledge, and knowledge requires (at least) truth and justification. Since Einstein's information on Brownian motion was lacking, Einstein clearly did not have "the epistemic justification required to know that the movements in question were in fact real. Thus, the *explanandum* in Einstein's explanation of Brownian motion did not constitute knowledge for Einstein at the time" (Dellsén 2016a, p. 76). Moreover, the kinetic theory on which Einstein's paper rests was in 1905 a disputed theory, and many reputable scientists did not accept the existence of atoms at that time (Perrin's decisive results were published starting from 1908, see Perrin 1908). So, also the *explanans* in Einstein's explanation of Brownian motion did not constitute a clear case of knowledge when Einstein published his paper.

Given that according to the epistemic account, scientific progress occurs when an increase in knowledge occurs, since in this case it cannot be affirmed that an increase in knowledge occurred, because neither the *explanans* nor the *explanandum* were known at the time, we should conclude that Einstein's contribution to science did not constitute progress in 1905.

According to Dellsén, this way of evaluating this historical episode is unacceptable, since it is unable to accommodate our common intuitions on what constitutes scientific progress. For Dellsén, this case underlines the inadequacy of the epistemic account in dealing with those cases in which progress occurs even if this progress cannot be said to constitute an increase in knowledge at the time when it occurs.

On the contrary, according to the noetic account, in 1905 a remarkable "cognitive" progress occurred, even if there wasn't a simultaneous increase in scientific "knowledge", since an explanation of an (until then) unexplained phenomenon was proposed. And Einstein's explanation was able to integrate such a phenomenon into a wider set of already explained phenomena, making it coherent with background knowledge, and so increasing the intelligibility of the phenomenon. In other words, an increase in the *understanding* of Brownian motion occurred in 1905.

In order to defend the epistemic account, it may be objected that Einstein's contribution to scientific progress consisted in gaining the knowledge that the kinetic theory would explain Brownian motion, if the kinetic theory is true. In this view, "Einstein's achievement amounts to gaining a kind of *hypothetical explanatory knowledge*—knowledge of how a potential *explanans* would explain a potential *explanandum* if the *explanans* and *explanandum* are both true" (Dellsén 2016a, p. 77). Unfortunately, this option seems to be unavailable for the supporter of the epistemic account. Indeed, if achieving some hypothetical explanatory knowledge may constitute genuine progress for the epistemic account, then this account is no more able to rule out Oresme's solution to the problem of splitting diamonds. In this view, to make Oresme's solution an acceptable solution, it would have been sufficient for its supporters to provide some theory about the supposed relation between hot goat's blood and diamonds, arranged in such a way that if this theory and Oresme's solution are true, then such theory would explain the splitting of diamonds through hot goat's blood. This theory would be an instance of hypothetical explanatory knowledge, and so it should be taken to constitute progress. But Bird, as we have seen, explicitly denies that Oresme's solution may constitute an instance of scientific progress. It is important to stress that, in order to discriminate between Oresme's and Einstein's cases, taking into account whether a hypothetical explanatory knowledge has later been confirmed is not a workable criterion. Indeed, if hypothetical explanatory knowledge has to wait until it is confirmed to be regarded as a genuine instance of progress, then it is not distinguishable from ordinary knowledge, and we should maintain that progress occurs when knowledge is acquired, i.e. when confirmation occurs. But if this is the case, then the epistemic account would again be unable to claim that genuine progress occurred in 1905 thanks to Einstein's work. So, the supporter of the epistemic account faces a dilemma: either she accepts that hypothetical explanatory knowledge may constitute an instance of genuine progress, and then she has to accept that Oresme's solution may have constituted progress, or she denies that hypothetical explanatory knowledge may constitute an instance of genuine progress, and then she has to accept that she is unable to assert that Einstein's work constituted progress in 1905. If she takes the first horn and accepts hypothetical explanatory knowledge, she has also to accept that an instance of progress may be considered as such independently of its confirmation. But, as we have seen above (Sect. 2.2), Bird explicitly claims that whether a belief (or a theory, a solution, etc.) constitutes genuine knowledge has to be determined with respect to our *current* knowledge, i.e. from a point of view from which we can assess whether that belief has been confirmed or not. So, the supporter of the epistemic account has to take the second horn of the dilemma, and conclude that her account of progress is unable to assert that Einstein's work constituted progress in 1905.

Thus, it seems fair to conclude that accepting hypothetical explanatory knowledge is not a viable proposal for the epistemic account, and that this account is unable to account for those cases that are equivalent to Einstein's case, i.e. cases in which an increase in the understanding of a phenomenon occurs, and we are justified in claiming that it in fact occurred, even if there was not a simultaneous increase in knowledge at the time it occurred.

It is worth underlining here that if, as Dellsén claims, the noetic account fares better than the epistemic account when dealing with this kind of case, this means that in Dellsén's view, whether a theory constitutes an increase in the understanding of a phenomenon at the time it is proposed, and so whether a theory constitutes progress, is independent of its later confirmation or rebuttal. In other words, the verdict on whether a theory provides an increase in the understanding of a phenomenon at time t_x, cannot be dependent on whether such theory will be confirmed and judged to be true in a specified later historical context t_{x+n} (for example: "according to our best current knowledge", as in the case of Bird's evaluation of Oresme's solution), or in some unspecified later historical context (for example: "if the theory will be confirmed in the future", as in the case of hypothetical explanatory knowledge). The verdict on the increase of understanding (i.e. the capacity to explain or predict some aspect of the target phenomenon) has to be provided with respect to the context in which it occurred.

3.2 Knowledge Increase Without Progress

Dellsén elaborates his proposal, and moves his criticisms to Bird's view, starting from a scenario which was proposed by Bird himself in his 2007 paper, and which was devoted to showing how an increase in scientific knowledge does not constitute progress in all circumstances. Bird describes this scenario as follows:

> imagine a team of researchers engaged in the process of counting, measuring, and classifying geologically the billions of grains of sand on a beach between two points. Grant that this may add to scientific knowledge. But it does not add much to understanding. Correspondingly it adds little to scientific progress. (Bird 2007, p. 84)

It is easy to agree with Bird. This classificatory activity represents a genuine instance of accumulation of scientific knowledge, since the research produces true beliefs relative to the state of the world that are justified in a non-accidental way according to scientific standards and methods. And certainly, it is difficult to figure out how this activity may increase our comprehension of the world. Nevertheless, this apparently innocent consideration constitutes, according to Dellsén, an insurmountable difficulty for the supporters of the epistemic account, because they are unable to accommodate this consideration with the very qualifying claim of the epistemic account, i.e. that the trend of knowledge accumulation matches the trend of scientific progress. If this is the case, then for every instance of accumulated knowledge, however insignificant, we should identify a correspondent (i.e. equivalent) instance of scientific progress. If, on the contrary, an increase in scientific knowledge can occur without being correlated to an equivalent increase in progress, then we should deny that scientific progress has to be identified with knowledge accumulation, as the supporters of the epistemic account maintain.

It is interesting to note that Bird himself suggests that the discrepancy between knowledge accumulation and progress growth is due to whether *understanding* increases. Dellsén elaborates exactly on this suggestion to develop his noetic account, according to which scientific progress has to be identified with increase in understanding.

Examples equivalent to Bird's "counting grains of sand" scenario may be multiplied. Consider all those cases in which spurious correlations are identified and correctly deemed as such.[7] Dellsén reports an interesting case:

> it happens to be true that there is a strong correlation between increases in childbirth rates outside of Berlin city hospitals and increases in stork populations around the city. When this piece of information was published in a[n] article that warned against coincidental statistical associations [...] there was an accumulation of knowledge that would have to count as scientific progress on the epistemic account. However, this information provides no understanding since it does not enable us to correctly explain or predict any aspect of childbirth rates or stork populations [...]. Intuitively, this is not a case of scientific progress. (Dellsén 2016a, p. 78)

To make Dellsén's remark more general, it may be useful to consider a paper recently published by Calude and Longo (2016). They base their argument, among other things, on Ramsey theory, i.e. the branch of combinatorics which investigates the conditions under which order must appear. If we restrict our attention to mathematical series, more precisely to arithmetic progressions, Ramsey theory investigates the conditions under which an arithmetic progression must appear in a string of numbers.

---

[7]On what spurious correlations are, cf. Dellsén (2016a, p. 78): "Suppose we have two variables V_1 and V_2 that are known on independent grounds to be unrelated, causally and nomologically. Let us further suppose that we learn, i.e. come to know, that there is some specific statistical correlation between V_1 and V_2—e.g. such that a greater value for V_1 is correlated with a greater value for V_2." This correlation represents an instance of spurious correlation, i.e. a correlation between two variables which is not due to any real relation between them. In these cases, such a correlation does not convey any information on the correlated variables, nor on some other relevant aspect of the world, so it is useless, irrelevant, or worse, it may lead us astray, if we do not identify it as spurious.

Calude's and Longo's analysis hinges on Van der Waerden's theorem, according to which for any "positive integers k and c there is a positive integer γ such that every string, made out of c digits or colours, of length more than γ contains an arithmetic progression with k occurrences of the same digit or colour, i.e. a monochromatic arithmetic progression of length k" (Calude and Longo 2016, p. 11).

For example, if we take a binary string of x digits, digits can be either '0' or '1'. Take '0' and '1' to be the possible colours of those x digits, i.e. $c = 2$. From Ramsey theory, we know that there will be a number γ such that, if x is bigger than γ, that string will contain an arithmetic progression of length k such that all k digits of that progression are of the same colour, i.e. either all the k digits are '0' or all the k digits are '1'.[8]

Consider now a database D, where some kind of acquired information about some phenomenon P is stored. We want to investigate the correlations among the data stored in D in order to increase our knowledge of P:

> In full generality, we may consider that *a correlation of variables in D is a set B of size b whose sets of n elements form the correlation* [...]. In other words, when a correlation function [...] selects a set of *n*-sets, whose elements form a set of cardinality b, then they become correlated. Thus, the process of selection may be viewed as a colouring of the chosen set of b elements with the same colour—out of c possible ones. [...]. Then Ramsey theorem shows that, given any correlation function and any b, n and c, there always exists a large enough number γ such that any set A of size greater than γ contains a set B of size b whose subsets of n elements are all correlated. (Calude, Longo 2016, p. 12)[9]

Calude and Longo prove that the larger D is, the more *spurious* correlations will be found in it. In other words, when our stock of available data increases, most of the correlations that we can identify in it are spurious. Since large databases *have to* contain arbitrary correlations, owing to the size of data, *not* to the nature of data, the larger the databases are, the more the correlations in such databases are spurious. Thus, the more data we have, the more difficult is to extract meaningful knowledge from them.[10]

---

[8]In this case (i.e. $c = 2$), if we have $k = 3$, then $\gamma = 8$. To see this, consider the following sequence of binary digits of length 8: 01100110. This string contains no arithmetic progression of length 3, because the positions 1, 4, 5, 8 (which are all '0') and 2, 3, 6, 7 (which are all '1') do not contain an arithmetic progression of length 3. However, if we add just one bit more to that string (i.e. if we add either '1' or '0'), we obtain the following two strings: 011001100 and 011001101. Both these strings contain a monochromatic arithmetic progression of length 3. Consider 011001100: positions 1, 5, 9 are all '0'. Consider 011001101: positions 3, 6, 9 are all '1'. More generally, it can be proved that if a string contains more than 8 digits, it will contain a monochromatic arithmetic progression of length 3. And in fact, all the 512 possible binary strings of length 9 contain a monochromatic arithmetic progression of length 3.

[9]It is important to stress that the nature of the *correlation function* is irrelevant: it can be completely arbitrary, i.e. in no way related to the nature of the data stored in the database.

[10]Cf. Calude and Longo (2016, p. 6): "it is exactly the size of the data that allows our result: the more data, the more arbitrary, meaningless and useless (for future action) correlations will be found in them." It may be interesting to note that, in order to derive their result, Calude and Longo

This result generalizes the difficulty deriving from the claim that scientific progress has to be identified with knowledge accumulation. Indeed, the search for spurious correlations within a stock of knowledge produces itself instances of knowledge, and so it contributes to the increase of the size of that stock of knowledge. But, as Calude and Longo demonstrate, the mere increasing of the stock of knowledge increases the number of spurious correlations that it is possible to find within it. Since, if our stock is sufficiently large, we can expect that *most* of the correlations that we can identify in it are spurious, we can also expect that the identification and the correct classification of these correlations as 'spurious correlations', while representing a genuine instance of knowledge, will hardly represent an instance of progress. So, the findings of spurious correlations are (almost) all cases in which knowledge increases while progress does not. And they are also cases that contribute to the increase of the number of possible spurious correlations, since they increase the stock of knowledge. Since the more spurious correlations are possible within a stock of knowledge, the more increasing our understanding becomes difficult,[11] we may even hypothesize that, contrary to Bird's view, in certain circumstances knowledge accumulation may 'negatively' contribute to an increase in the understanding of relevant phenomena.[12]

According to Dellsén, the epistemic account is unable to account for all those relevant cases of knowledge production that do not imply an increase in progress, since it takes scientific progress to amount to knowledge accumulation. On the contrary, the noetic account is able to account for those cases, because while it acknowledges that those cases represent genuine instances of knowledge, it denies that they also constitute genuine instances of progress, given that they do not increase our understanding of the world, i.e. they do not enable us to correctly explain or predict any relevant aspect of the world.

4 Knowledge and Understanding: The Problem of the Factivity of Knowledge

We have seen that, according to Dellsén, the superiority of the noetic account with respect to the epistemic account is due to the fact that the noetic account is able to adequately account for some kinds of cases that the epistemic account is not able to

---

define "spurious" in a more restrictive way than Dellsén. According to them, "a correlation is spurious if it appears in a 'randomly' generated database" (p. 13). Details can be found in Calude and Longo (2016). In any case, this does not impinge on the considerations that follow.

[11]Think of the increase in the understanding of some phenomenon X that may be derived by the findings of relevant (i.e. not-spurious) correlations among X and other phenomena Y and Z: if the number of spurious correlations increases, the number of correlations that we have to discard before finding the relevant ones increases too. Thus, increasing the understanding becomes more difficult when the number of spurious correlations increases.

[12]On a similar point, see Rancourt (2015).

adequately account for. In Dellsén's view, the inadequacy of the epistemic account in accounting for such cases is due to some qualifying features of the concept on which the epistemic account rests, i.e. "knowledge". In this perspective, the superiority of the noetic account is due to the fact that it rests on the concept of "understanding". In other words, it would be thanks to some features of the concept of understanding that the noetic account fares better than the epistemic account. The question then arises: What are the differences between the concept of "knowledge" and the concept of "understanding" that allow "understanding" to perform better than "knowledge" when we deal with scientific progress?

Before trying to answer this question, we have to keep in mind that we are dealing here with a *realist* perspective on science and scientific progress. The confrontation between Bird and Dellsén takes place in a shared realist framework. Dellsén explicitly affirms that he considers his proposal to be a realist account of progress,[13] and it is exactly for this reason that he takes the noetic account to be a valid alternative to Bird's epistemic account, which is a full-blooded realist view of progress. Dellsén sees the noetic account as a sort of amelioration of the epistemic account. In his view, the noetic account is a more adequate (and nuanced) realist account of progress than the epistemic one, since it is able to satisfyingly account for more kinds of cases.

Turning back to the differences between knowledge and understanding, one of the most salient is the strength of the truth and justification requirements: for a belief to be knowledge, it must be (at least) true and justified. On the contrary, in order to have some understanding of a phenomenon, what is required is the ability to make previsions or provide explanations relative to such phenomenon.[14] But the history of science shows us that the capacity for providing explanations or making predictions may be due to beliefs or theories that are incomplete, (partially) false, or not yet empirically confirmed. So, the question is: Can this ability of making predictions and providing explanations, i.e. understanding, really be decoupled from the truth requirement?

There is a bifurcation here: indeed, it is very possible to give a realist interpretation of the concept of understanding, according to which the ability to make predictions and provide explanations depends, in the ultimate analysis, on the truth of the theory we are dealing with; but another interpretation of the concept of

---

[13]Cf. Dellsén (2016, p. 73, fn. 6): "the noetic account amounts to a moderately realist view of the aim of science."

[14]Several very different views have been advanced on the requirements that have to be fulfilled in order to have understanding (see for a survey Baumberger et al. 2017). The main distinction is between those authors who think that understanding is just a species of knowledge (and so there is not a real distinction between these two concepts, see Grimm 2006), and those who, on the contrary, think that understanding is not just a species of knowledge (see Dellsén 2016b). Those who belong to this latter group have different ideas on how exactly understanding differs from knowledge. They usually claim that understanding lacks one or more of the traditional knowledge requirements, i.e. truth, justification, and some anti-luck condition. Here we will follow Dellsén's characterization of understanding, and assume, at least for the sake of the argument, that understanding is not just a species of knowledge.

understanding is also possible, which, since it is opposed to the previous one, can be labeled "anti-realist", according to which the abilities of making predictions and providing explanations are not necessarily related to the truth of the theory we are dealing with. In other words, according to the realists, scientific progress cannot be due to anything other than an approximately true theory, while according to the anti-realists even a false theory may constitute a genuine scientific progress.[15] Thus, a realist interpretation of the concept of understanding gives rise to a realist account of scientific progress, while an anti-realist interpretation of the concept of under-standing gives rise to an anti-realist account of scientific progress (see e.g. Khalifa 2011; de Regt and Gijsbers 2017). Both those interpretations have been advanced and defended in the last years.[16]

Consider now that we are dealing here with the noetic account, and that, as we have already noted above, Dellsén takes it to be a *realist* account of progress. So, we can conclude that Dellsén adopts a realist stance on understanding, and thinks that, in the ultimate analysis, understanding depends on the truth of the theory we are dealing with. But if understanding depends on truth as knowledge does, how is it possible that it differs from knowledge in such a way that the noetic account is able to account for those cases that the epistemic account is not able to account for precisely because of the rigidity of its truth requirement?

Dellsén, in order to maintain both the claim that there is a relevant difference between understanding and knowledge with respect to the truth requirement, and the claim that the noetic account, which is based on the concept of understanding, is a realist account of progress, adopts a line of reasoning developed by some epis-temologists in recent years, and takes understanding to be a *quasi-factive* concept.[17]

Knowledge is usually considered to be factive: if a person has the belief that p, she does not know that p unless 'p' is true.[18] Understanding would instead be quasi-factive, in the sense that it is not necessary that each component of a theory be true to make this theory able to increase understanding. It is sufficient that "the explanatorily/predictively essential elements of a theory" be "true in order for the theory to provide grounds for understanding" (Dellsén 2016a, p. 73, fn. 6). On this view, at the "periphery" of a theory there well may be inaccuracies, falsities,

---

[15]The idea that empirical success is a good indicator of truth is a pillar of scientific realism (see e.g. Wray 2013). Thus, a realist view of scientific progress cannot *completely* sever the link between the empirical success of a theory and its truth.

[16]See e.g. Kvanvig (2003), who maintains that the truth requirement is necessary for under-standing; Elgin (2007), who maintains that we may have understanding even through falsities; Rancourt (2015), who maintains that, in certain circumstances, an increase in truth-content may even lead to a decrease in understanding.

[17]On the quasi-factivity of understanding, see Kvanvig (2009), and Mizrahi (2012). For some criticisms of this view, see Elgin (2009), and de Regt and Gijsbers (2017).

[18]Many authors argue that knowledge is a *factive* propositional attitude. To say that a propositional attitude is factive is to say that "it is impossible for you to have that attitude towards anything other than a true proposition" (Pritchard and Turri 2014), where "true" has to be intended in the sense of "corresponding to facts". Williamson (2000) argues that knowledge is *the most general factive* propositional attitude. For a radically different view on knowledge, see Cellucci (2017).

abstractions, idealizations, i.e. all those elements that prevent us from baldly claiming that a theory is straightforwardly "true" and that everything it states constitutes "knowledge". And those elements may even contribute in some way and to some extent to the understanding of the target phenomenon. But what, in the ultimate analysis, grounds our understanding of a phenomenon is the *truth* of the "central propositions" of the theory we are dealing with (Kvanvig 2009; Mizrahi 2012).

In this way, the abilities of making predictions and providing explanations of some phenomenon are not completely severed by the possession of some relevant *truths* about this phenomenon. And according to Dellsén, this would be sufficient both (1) to consider the noetic account a *realist* account of progress, and (2) to consider genuine contributions to progress those cases in which understanding increases thanks to a theory replete with elements that are not strictly true, i.e. elements whose presence prevents us to claim that knowledge also increases, since knowledge is a more demanding concept.

If we conceive of understanding in this way, understanding admits of degrees, while knowledge is an all or nothing matter. There is no such thing as degrees of knowing: either you know or you don't. This is mainly due to the fact that truth is an all or nothing matter. Since knowledge requires truth, and truth does not admit of degrees, knowledge does not admit of degrees in its turn. Thus, understanding may increase even if there is not an increase in the truth content of the theory we are dealing with, if some, strictly speaking, false elements of a theory allow us to make better predictions or provide better explanations of the target phenomenon. From this, it follows that while it makes no sense to claim that a true theory is more true than another true theory, it makes perfect sense to claim that a theory provides a better understanding of a phenomenon than another theory. Since there may be an increase in understanding even if there is not an increase in the truth content of a theory, there may be an increase in the understanding of a phenomenon even if there is not an increase in our knowledge relative to such phenomenon.

According to Dellsén, those features of the concept of understanding make it perfectly suited to account for scientific progress, and for other relevant aspects of scientific practice, such as the use of highly idealized models in science.

5   The Noetic Conception as a Realist Conception of Progress

In order to assess Dellsén's claim that the noetic account is a more adequate realist account of progress than the epistemic account, we will leave aside many of the criticisms that have been raised against the quasi-factive view of understanding.[19]

---

[19]The main criticism to the quasi-factive view of understanding is the one developed by Elgin (2007, 2009), who shows how in many cases idealizations and other 'untrue' elements of a scientific theory are "central terms" of that theory, not peripheral. They are "essential elements of a theory" that cannot be replaced, nor are expected to be replaced in the future by scientists.

We will instead try to develop an argument elaborating on some works of Barrett (2003, 2008), Stanford (2003), and Saatsi (2016). This argument aims at showing that, if understanding is quasi-factive, the noetic account may face some difficulties in accounting for the case of the incompatibility of Quantum Mechanics (QM) and General Theory of Relativity (GTR), and that this impinges on Dellsén's claim that the noetic account is an adequate realist view of progress.

QM and GTR are certainly our current best scientific theories, in any sense in which 'best' is usually understood by philosophers of science. In a realist perspective, since scientific realism claims that our best theories are true (Chakravartty 2015), both QM and GTR should be regarded as true or, at least, approximately true. Thus, a realist view of scientific progress should account for QM and GTR accordingly.

But there is a problem. Roughly, we know that QM and GTR are incompatible, in the sense that, due to the different fundamental theoretical assumptions they make, they provide two contradictory descriptions of the world (see e.g. Macías and Camacho 2008). And to the extent that two incompatible theories cannot both be true, we know that QM and GTR cannot both be true (Barrett 2003; Bueno 2017).[20]

Let's try to consider how Dellsén's proposal would account for this case. Since the noetic account rests on the concept of understanding, and conceives of understanding as our ability in making predictions or providing explanations, if we adopt the noetic view, we should admit that QM and GTR constitute genuine cases of scientific progress. Indeed, since they are the most powerful theories we have, we can fairly affirm that they allow us to make accurate predictions and provide deep explanations of a large class of phenomena.

But recall that according to Dellsén, the noetic account rests on a quasi-factive conception of understanding. In this view, the abilities of making predictions and providing explanations that QM and GTR give us, have to be due, in the ultimate analysis, to the truth of the "essential elements" of those theories. If this is the case, then QM and GTR may be deemed to be (approximately) true, at least in their essential theoretical 'core'. But if QM and GTR are both (approximately) true (at least) in their essential theoretical 'core', they should not be radically incompatible, at least in principle. Nevertheless, there is a wide consensus among scientists and philosophers of science on the claim that QM and GTR are in fact *radically* incompatible.

There seems to be a tension here. As we will see below, the supporter of the noetic account may try to solve this difficulty either by accepting the claim that QM and GTR are incompatible, and so that one of those theories is false (Sect. 5.1), or by denying this claim, and so maintaining that both those theories are true (Sect. 5.2).

---

[20]It is worth noticing that this argument does not rest on a sort of pessimistic induction over past science (Laudan 1981), as many anti-realist arguments do. And so it does not display the weakness that is usually thought to afflict that kind of arguments, i.e. their inductive character (Barrett 2003). Indeed, the argument we are dealing with here is based on the *theoretical* impossibility of reconciling the images of the world provided by our current most powerful scientific theories.

5.1   First Option: Affirming the Incompatibility
of QM and GTR

Let's consider the hypothesis that QM and GTR are really incompatible, and so
they cannot both be true. This would mean that even the 'essential elements' of at
least one of QM and GTR are not true. Since, as we have seen, the supporter of the
noetic account should consider both QM and GTR to be cases of progress, this
leads her to a dilemma: (1) either she has to admit that the noetic view is not able to
account for the fact that a theory whose 'essential elements' are false may nevertheless increase our understanding; (2) or she has to claim that such a theory has
never really increased our understanding, given that its 'essential elements' were
false.

Both the horns of this dilemma are problematic to take. Indeed, if one takes (1),
one has to admit that the noetic account is an inadequate account of scientific
progress, because it is not able to account for such a relevant case, i.e. that a
radically false theory may constitute progress. If, on the other hand, the supporter of
the noetic account takes (2), she finds herself in a position that is in some way
similar to the one in which the supporter of the epistemic account finds herself if she
tries to consider Einstein's work a case of progress (see above, Sect. 3.1).

Let's try to see why. As we have seen, according to Dellsén's criteria, both QM
and GTR are cases of progress, since they increase our understanding. But
according to the quasi-factive view of understanding, if QM and GTR are radically
incompatible, they cannot both be genuine cases of progress. If we want to maintain
the quasi-factive view, we have to deny that one of QM and GTR really provides
understanding. But at the moment we are not able to assess, between QM and GTR,
which is the approximately true theory and which is the one that will be discarded
because its essential elements will be proven false. According to our current
standards, they are both extremely empirically successful.

To face this difficulty, we may be tempted to claim that QM and GTR provide us
just a sort of *hypothetical understanding*, i.e. the understanding that a theory would
provide if its essential elements are true. In this view, when in the future the theory
whose essential elements are true will be identified, we will be able to assess which
theory between QM and GTR *really* increases our understanding, and so constitutes
a genuine case of progress. But if the verdict on the progressiveness of QM and
GTR is made dependent on (an eventual) future confirmation of the truth of their
essential elements, we would not be really able to claim *now* that they constituted
genuine cases of progress when they were formulated, exactly as the supporter of
the epistemic account was unable to claim that Einstein's paper constituted a
genuine progress when it appeared in 1905.

But the idea of making our judgment over the progressiveness of a case
dependent on future confirmation is exactly what Dellsén deemed unacceptable in
considering the attempt of rescuing the epistemic account with respect to Einstein's
work by considering it a case of *hypothetical explanatory knowledge*. Now, as we
have seen, the epistemic account faces a dilemma: either it is not able to consider

Einstein's work a case of progress, or, if it tries to rely on the idea of *hypothetical explanatory knowledge* to account for such a case, it is no more able to rule out Oresme's solution from the set of genuine cases of progress. In a similar way, the noetic account faces a dilemma with regard to QM and GTR: either it is not able to rule out the false theory between QM and GTR, or if it tries to rely on the idea of *hypothetical understanding* to discriminate between these two cases, it is no more able to claim *now* whether they are genuine cases of progress or not.

Moreover, the strategy of making our judgment over the progressiveness of a case dependent on future confirmation faces two more general problems. The first is the risk of a regress. If at any time t_x, the supporter of the noetic account has to wait for the truth of the essential elements of our current best theory to be confirmed by future science at time t_{x+n} in order to judge its progressiveness, then she risks being able to correctly determine what episodes in the history of science constituted genuine cases of progress only at the end of time. Indeed, the *truth* of the essential elements of any past theory can be determined only in the light of our best current theory. But the truth of the essential elements of our best current theory is determinable only by future science (Barrett 2008; Stanford 2003; more on this below). It is also conceivable that in the light of future science, say at time t_{x+n}, the essential elements of some past theory T that we now, at time t_x, deem true in the light of our current science will be discarded and regarded as false. So, there is the risk that we may have to modify our verdicts on the progressiveness of any past episode at any future time.

The second general problem that this strategy faces is that, if we have to wait for some future confirmation to judge over the progressiveness of an episode, the distinction between the noetic account and the epistemic account fades. Indeed, Dellsén explicitly argues (see above, Sect. 3.1) that these accounts are distinct because the noetic account is able to determine whether a theory is progressive by focusing on the historical context in which it is proposed, i.e. on the contribution to the understanding of a phenomenon that it provides at the time of its formulation, independently of whether this theory will later be confirmed or not, while the epistemic account, since it rests on the concept of knowledge, has to wait for a theory to be confirmed to regard it as an instance of progress. If confirmation is also required by the noetic account, the noetic account becomes equivalent to the epistemic account.[21]

---

[21]There are some additional difficulties for Dellsén's account worth being pointed out: the first is that it is true that if QM and GTR are incompatible, at least one of them cannot be true. But we cannot exclude that neither is. How would the noetic view account for the possibility that, in light of future science, the essential elements of both QM and GTR will be deemed untrue? Shall we then claim that neither QM nor GTR were cases of progress when they were formulated and applied? In the same vein: even if we concede that future confirmation will allow us to determine that one theory between QM and GTR really increases our understanding because of the truth of its essential elements, there will still be the difficulty for the supporter of the noetic account *to explain* the empirical success enjoyed by the other radically 'false' theory, given that it is uncontroversial that both QM and GTR are extremely successful in dealing with the world, and that the noetic account claims to be a *realist* account of progress.

To conclude, if we take (2), i.e. we claim that one of QM and GTR has never really increased our understanding, since its 'essential elements' are false, this would amount to admit that there may be cases in which an increase in our ability to make predictions and provide explanations increases while our understanding does not increase. This kind of case could not be accounted for by the noetic view, since these cases contradict the qualifying claim of the noetic account, namely that scientific progress has to be identified with an increase in understanding.

Thus, since neither (1), nor (2) seem to be a viable route to take for the supporter of the noetic account, it is fair to conclude that the first option we considered, i.e. affirming that QM and GTR are really incompatible, is not a promising way to solve the problem of accounting for the incompatibility of QM and GTR if we adopt the noetic account.

5.2   Second Option: Denying the Incompatibility of QM and GTR

The supporter of the noetic account may try to defend her view by claiming that QM and GTR are not really incompatible, that they are both approximately true theories, and that their now *seeming* incompatibility derives exclusively from some of their non-essential elements, which could be amended in the future. If this is the case, both QM and GTR can be regarded as genuine cases of progress in light of the noetic account.

There are two main and related problems with this line of reasoning. The first is that the claim that QM and GTR are not really incompatible contradicts the 'received view' among scientists and philosophers on this issue. Obviously, this is not a decisive rejoinder, but it is a reminder of the fact that the burden of proof of the claim that such a deep incompatibility as the one that obtains between QM and GTR may be due to the incompatibility of just some peripheral elements of those theories, is on the supporter of this line of defense. And proving such a claim is not an easy task (Barrett 2008).

The second, and more difficult, problem is that this defense of the noetic account postpones the ascertainment of the (approximate) truth of QM and GTR to some future time, and this makes this line of defense ineffective.

Indeed, QM and GTR are our current *best* theories. From a realist point of view, we can at most hypothesize that they are (approximately) true, since they are actually enjoying huge empirical success. Nevertheless, as realists themselves usually acknowledge, this kind of inference "from the success to the truth" of a theory is insufficient to establish with certainty that a theory is actually true. There is always the possibility that a theory, even if by our current standards empirically successful, will be dismissed and deemed to be false in light of future science. Thus, we cannot, at the moment, estimate the degree of approximation to the truth of QM and GTR, nor can we, strictly speaking, even really claim that they are "approximately" true in any meaningful sense, since we cannot assess their approximation to the truth in the same way we usually do in order to maintain that a past theory is approximately true. Indeed, we can claim that a past theory T was approximately true, only by comparing it with our best (and more complete) current theory T^* of the same domain (Stanford 2003). In the case of QM and GTR, since they are our *current* best theories, we cannot estimate their approximation to the truth by comparing them to a more complete theory that we deem to be true (Barrett 2008).

To better see this point, consider Newtonian Gravitation (NG): realists may claim that NG is approximately true, since they are able to compare NG to GTR, and thus show that the former can be derived as a special case of the latter (Saatsi 2016; Barrett 2008). But it would be pointless to claim that GTR is "approximately true", since we do not know *what* it may approximate. Since GTR is so successful, we use it to estimate the degree of "approximation" to the truth of past theories, but we cannot say that GTR is approximating some more true theory. Nor can we claim that GTR is the true and definite theory, since we know that it is theoretically unable to explain the phenomena pertaining to a relevant domain of nature (the domain of QM), and that it is incompatible with QM. So, even if we think that QM and GTR are not really radically incompatible, and that they are both approximately true, this very fact means that none of them can be the *definite* theory. At most, it may be claimed that they both approximate the same future definite theory.

It may be claimed that in the future there will certainly be developed a theory TX, and that it will be possible to show that GTR approximates TX, in the same way in which it has been possible to show that NG approximates GTR. But this claim is just as compelling as any other induction is, since it rests on our hope that the future will resemble the past. We have no effective way to support the claim that GTR will be proven to approximate TX, since we have no way to know *what* TX may look like. Since we are unable to compare TX and GTR, we are unable to assess whether GTR approximates TX. We cannot rule out the possibility that GTR and TX will be radically distinct, and that the essential elements of GTR will be deemed to be false in the light of TX. If this is the case, GTR would be a false but empirically successful theory, and so it cannot be claimed to be approximately true. Given that QM and GTR are incompatible, and that this *may* imply that GTR is in reality radically false (since, even if we are now considering the possibility that QM and GTR are *not* radically incompatible, we have not yet succeeded in showing that in fact *they are compatible*), this scenario is not implausible, and so cannot be easily dismissed. This argument may be developed, *mutatis mutandis*, with regard to the claim that QM will approximate TX.

It may be objected that if QM and GTR are not really incompatible, i.e. their incompatibility is due just to some of their peripheral elements, then it is not true that TX is completely undetermined. We have some clues of how TX *should* be: if we take that GTR's essential elements are true, since GTR constitutes progress and we adopt a realist view of progress, then TX should retain GTR's essential elements and dismiss just the false, peripheral elements of GTR. Moreover, because of this continuity between GTR and TX, TX should be such that GTR may be derived from it as a special case, in the same way in which NG is derivable from GTR.

The problem is that, even if we concede that TX will have to meet these constraints, we are nevertheless still unable to effectively determine whether GTR *really* approximates TX. The fact is that we have no idea *how* GTR should be modified in order to be improved.[22] We do not know *now* what elements will be deemed central and what elements will be deemed peripheral, and so dismissed, in the future in light of TX (Barrett 2003).[23] Such a distinction between central and peripheral elements will be possible only *after* TX will have been elaborated and confirmed. Think again of NG. If we accept the line of reasoning we are analyzing, we should maintain that, by reasoning alone, at Newton's time, it would have been possible to determine (at least to some extent) the form that GTR should have taken, and what elements of NG would have consequently been deemed essentials. If this were the case, it would have been possible to compare NG and GTR (before this last one had even been actually elaborated), and safely claim that NG approximates GTR. But things go the other way around. It is starting from GTR that we can now claim what elements of NG are essential and what "errors" or "falsities" are embedded in NG. And it is in large part due to the post-GTR development of a geometrized version of NG, which is empirically equivalent to NG, "but absorbs all the information about gravitational forces into information about spacetime curvature so that gravitation is rendered a manifestation of spacetime curvature" (Saatsi 2016, p. 9), that we are now able to understand *how* NG may be shown to be a special case of GTR, and thus *why* we can safely claim that NG approximates GTR.[24]

To sum up, if we try to defend the noetic account maintaining that QM and GTR are not really incompatible and are both approximately true, in order to make our defense effective, we should be able to compare QM and GTR to the theories (or the theory) that will replace them in the future. But, since we have no idea about what such future theories may look like, we are not able to show that QM and GTR approximate them, and so we cannot claim that QM and GTR are approximately true. So, it does not seem easy to find a way to defend the claim that the incompatibility between QM and GTR is due just to some of their peripheral elements. Thus, taking the second option, i.e. maintaining that QM and GTR are not really incompatible, does not seem to be a promising way to defend the noetic account.

---

[22]Cf. Barrett (2003, p. 1216): "While we do have a vague commitment that our future physical theories will somehow be better than our current physical theories, we do not now know how they will be better. If we did, we would immediately incorporate this insight into our current physical theories."

[23]Cf. Barrett (2003, p. 1216): "insofar as we expect surprising innovations in the construction of future theories […], we cannot now know even what the structure of the space of possible options for refining our current theories will prove to be." This point cannot be developed here, but it is worth underlining that this line of reasoning is analogous to the *unconceived alternatives* argument developed by Stanford (2006), who elaborates on Sklar (1981).

[24]See also Barrett (2008). For more details, see Malament (1986a, b).

Since neither the first option nor the second option we considered seems to provide a viable route to take for the supporter of the noetic account, we can fairly conclude that the problem of the incompatibility between QM and GTR represents a serious challenge for the noetic account.

6 Factivity and Modality

From what we have said so far, it seems that the supporter of the noetic account faces a dilemma: either (1) she dismisses the noetic account as inadequate, or (2) she dismisses her quasi-factive view of understanding. Indeed, if she tries to rescue the realist construal of the noetic account by adopting a full-blooded factive view of understanding, she risks being unable to distinguish understanding from knowledge, and so to let the noetic account become equivalent to the epistemic view. But if she discards the quasi-factive view and adopt a non-factive view of understanding, she has to dismiss the claim that the noetic account is a realist account of progress, since adopting a non-factive view would amount to admitting that even a radically false theory may constitute progress, a claim that realists usually deny.

But there may still be another option for the supporter of the noetic account who wants to maintain the claim that the noetic account is a *realist* account of progress. Rice (2016) has recently claimed that we can construe the idealized models commonly used in science in a realist fashion, even if we concede that they do not accurately represent the way the world really is, and that they indispensably rest on idealized assumptions, i.e. even if some of their *essential* elements are known to be false. Rice's aim is to solve the difficulty that the realist has to face when she tries to account for the role that idealizations play in science, a theme deeply connected with the debate on the concept of understanding (Elgin 2009; Saatsi 2016). In a nutshell, Rice suggests that highly idealized models may nevertheless provide *factive* scientific understanding, since they give us *true modal information* about the counterfactual relevance and irrelevance of various features of the target system. So, despite their representations being inaccurate and many of their assumptions being false, those models may increase our understanding. This view of understanding seems to be compatible with the noetic account, because, despite the fact that Rice names his view of understanding 'factive', he does not construe 'factive' in a too demanding way as necessarily requiring truth: "my view is that scientific understanding is factive because in order to genuinely understand a natural phenomenon *most* of what one believes about that phenomenon [...] must be true" (Rice 2016, p. 86). This may be sufficient for distinguishing the concept of 'understanding' from the concept of 'knowledge'. Thus, it seems, at least *prima facie*, that Rice's proposal may be compatible with Dellsén's account of progress.[25]

Relying on Rice's proposal, the supporter of the noetic account may try to account in a realist fashion for those cases in which a theory provides an increase in our understanding even if its essential elements have to be considered, strictly speaking, false. This kind of case is exactly the one we analysed above (Sect. 5) with respect to the incompatibility between QM and GTR. By adopting Rice's view, the supporter of the noetic account may claim that, even if in the future one of QM and GTR (or both of them) will be deemed to be strictly speaking false, this false theory is nevertheless really increasing our understanding, and so it constitutes a genuine instance of progress. She may also maintain that the previous claim is compatible with the idea that the noetic account is a realist conception of scientific progress, arguing that such a false theory, despite the falsity of its essential elements, provides us *true* modal information on some relevant aspect of the pertinent phenomena, and it is this true modal information that is responsible for the increase of our understanding. Since in this view the truth requirement is, at least to some extent, fulfilled by the modal essential element of the theory, this may well be claimed to be a realist view of progress: "it is the model's ability to provide true modal information about the space of possibilities that enables the model to produce" scientific understanding (Rice 2016, p. 92).

The main problem with this approach is that, in order to salvage the claim that only true elements may be responsible for an increase in understanding, it assumes that a false theory may convey *true modal knowledge*.

Now, let's concede, for the sake of the argument, that it may be possible to provide a coherent realist notion of "a false theory that is able to convey true modal information", and that it may also be possible to provide an account of *how* a theory such as QM or GTR may be considered an instance of this kind of false theory. The problem on which we will focus is that this 'move' amounts to embracing modal realism. In other words, if the supporter of the realist account tries to salvage her realist view of progress by relying on Rice's proposal, she commits herself to modal realism. But this commitment may be challenging for her.

Indeed, if a theory is supposed to be able to provide *true* modal knowledge, this means that we assume that modal knowledge is possible, i.e. that we are able to know what is necessary and what is possible. Since we are dealing here with a realist conception of understanding, and since the theory we are considering is supposed to be false, the element that has to be true, in order to fulfill the realist truth requirement, is the modal information provided by the theory. Indeed, if we take an anti-realist stance on modality, nothing is left in the theory which can be regarded as 'true' from a realist point of view, and so Rice's conception of

---

[25]Some work would be necessary to generalize Rice's proposal and make it suitable to account for the case of the incompatibility between QM and GTR, since Rice (2016) refers just to some kinds of models, and especially focuses on some optimality models used in biology, while QM and GTR are theories. This point cannot be developed here, but a promising route may be to adopt the semantic view of theories, according to which a theory is the class of its models.

understanding cannot be applied, nor can the noetic account be claimed to be a realist account of progress anymore. So, if we adopt Rice's proposal to support the noetic account, we have to take the modal information provided by the theory as true, and we have to intend 'true' in a realist sense. In the context of modal epistemology, this usually amounts to adopting some formulation of 'possible worlds' modal realism, a view which has been fiercely disputed and it is certainly not easy to defend (Vaidya 2016; Bueno 2017).

It may be objected that the true modal information provided by the false theory refers only to some actual features of the relevant phenomena, and that this fact exonerates us from committing ourselves to some possible-worlds construal of modality, since in this case we do not have to account for non-actual possible features of the relevant phenomena, and so we can avoid making reference to non-actual possibilities in terms of possible worlds, which are ontologically dubious and epistemically inaccessible entities. Indeed, if the theory tells us what is actual, from what is actual we can safely infer what is *actually possible* without making reference to any world other than ours (Bueno 2017).

But, even granting for the moment the 'soundness' of this objection, it seems to be inadequate. Let's try to see why. Rice (2016) distinguishes between system-specific models and hypothetical models. System-specific models provide accurate information about the counterfactual relevance and irrelevance of salient features of the target system. On the contrary, hypothetical models are not intended to accurately represent any particular features of a real-world system, i.e. "the model has no real-world 'target system' whose (difference-making) features it aims to accurately represent" (Rice 2016, p. 91). To sum up, "system-specific models aim to provide 'how actually' information while hypothetical models typically aim to provide 'how possibly' information" (Rice 2016, p. 92). Now, if we try to adapt Rice's proposal to the noetic perspective on the falsity of one of QM and GTR, we have to keep in mind that here we are dealing with a theory that is, by hypothesis, *radically* false, i.e. whose *essential* elements are *false*. So, it is reasonable to think that such a theory would be considered, in the modified noetic account that we have assumed it is possible to elaborate according to Rice's proposal, as being analogous to Rice's hypothetical models, and not to Rice's system-specific models. Indeed, if such a theory is radically false, this means that it does not accurately tell us *anything* about the *actual* target system, otherwise some of its essential elements would have been true. If this is the case, the objection mentioned above does not hold, since hypothetical models do not provide us with modal information about actual features of the target phenomenon; they instead provide information about non-actual possibilities of the target phenomenon: they "explore the possibility space in which features differ [...] dramatically from those of the actual system" (Rice 2016, p. 92, fn. 12). Thus, this is not a viable route for the supporter of the noetic account to avoid the commitment to some formulation of possible-worlds modal realism.

7   Understanding and Modalism

There are several difficulties that arise for the supporter of the noetic account from her commitment to such a disputed view as possible-worlds modal realism.[26] Perhaps the main difficulty comes from the fact that the noetic account aims to be a realist account of scientific progress. Many scientific realists think of themselves as naturalists, i.e. they adopt a naturalist stance (Morganti 2016). But it is at least very controversial whether possible-worlds modal realism may be compatible with a naturalist stance. Indeed, modal realists claim that, for every way the world could be, there is a world that is that way (Lewis 1986). This means to assume that if something is impossible in our world but it is conceivable, it is true in some other possible world *causally isolated* from ours. This contrasts with the way 'naturalism' is usually construed, i.e. as implying that natural entities are all there is, and that for an entity to be regarded as natural, it has to not be *in principle* spatiotemporally and causally isolated from our world (Papineau 2016).[27]

In any case, we will put aside these difficulties here, and we will grant that there is an escape route for the supporter of the noetic account, i.e. another promising realist view of modality to take, which does not rest on possible worlds, i.e. modalism, according to which modality is a primitive notion (Bueno 2017; Bueno and Shalkowski 2004).

Let's grant, for argument's sake at least, that modalism is able to avoid or solve many of the problems that afflict possible-worlds modal realism (Vaidya 2016, Sect. 1.2.3). The problem that we will try to point out is that this account of modality seems to be in conflict with a (quasi-)factive view of understanding, i.e. the view of understanding advocated by Dellsén.

Modalism draws on an analogy between modal and mathematical knowledge originally developed by Lewis (1986), but it reaches different conclusions:

> the key idea is that we have mathematical knowledge by drawing (truth-preserving) con-
> sequences from (true) mathematical principles. And we have modal knowledge by drawing
> (truth-preserving) consequences from (true) modal principles. (Bueno, Shalkowski 2004,
> p. 97)

According to this view, to know that *P* is possible (or necessary) means to derive "it is possible that *P*" from particular assumptions. More precisely, to know that *P* is possible amounts to being entitled to introduce a possibility operator: *it is possible that P*. In some cases, it is easy to do so: we know that *P* is actual, and therefore possible. But in many cases, *P* is not actual. So, in these cases, when we claim that

---

[26]For a survey of the problems afflicting possible-worlds modal realism, see Vaidya (2016), Bueno (2017), Bueno and Shalkowski (2004).

[27]The adoption of possible-worlds modal realism amounts to assuming that there is something "like a realm of metaphysical possibility and necessity that outstrips the possibility and necessity that science deals with, but this is exactly what naturalists should not be willing to concede" (Morganti 2016, p. 87).

'we know that P is possible', what we are really doing is deriving *it is possible that P* from some particular assumptions.

Here the different way in which Bueno and Shalkowski interpret the analogy between modality and mathematics marks their distance from Lewis's approach. In a nutshell, they develop this analogy as implying that in those cases in which we cannot know (by empirical means) that P is actual, the modal knowledge that we can at most reach is *conditional*, i.e. it is of the form: « '*it is possible that P*' is true, given that the assumptions on which we rest to derive '*it is possible that P*' are true ».

As in mathematics, due to Gödel's results, we are generally unable to prove with certainty that the axioms of the theory we use to derive a theorem *T* are 'true', and we take our knowledge of such theorem to be of the form: «the theorem *T* is true, if the axioms from which it is derived are true»; when dealing with modality we are unable to prove that the modal principles that we choose in order to derive the target conclusion are true.[28] Indeed, in many cases the possibility of determining whether something is possible or not will depend on controversial assumptions. There are several incompatible and competing assumptions available to be taken as the starting point from which we derive our target conclusions on what is possible, and there is not a way of proving that such 'first assumptions' are at their turn 'true' without ending in an infinite regress or committing a *petitio principii*.

According to modalism, we have to accept that with respect to cases involving non-actual possibilities, "instead of having *categorical* modal knowledge [...] (that is, knowledge of what is possible or necessary, independently of particular assumptions)," we can at most "have *conditional* modal knowledge [...] (that is, knowledge of what is possible or necessary *given* certain philosophical assumptions)" (Bueno 2017, p. 80).

Now, in the context we are dealing with, as we have already seen above, the *radically false* theory between QM and GTR, which is supposed to be able to give us true modal knowledge, is supposed to convey modal information relative to *non-actual* possibilities. Thus, if we adopt modalism to escape the conundrums deriving

---

[28]On the consequences of Gödel's results for how mathematical knowledge should be conceived, see Cellucci (2013, 2017). On how modalism construes the analogy between modality and mathematics, cf. Bueno and Shalkowski (2004, pp. 97–98): "If the analogy with mathematics is taken seriously, it may actually provide a reason to *doubt* that we have any knowledge of modality. One of the main challenges for platonism about mathematics comes from the epistemological front, given that we have no access to mathematical entities—and so it's difficult to explain the reliability of our mathematical beliefs. The same difficulty emerges for modal realism, of course. After all, despite the fact that, on Lewis' account, possible worlds are *concrete* objects, rather than abstract ones, we have *no access* to them. Reasons to be skeptical about a priori knowledge regarding mathematics can be easily 'transferred' to the modal case, in the sense that difficulties we may have to *establish* a given mathematical statement may have a counterpart in establishing certain modal claims. For example, how can we know that a mathematical theory, say ZFC, *is consistent*? Well, we can't know that in general; we have, at best, relative consistency proofs. And the consistency of the set theories in which such proofs are carried out is far more controversial than the consistency of ZCF itself, given that such theories need to postulate the existence of inaccessible cardinals and other objects of this sort."

from possible-worlds modal realism, we have to regard the modal knowledge provided by this false theory as an instance of *conditional knowledge*. The problem is that *conditional knowledge* is unable to fulfill the realist requirement that is necessary to claim that the noetic account is a *realist* account of scientific progress. Indeed, if we adopt Dellsén's and Rice's (quasi-)factive view of understanding, instances of *conditional modal knowledge* cannot be considered to be able to increase our understanding, since we cannot assess whether they are *really* true.

Thus, modalism is not really a viable route to take for the supporter of the noetic account: if she tries to rely on Rice's proposal and 'go modal' to face the challenge deriving from the incompatibility of QM and GTR, she seems unable to avoid committing herself to possible-worlds modal realism.

8 Conclusion

In this article, we tried to assess whether the noetic account is a more adequate realist account of progress than the epistemic account. We elaborated an argument that aims to show how the quasi-factive view of understanding that Dellsén adopts in order to maintain that the noetic view is a realist account of progress is in tension with Dellsén's definition of understanding. We examined a possible way out for Dellsén, the adoption of Rice's proposal that highly idealized models may nevertheless provide us factive scientific understanding by giving us true modal information about pertinent phenomena. But this shift to modality comes with a cost for the supporter of the noetic account: it implies that she has to commit herself to possible-worlds modal realism, an option that may be unpalatable for many scientific realists. Finally, we have proposed another way out for Dellsén, i.e. the adoption of modalism. But we showed that modalism is not able to support a realist view of understanding.

To sum up: if the supporter of the noetic account wants to maintain the standard way to conceive of knowledge and understanding, she faces the following dilemma: either (1) she dismisses the noetic account as an inadequate realist account, or (2) she dismisses her quasi-factive view of understanding. If she tries to escape this dilemma by 'going modal', she faces a new dilemma: either (1) she commits herself to possible-worlds modal realism, or (2) she dismisses her quasi-factive view of understanding.

In both these dilemmas, to take option (1) is very difficult: in the first case, it amounts to admitting that the noetic account does not really fare better than Bird's view, while in the second case, it implies adopting such a controversial perspective on modality that many scientific realists may tend to prefer Bird's view in any case. If one of the main rationales for develop the noetic account was advancing the epistemic view, this seems not to be a good result. We think that option (2) is the more promising one. Obviously, taking (2) amounts to dismissing the idea that the noetic account is a *realist* account of progress, and this may seem an even worse result than the one that can be achieved by choosing (1). But we instead think that this option is worth further investigation, since developing an anti-realist noetic account of scientific progress, which relies on a well-defended anti-realist view of understanding, as e.g. the one recently provided by de Regt and Gijsbers (2017),[29] may be the best way to emphasize the several interesting features that the noetic account displays, and let it spread all its theoretical fertility without having to struggle with the conundrums arising from the adoption of possible-worlds modal realism.

References

Baumberger, C., Beisbart, C., & Brun, G. (2017). What is understanding? An overview of recent debates in epistemology and philosophy of science. In S. Grimm & C. Ammon (Eds.), *Explaining understanding: New perspectives from epistemology and philosophy of science* (pp. 1–34). New York: Routledge.

Barrett, J. A. (2003). Are our best physical theories probably and/or approximately true? *Philosophy of Science, 70,* 1206–1218.

Barrett, J. A. (2008). Approximate truth and descriptive nesting. *Erkenntnis, 68,* 213–224.

Bird, A. (2007). What is scientific progress? *Noûs, 41,* 64–89.

Bueno, O. (2017). The epistemology of modality and the epistemology of mathematics. In B. Fischer & F. Leon (Eds.), *Modal epistemology after rationalism* (pp. 67–83). Cham: Springer.

Bueno, O., & Shalkowski, S. (2004). Modal realism and modal epistemology: A huge gap. In E. Weber & T. De Mey (Eds.), *Modal epistemology* (pp. 93–106). Brussels: Koninklijke Vlaamse Academie van Belgie.

Calude, C. S., & Longo, G. (2016). The deluge of spurious correlations in big data. *Foundations of Science.* https://doi.org/10.1007/s10699-016-9489-4.

Cellucci, C. (2013). *Rethinking logic: Logic in relation to mathematics, evolution, and method.* Dordrecht: Springer.

Cellucci, C. (2017). *Rethinking knowledge: The heuristic view.* Dordrecht: Springer.

Chakravartty, A. (2015). Scientific realism. In: E. N. Zalta (Ed.), *The stanford encyclopedia of philosophy*, URL: http://plato.stanford.edu/archives/fall2015/entries/scientific-realism/.

Chang, H. (2007). Scientific progress: Beyond foundationalism and coherentism. *Royal Institute of Philosophy Supplement, 61,* 1–20.

Dellsén, F. (2016a). Scientific progress: knowledge versus understanding. *Studies in History and Philosophy of Science, 56,* 72–83.

Dellsén, F. (2016b). Understanding without justification or belief. *Ratio.* https://doi.org/10.1111/rati.12134.

de Regt, H. W. (2009). Understanding and scientific explanation. In H. W. de Regt, S. Leonelli, & K. Eigner (Eds.), *Scientific understanding: Philosophical perspectives* (pp. 21–42). Pittsburgh: Pittsburgh University Press.

de Regt, H. W. (2015). Scientific understanding: Truth or dare? *Synthese, 192*(12), 3781–3797.

---

[29]There is no space here to adequately argue for this claim, but we think that the noetic account may well be construed in anti-realist terms, and that a promising way for developing this conception is relying on the notion of 'effectiveness' proposed by de Regt and Gijsbers (2017). They propose to replace the usual truth requirement for understanding with an effectiveness condition on understanding, according to which understanding requires representational devices that are scientifically effective, where being 'scientifically effective' means being able to produce useful scientific outcomes such as correct predictions, successful applications and fertile ideas.

de Regt, H. W., Leonelli, S., & Eigner, K. (2009). Focusing on scientific understanding. In H. W. de Regt, S. Leonelli, & K. Eigner (Eds.), *Scientific understanding: Philosophical perspectives* (pp. 21–42). Pittsburgh: Pittsburgh University Press.

de Regt, H. W., & Gijsbers, V. (2017). How false theories can yield genuine understanding. In S. R. Grimm, C. Baumberger, & S. Ammon (Eds.), *Explaining understanding: New perspectives from epistemology and philosophy of science* (pp. 50–75). New York: Routledge.

Einstein, A. (1956). On the movement of small particles suspended in a stationary liquid demanded by the molecular kinetic-theory of heat. In: Idem, *Investigations on the theory of the brownian movement*. New York: Dover: 1–18. Original Edition: Idem, (1905). Über die von der molekularkinetischen Theorie der Wärme geforderte Bewegung von in ruhenden Flüssigkeiten suspendierten Teilchen. *Annalen der Physik, 322*: 549–560.

Elgin, C. (2007). Understanding and the facts. *Philosophical Studies, 132*, 33–42.

Elgin, C. (2009). Is understanding factive? In A. Haddock, A. Millar, & D. Pritchard (Eds.), *Epistemic value* (pp. 322–330). New York: Oxford University Press.

Grimm, S. R. (2006). Is understanding a species of knowledge? *The British Journal for the Philosophy of Science, 57*, 515–535.

Ippoliti, E. (2014). Generation of hypotheses by ampliation of data. In L. Magnani (Ed.), *Model-based reasoning in science and technology: Theoretical and cognitive issues* (pp. 247–262). Berlin: Springer.

Ippoliti, E., & Cellucci, C. (2016). *Logica*. Milano: Egea.

Khalifa, K. (2011). Understanding, knowledge, and scientific antirealism. *Grazer Philosophische Studien, 83*, 93–112.

Kuhn, T. S. (1970). *The structure of scientific revolutions* (2nd ed.). Chicago: University of Chicago Press.

Kvanvig, J. L. (2003). *The value of knowledge and the pursuit of understanding*. New York: Cambridge University Press.

Kvanvig, J. L. (2009). Responses to critics. In A. Haddock, A. Millar, & D. Pritchard (Eds.), *Epistemic value* (pp. 339–353). New York: Oxford University Press.

Laudan, L. (1977). *Progress and its problems*. London: Routledge.

Laudan, L. (1981). A confutation of convergent realism. *Philosophy of Science, 48*, 19–48.

Lewis, D. (1986). *On the plurality of worlds*. Oxford: Blackwell.

Macías, A., & Camacho, A. (2008). On the incompatibility between quantum theory and general relativity. *Physics Letters B, 663*, 99–102.

Malament, D. (1986a). Gravity and spatial geometry. In R. Barcan Marcus, G. J. W. Dorn, & P. Weingartner (Eds.), *Logic, methodology and philosophy of science VII* (pp. 405–411). Amsterdam: Elsevier.

Malament, D. (1986b). Newtonian gravity, limits, and the geometry of space. In R. Colodny (Ed.), *From quarks to quasars: Philosophical problems of modern physics* (pp. 181–201). Pittsburgh: University of Pittsburgh Press.

Mizrahi, M. (2012). Idealizations and scientific understanding. *Philosophical Studies, 160*, 237–252.

Morganti, M. (2016). Naturalism and realism in the philosophy science. In K. J. Clark (Ed.), *The Blackwell companion to naturalism* (pp. 75–90). Oxford: Blackwell.

Niiniluoto, I. (2015a). Optimistic realism about scientific progress. *Synthese*. https://doi.org/10.1007/s11229-015-0974-z.

Niiniluoto, I. (2015b). Scientific progress. In E. N. Zalta (Ed.), *The stanford encyclopedia of philosophy*, URL: http://plato.stanford.edu/archives/sum2015/entries/scientific-progress/.

Papineau, D. (2016). Naturalism. In E. N. Zalta (Ed.), *The stanford encyclopedia of philosophy*, URL: https://plato.stanford.edu/archives/win2016/entries/naturalism/.

Perrin, J. (1908). Agitation moléculaire et mouvement brownien. *Comptes Rendus de l'Académie des Sciences, 146*, 967–970.

Popper, K. R. (1963). *Conjectures and refutations*. London: Routledge and Kegan Paul.

Pritchard, D., Turri, J. (2014). The value of knowledge. In E. N. Zalta (Ed.), *The stanford encyclopedia of philosophy*, URL: https://plato.stanford.edu/archives/spr2014/entries/knowledge-value/.

Rancourt, B. T. (2015). Better understanding through falsehood. *Pacific Philosophical Quarterly.* https://doi.org/10.1111/papq.12134.

Rice, C. C. (2016). Factive scientific understanding without accurate representation. *Biology and Philosophy, 31,* 81–102.

Rowbottom, D. P. (2010). What scientific progress is not: Against bird's epistemic view. *International Studies in the Philosophy of Science, 24,* 241–255.

Saatsi, J. (2016). What is theoretical progress of science? *Synthese.* https://doi.org/10.1007/s11229-016-1118-9.

Sklar, L. (1981). Do unborn hypotheses have rights? *Pacific Philosophical Quarterly, 62,* 17–29.

Stanford, K. (2003). Pyrrhic victories for scientific realism. *Journal of Philosophy, 100,* 553–572.

Stanford, P. K. (2006). *Exceeding our grasp.* Oxford: Oxford University Press.

Vaidya, A. (2016). The epistemology of modality. In E. N. Zalta (Ed.), *The stanford encyclopedia of philosophy,* URL: https://plato.stanford.edu/archives/win2016/entries/modality-epistemology/.

Williamson, T. (2000). *Knowledge and its limits.* Oxford: Oxford University Press.

Wray, K. B. (2013). Success and truth in the realism/anti-realism debate. *Synthese, 190,* 1719–1729.

Zagzebski, L. (2001). Recovering understanding. In M. Steup (Ed.), *Knowledge, truth, and duty* (pp. 235–251). New York: Oxford University Press.

Rhythms, Retention and Protention: Philosophical Reflections on Geometrical Schemata for Biological Time

Giuseppe Longo and Nicole Perret

Abstract In this paper, following the technical approach to biological time, rhythms and retention/protention in Longo and Montévil (Perspectives on organisms: Biological time, symmetries and singularities. Springer, Berlin, 2014), we develop a philosophical frame for the proposed dimensions and mathematical structure of biological time, as a working example of "theory building". We first introduce what "theory building" means to our perspective, in order to make explicit our theoretical tools and discuss the general epistemological issue. Then, through a conceptual articulation between physics and biology, we introduce protention (anticipation) and retention (memory), as proper biological observables. This theoretical articulation, which we consider at the core of moving from physical to biological theorizing, allows us to use some of the properties of these observables as principles around which it is possible to outline a proper geometrical schema for biological time. We then philosophically motivate the analysis of "time" as an operator that acts in biological dynamics in a constitutive way. In other words, space and time become specials concepts of order, actively involved in the theoretical organization of biology, in contrast to existing theories in physics where they appear as parameters. In this approach, we first consider the usual dimension of an

Articles (co-)authored by Longo are downloadable from http://www.di.ens.fr/users/longo/.

G. Longo (✉)
Centre Cavaillès, République des Savoirs, CNRS, Collège de France
et Ecole Normale Supérieure, 29, Rue d'Ulm, 75005 Paris, France
e-mail: giuseppe.longo@ens.fr
URL: http://www.di.ens.fr/users/longo

G. Longo
Department of Integrative Physiology and Pathobiology,
Tufts University School of Medicine, Boston, USA

N. Perret
Centre Cavaillès, République des Savoirs, Institut d'Etudes
Avancées de Nantes, et Ecole Normale Supérieure de Paris, 5,
Allée Jacques Berque, 44000 Nantes, France
e-mail: nic.perret@gmail.com

© Springer International Publishing AG 2018
D. Danks and E. Ippoliti (eds.), *Building Theories*, Studies in Applied Philosophy,
Epistemology and Rational Ethics 41, https://doi.org/10.1007/978-3-319-72787-5_12

irreversible physical time. We then add to it a dimension specific to the internal rhythms of organisms. We motivate this dimensional extension by the relative autonomy of biological rhythms with respect to physical time. This second dimension of time is "compactified" in a simple but rigorous mathematical sense. In short, as soon as there are life phenomena, their rhythms scan biological time. We will consider such a statement as a starting point for an original notion of biological inertia.

Keywords Biological inertia · Biological time · Geometrical schema Protention · Retention · Rhythms

1 Introduction

In this paper, we provide a philosophical account to an original approach on biological time, developed in Longo and Montévil (2014). This approach is a working example of "theory building". Before going into more detail, we aim to state some philosophical considerations on the practice of "theory building", which we understand with the meaning of "constituting objectivity" [see Bitbol, Kerszberg and Petitot (2009)].

Our constructivist approach is primarily built on the philosophical belief that science is not the mere description of phenomena, nor are theories the plain representation of reality. The objectivity of the *laws of nature* does not belong to nature itself, as an intrinsic truth. Rather, it results from scientific activity, in interaction with the perception of the world. That is, the scientific object differs from simple perception. Its configuration is related to the conceptual and practical tools used to access the phenomena and it is part of a *construction of objectivity* that is proper to scientific knowledge.

It should be clear that, for us, theory building is not an arbitrary speculation on "reality", whatever this word may mean. Instead, it begins with the active access to the world by our sensitive and scientific tools for "measuring", in the broadest sense. But these tools, starting with our physiological sensitivity, are biased by our actual being in the world: we may be insensitive to a visual perception or even to pain, say, by ongoing forms of conscious or preconscious being. Similarly, the construction of measurement tools in science is biased by strong theoretical commitment, such as the construction of the instruments of microphysics. These depend on the choice of the observables to be measured, for example, the physical quantities expressed by the bounded operators in Schrödinger's equations (Bitbol 1996), as well as on countless technical details all depending on theoretical and engineering frames. Thus, they depend on both previously theorized experiences (the search for the spin of a particle by these or that tools, say) as well as on their mathematical expression (Schrödinger's approach, typically). The choice of the observables and measurement contexts and tools is just as important in biology, as our working example will show.

In other words, the process of building theories happens in a halfway place between thinking and reality. The convergence of these two *ways of access* to phenomena is the place where concepts appear and stabilize. According to this perspective, a main difficulty of scientific activity has been to take the conceptual elements of this intermediate world as reality itself. It is typically the case of categories such as causality, of numbers, of symbolic structures such as time and space, and so on. In the particular case we account for, we make extensive use of the conceptual construction of time and geometry.

In the approach to biological time presented here, we recognize two different forms of observable time in the same dimension, the irreversible time of thermodynamics and the time of the developmental process in biology (embryo/ontogenesis). Then, we also distinguish between the dimension of these two forms of time and a new one, where it is possible to represent biological rhythms. The choice of this new observable is far from being arbitrary as it is motivated by the relevant empirical role of rhythms from the analyses of minimal forms of metabolism to the clinical relevance (and extensive literature) on the heart and the respiratory rhythms (see below and the references). Our philosophical bias should be clear: we noticed a different practical role given to rhythms in concrete actions on living beings, from investigations to therapies. Historically, this role has been made possible by the relatively early invention of a variety of tools for their measurement. We then integrate this knowledge and practical activities into a basic, schematic representation which may suggest both further theoretical developments and new tools for measurements.

2 Philosophy and Geometrical Schemata for Biological Time

Contemporary studies on the temporal orientation of consciousness insist on the importance of anticipation and memory. The role played by these two temporal elements has been largely explored through a long philosophical and phenomenological tradition, among others [see for example Depraz (2001), Petitot, Varela and Pachoud (1999) for recent approaches and syntheses]. Husserl designated as "protention" the particular extension of the present towards the future by anticipation. Moreover, he proposed to link this aspect to "retention" as the extension of the present towards the past by memory (Husserl 1964). Recent works, directly inspired by the phenomenological approach, formalize this aspect in an increasingly precise way [see for example Gallagher and Varela (2003), Vogeley and Kupke (2007)]. This philosophical background is an essential component of the theory building introduced here. Too often the scientific work hides philosophical biases, Perret et al. (2017); by our example we will also consider the implicit consequences of an intended philosophy.

In our theoretical approach, we propose to see protention and retention as the starting point to exhibit a specific temporal structure of the *preconscious* living systems, as a compatible extension of the phenomenological view. Through a conceptual articulation between the role of observables and parameters in physics and in biology, we argue that these two elements are specifically observable for biology. Accordingly, it is possible to outline a geometry of the temporal structure specific to living organisms (Bailly, Longo and Montevil 2011). This is a broader theoretical approach that seeks to propose specific principles for a conceptual organization of the living. This global perspective is grounded on the new theoretical framework of "extended criticality" (Bailly and Longo 2008; Longo and Montévil 2011a, b).

2.1   Constitutive Space and Time

In this paper, we invoke a special constitution of time for the living. Inspired by the philosopher Kant (2000), we consider that reflecting on time and space means reflecting on the deep conceptual conditions of a scientific domain as a priori forms. That is to say, we are not invoking a description of some real temporal properties that would be present within biological objects, or even a measurement of quantities, but more a mathematical conceptualization of some temporal specificities that we recognize as observables in the living. To do this operation, we have to clarify the theoretical role of mathematics in our perspective. According to Kant in the *Critique*, the a priori forms of space and time receive and shape phenomena. Mathematics, because of its *synthetic* and a priori character, organizes the content of the a priori forms by a conceptualisation. This amounts, therefore, to a constitutive role of mathematics in the construction of scientific concepts (Cassirer 2004). Space and time, then, are the *conditions of possibility* of constituting objectivity operated by mathematics (Kant 2000).

Now, through a further abstraction of the a priori transcendental principles,[1] it is possible to overcome the Kantian dualism between, on the one hand, the pure forms of the a priori intuition of space and time, and concepts, on the other hand.

---

[1] The process of relativisation of the Kantian a priori comes from the neo-Kantian School of Marburg and especially from Cassirer. With non-Euclidean geometries, a priori forms of intuition of space and time (which, for Kant, had the form of Euclidean geometry) could no longer constitute a scientific foundation for localisation. Moreover, after the formulation of the theory of relativity (restrained and general, both basing themselves on non-Euclidean spaces), the very concept of an object and its relationship to space was no longer immediate in intuition. More specifically, in classical mechanics, the dependency of the notion of "object" upon a complex of universal laws was founded on the laws of geometry. In relativity theory, instead, the localisation of an object takes place through operations that enable a transition from one reference system to another. It is the invariants of such transformations that may be deemed "objects". We refer here to Cassirer (2004), for broad overviews of the possible modulations of the a priori we refer to Kauark-Leite (2012), Lassègue (2015).

In particular, time becomes an *operator*. It participates in the organizing and constitutive activity of mathematics rather than being a set framework making such activity possible. In other words, space and time become true *concepts of order* actively involved in the conceptual organization of a science. As for our perspective, this is a major methodological shift when moving from physical to biological theorizing.

In this context, the project of making the geometrical complexity of biological temporality mathematically intelligible corresponds to the construction of a geometrical schema specific to living phenomena, which is characterized, in a broadened Kantian sense, by a set of principles of order. Indeed, in the process of constituting scientific objects, the determination of sensory perception through concepts provides not only the scientific object, but also the rule of construction of this object. We call this rule, in a Kantian sense, a *schema*. That is, the process of determination assumes genericity, because it comes with the exhibition of a general process of construction. Therefore, providing a geometrical schema for biological time means operating a geometrical determination on the specificity of time shown by the living, and constructing, through this determination, a general theory. Now, a biological object as a *physical singularity* (Bailly and Longo 2011) presents specific temporal characteristics, as described in Longo and Montévil (2014), which we will survey and discuss here.

Note that, in general, as in any scientific theorizing, a sound theory of life phenomena must stem from its specificities in order to construct a conceptual organization adapted to the biological. The physicalistic reduction of the biological constitutes an illegitimate theoretical operation, based upon a realist prejudice according to which the laws of physics represent real properties of phenomena. Instead, physical theories are also conceptual organizations of "reality" constructed from transcendental principles (Bitbol 1998, 2000). Furthermore, they propose various notions of a *causal field,* which are not even reducible to one another, as, for example, in classical and relativistic versus quantum physics. So, to which of these fields should we reduce life phenomena? Note that both classical and quantum phenomena are relevant in biology, as they may both have phenotypic effects, from cells to organisms (Buiatti and Longo 2013); not to mention the hydrodynamics of fluids, a field far from being understood in terms of quantum (particle) physics (Chibbarro et al. 2014).

It seems then more pertinent to construct a new "causal field" for the biological, which is founded upon its own specific principles, the same as the physical causal fields within their own theoretical domains. Later on, one may better establish a unification project, as it is also considered in physics (as, for example, in the ongoing tentative unifications of relativistic/quantum physics: what would there be to unify if there were not two theories?). Now, for biology we suggest to ground the causal field on a geometrical schema for biological time; as a part of the theory of organisms we are working at, see Soto and Longo (2016).

The temporality of the living organisms is very specific compared to the physical treatment of time. Development, aging, biological rhythms, evolution and metabolic cycles attest to this peculiarity (Chaline 1999). Here, protention and retention will play a constitutive role.

We propose first to take minimal protentional behaviors as mathematically quantifiable observables, without requiring the production of a physical theory of teleonomy. On the contrary, the teleonomic dynamic becomes a simple principle, founded upon the intuition of a movement of retention/protention, which is an observable.

The protentional act indeed represents a temporal specificity, which we observe in the simplest living forms (for example, paramecia, Misslin (2003)). It is from the internal rhythms of life phenomena, which we will address, and from this act that we can establish an autonomy of biological time in contrast with physical time. This demonstrates the need for an autonomy of the transcendental principles of biology. It will, therefore, be entirely legitimate to add specific observables as well as a second temporal dimension specific to biological rhythms. We will then construct a geometrical schema for biological time on two dimensions: one to represent the rhythms specific to life phenomena, the other to quantify the usual dimension of the irreversible time of physics. In this dimension, a new observable of time appears to be relevant: the irreversibility specific to biological processes, in particular the setting up and renewal of organization (Longo and Montévil 2014).

2.2 Dichotomous External/Internal Reference Systems

The transcendental role in the geometric construction of biological time manifests even more radically in the way in which a biological object's two internal/external poles of temporal reference are articulated. Indeed, due to the relativization of the Kantian a priori, it is legitimate to consider abstract notions of space and time as able to relate to the mathematical structures of group and semigroup, respectively. In particular, the determination of displacement groups (reversible) is involved in formalizing the abstract notion of space. Analogously, the characteristics of semigroups participate in formalizing the abstract notion of time and, namely, of the properties of compositionality and irreversibility of the flow of time (Bailly and Longo 2011, p. 169).

In short, we consider, first, physical space, where displacements (reversible, group transformations) are possible, and *within which* we can describe the internal/external spaces for each organism and, second, an irreversible physical time (whose transformations form a semigroup). More generally, by an extension of this correspondence to logic,[2] we can see the outline of a dichotomic structure of constitution taking another step towards abstraction. We have, on the one hand, the *space,*

---

[2]The notion of group can be put into correspondence with the logical relationship of equivalence, and the notion of semi-group has the same form of ordered relation, (Bailly and Longo 2011, p. 163).

group structure, as the *equivalence relation* pole, and, on the other hand, the *time, semigroup structure,* as the *order relation* pole.

To this ordered time line, *we add* a second dimension specific to the internal time of organisms, the time of biological rhythms. This dimensional extension will be motivated by the relative autonomy of biological rhythms with respect to physical time. They present themselves in fact as "pure numbers", that is, as invariants common to different biological species. In short, to irreversible physical time (the thermodynamic straight line or, algebraically, a semigroup), we add an orthogonal dimension represented by a compactified straight line (a circle, a "compactification method" that has already been exploited in physics for space by Kaluza and Klein, see Bailly and Longo (2011)). It intertwines with it as an iterative and circular observable, that of biological rhythms, which we will address.

Now, these two dimensions articulate with one another through a dichotomy of the internal/external type, which participates, constitutively, in a new conceptual organization of biology. This also comes down to the constitution of a causal field specific to life phenomena, because we will correlate protention with the setting of these internal rhythms, enabling us to conceptualize a form of teleonomy without, nevertheless, referring to a retrograde causality.

To return to our Kantian considerations, the space and time of the *Critique* (2000) were in opposition and, more precisely, they assumed, within the subject, the a priori form of the external sense and the a priori form of the internal sense, respectively. Recall here the progressive rediscovery of the Leibnizian arguments defended by the later Kant of the *Opus Postumum* (1995), according to which space and time can no longer be in such opposition, but themselves possess intrinsic forms on the side of the object. We are then led to rediscover the legitimacy, at least a theoretical one, of the structuring of a proper internal temporal dimension for life phenomena, insofar as both internal/external poles must be found within the same object, the living organism. However, this does not mean that they constitute properties that are intrinsic to the objects, because we are still at an epistemic level of constituting objectivity. What we have, in particular, are forms of constituting the localization of objects coming from their actual determination. In other words, space and time become the active conditions of constituting the intelligibility of the object: some sort of *forms of sensible manifestation* (Bailly and Longo 2011, p. 153). The external sense, then, determines the form of the manifestation of the relations, and the internal sense governs the form of the manifestation of the identification of the objects. By means of this process, and in conjunction with relativisation of the a priori, a transformation of the abstract notions of space and time is operated. This transformation, in conclusion, comes down to justifying the epistemic role of the internal spatiotemporal dimensions specific to biology, governing the very conditions of the possibility for individuating the object.

Following Longo and Montévil (2014), we reconstruct and elaborate on this process through two movements. First, we identify a proper observable, the time of biological irreversibility, and we place it in the dimension of physical time (thermodynamic time, thus oriented), anchored upon an extended present (retention/ protention). Then, we add a second compactified dimension to this temporal dimension shared with physics. This dimension is supposed to be the proper dimension to describe biological rhythms. This geometrical schema constructs a new intelligibility using the internal constitutive property of the abstract notion of time.

3   Retention and Protention

Husserl undertakes a fundamental analysis of the temporality specific to consciousness, separated from objective time, based upon two opposing temporal directions: memory and anticipation. Memory is characterized as a reconstruction of a distant past and anticipation as the expectation of a possible future. Now, these two poles belong respectively to the past and to the future, but a tendency towards these two directions along the same axis takes place in the *present apprehension* of phenomena.

We will very generally address the movements of retention and protention, even in the absence of intentionality (so also for preconscious activities). Retention and protention are forms of the present: the present instant is therefore constituted as a dialectic situation, which is never simple or defined, a situation *that is not to be described as punctual.*

More specifically, in physics, one can conceive of a punctual (pointwise) present, a singular instant which is a number on Cantor's straight line of real numbers. The temporal singularity of the biological, instead, is *extended*: an extended transition from the past to the future, a union of minimal retention and of the corresponding protention. This change is fundamental and paradigmatic with respect to physics. With the invention of speed and acceleration as instantaneous values, the limits of a secant that becomes a tangent line (Newton) or of a ratio of which the numerator and denominator tend towards 0 (Leibniz), mathematics sets itself within modern physics. Then, by their punctual values, speed and acceleration also become functions of time.[3]

Now, in biology, in this case and others, the punctuality of a process is devoid of meaning: the snapshot loses what is most important, the functions and action of the living phenomenon, which *is never* devoid of activity. The instantaneous picture of a rock that is falling is identical to the picture of the rock when stationary, the rock being *inert* even during its *inertial* movement. Life can be understood only in its processes, which are constantly renewing and changing, from internal physiological activity to movement. Biological time is therefore not to be grasped based on a

---

[3]Note that H. Weyl, a major mathematician of relativity theory, while working on "Space, time and matter", a fundamental book for that theory, stresses the limits of the physical description of time. He does so in Weyl (1918), in reference to the non-pointwise experience of phenomenal time, where the knowing, living subject plays a role.

possible punctuality; and this will also apply, as far as we are concerned, to *all* biologically relevant parameters and observables. Even more strongly, life is not only a process, a dynamic; it is always (in) a "critical transition". We have rendered this analysis of the "extension" of biological observables and parameters by the notion of "extended criticality", which is specific to the living state of matter (Bailly and Longo 2008, 2011; Longo and Montévil 2011a), to be briefly hinted below.

In time, retention, directed towards an immediate past, and protention, directed towards the immediate future, constitute an extension of the present that distinguishes itself from the objective time of physics, all the while articulating itself with it. We refer to Longo and Montévil (2011b) for the mathematical analysis: retention is described by means of a relaxation function (an exponential that decreases in physical time), whereas protention is described by its symmetrical, corrected by a linear dependence of retention. The composition of these formal symmetrical exponentials formalizes the fact that there is no anticipation without memory, as advanced by Husserl and as confirmed by recent empirical evidence (Botzung, Denkova and Manning 2008) (for other works on conscious activity, see Nicolas (2006), Perfetti and Goldman (1976)). Protention is therefore mathematically dependent upon retention, an asymmetry that orients biological time. In short, we consider as if the organism, as elementary as it may be, were capable of protention. Such protention is able to govern the behavior of the organism in its present on the basis of prior experience. Even a paramecium manifests clear forms of protention and retention; see Misslin (2003).

To conclude, to this construction of objectivity specific to biological time, we added, taking Husserl as a starting point, a temporal observable that is specific to biology based on the interplay between retention and protention. This notion, albeit in the same mathematical dimension as the physical arrow of time, oriented by all irreversible phenomena (at least thermodynamically) does propose a new observable for us: the irreversibility specific to biological time, oriented by the mathematical asymmetry of retention/protention (Longo and Montévil 2011b).

Notice that within the same physical dimension we can have several observables: energy, for example, can be potential or kinetic. For us, the irreversibility specific to biological time adds itself to that of thermodynamic time. Its irreversibility is not only due to the dispersal of physical energy (entropy), but also to the establishment and maintenance of biological organization (which we have analysed in terms of anti-entropy, see Bailly and Longo (2009), Longo and Montévil (2012)). Evolution and embryogenesis (ontogenesis, in fact) have their own constitutive irreversibility, which adds itself to that of thermodynamic processes. This irreversibility is the observable of time specific to life phenomena; in Bailly and Longo (2009), it is considered, mathematically, as an *operator* and not as a parameter as is time in physics, because it operates and constitutes life phenomena, *which is always the result of a history*. In summary, the asymmetry of retention and protention contributes to this new irreversible observable time proper to biological objects and their determination.

4 Biological Inertia

The minimal protentional capacity of living organisms may be founded upon observing the propensity of any organism to simply extend a situation. This capacity may be more or less pronounced according to the level of evolution and the presence or absence of a nervous system. It is, first, the observation of an aptitude to adapt to a situation, by changing and through "self-preservation", that leads us to introduce a function of retention, a component of identity and "structural stability". This may be conceived as the possibility of registering a *morphological memory* at various levels, for example at the biochemical, immune, neural, vestibular or cerebral levels; however, its main biological purpose is precisely to enable protention. In other words, we consider the possibility for an organism to conserve a memory of a comparable previous situation, through learning, even at a very simple level of organization, as the precondition of an *adaptability through anticipation* of a similar situation. In this approach, the genome could be considered as the main retentional component specific to a species. As such, it would play as much of a constraining role with respect to the huge range of hypothetical possibilities as it would the role of an activator with respect to the development of such or such an organism belonging to a given species. This constraint would in a way "canalize" the possibilities of development as a function of the retentional heritage, that is, to the whole biochemical history of the species. The eventual "explosions" associated with the rupture in punctuated equilibria (c.f. the Burgess fauna as analysed by Gould (1989), for example) would then correspond to the lifting of entire classes of inhibitions with respect to the activating role of genomes. This representation would then correspond to the viewpoint according to which life phenomena, far from selecting singular and *specific* geodesic trajectories as in physics, would evolve within a very *generic* framework of possibilities. Among such possibilities some would be inhibited either by internal constraints (from the genome to the structure of the organism) or external constraints (the environment).

At the level of the organism, we can interpret protentional behavior as an anticipation played upon the activation of memory. Thus, the trace of experience also plays a role of constraint: some consequent reactions become plausible and then generate a related behaviour, even if it then proves to be poorly adapted, thus leading to further learning. Anticipation of this type becomes an instrument for interpreting the behaviour of the organism with respect to the unpredictability that it continuously faces. It can even be seen as a sort of instrument for continuous reorganization as a consequence of the impossibility of rendering explicit the whole field of possibilities. Thus, as clearly distinguished by Husserl, retention is not memory itself, but the process of memory activation in the present instant—*in view of action*, we emphasize. Likewise, protentional movement is not anticipation into the future, but the process of projecting the immediate possibilities of a previously lived, yet, in fact, reconstructed state.

By these movements of dynamic extension of the present, we have a sort of inertial principle of life phenomena, which we could call *biological inertia*.

In Longo and Montévil (2011b), this inertia is mathematically represented as the coefficient of protention: it gives it mathematical "weight", so to speak, depending on retention, in the same way as (inertial) mass, in physics, is the coefficient of acceleration in the presence of a force.

5 Biological Rhythms, A Geometrical Schema for Life Phenomena

Using the same process of mathematical objectification, a new dimension of time founded upon the consideration of rhythms that are internal to life phenomena may be added to the dimension of thermodynamics in which retention/protention also resides (Longo and Montévil 2014). This second dimension of time is compactified (a circle, a loop, instead of the usual straight line of the Cartesian plane), and is thus autonomous in an even more radical way with respect to physical time. In short, as soon as there are life phenomena, there is a rhythm that takes place within: the metabolic rhythm, at least, and then the cardiac, respiratory, and hormonal rhythms, among others. Observation proposes them to us as pure numbers: they give us the time of an organism (life-span, typically), by allometric coefficients, but they do not have the dimension of *time*. For example, the number of heartbeats of mammals is an a-dimensional invariant, a number (approximately 1.2×10^9) and, by a coefficient given by the mass, it gives the average life-span of the organism in question. Thus, a mouse and an elephant have average life-spans that differ by a factor of 50, but they have the same number of heartbeats, the frequency of heartbeats being 50 times higher in the mouse (refer to Bailly et al. (2011), Longo and Montévil (2014) for the technical details).

This second temporality contributes to establishing and justifying a specific causal field for life phenomena. Maybe it is this aspect that must in certain respects be interpreted as a *retrograde causality* but without constituting a temporal inversion. It is rather a circular movement which establishes itself and is also at the heart of the minimal retention/protention dynamic: the expectation of the return of a rhythm, as we will argue below.

From a mathematical standpoint, the introduction of a compactified dimension of time gives, for the topology of our biological schemata for time, $\mathbf{R} \times \mathbf{S_1}$ (a straight line times a circle, a cylinder). Of course, the compactification "radius" remains null in analyses of the inert.

This structure of time breaks certain classical causal aspects, as we were saying: through protention, there may be a change in the present following an *anticipation* of the future. However, the second compactified dimension is exclusively relative to the biological rhythms and fluxes of the very special component of "information" that is related to protention.

In these analyses, two types of biological rhythms are proposed:

1. "External" rhythms, directed by phenomena that are exterior to the organism, with a physical or physicochemical origin and that impose themselves upon the organism. These rhythms are the same for many species, independent of their size. They express themselves in terms of physical, hence dimensional, periods or frequencies (s, Hz) and the invariants are dimensional; they are described relative to the dimension of physical time (in exp(it)). Examples: seasonal rhythms, the circadian rhythm and all their harmonics and subharmonics, the rhythms of chemical reactions that oscillate at a given temperature, etc.
2. "Internal" rhythms, of an endogenous origin, specific to physiological functions of the organism that therefore depend on purely biological functional specifications. These rhythms are characterized by periods that scale as the 1/4th power of the organism's mass and are related to the life-span of the organism, which scales in the same way; they are expressed as pure numbers. For this reason, these invariants are *numerical*, in contrast with the great constants of physics, which have *dimensions*—acceleration, speed, action ... In our description, by a new compactified "temporal" dimension, the numerical values then correspond to a "number of turns", independent of the effective physical temporal extension (examples: heartbeats, respirations, cerebral frequencies, etc. See the graphical representation in Longo and Montévil (2014)).

In short, endogenous biological cycles, which do not depend directly on external physical rhythms that impose themselves, are those which:

1. Are determined by pure numbers (number of respirations or heartbeats over a lifetime) more than by dimensional magnitudes as in physics (seconds, Hertz...).
2. Scale with the size of the organism (frequencies brought to a power −1/4 of the mass, periods brought to a power 1/4), which is generally not the case with constraining external rhythms, which impose themselves upon all organisms (circadian rhythms, for example).
3. Can thereby be put into relation with an additional compactified "temporal" dimension (an angle, actually), in contrast with the usual temporal dimension (physical, thermodynamic, more specifically), non-compactified and endowed with dimensionality.

In this framework, the extended critical situation, corresponding to the self-referential and individuated, but changing, character of the organism, presents a topological temporality of the $\mathbf{R} \times \mathbf{S_1}$ type; whereas the externality of the organism (and the way in which this externality reacts with the organism) preserves its usual temporal topology of $\mathbf{R}$.

Without changing the basic question, we can present a somewhat different perspective: for a living organism, the extended critical situation would occupy a volume within an n-dimension space, with n $\geq$ 5. Among these n dimensions we would distinguish the three dimensions of classical physical space ($\mathbf{R^3}$ topology) and the two dimensions of biological time ($\mathbf{R} \times \mathbf{S_1}$ topology), of which the compactified dimension would have a null radius beyond this volume. The remaining n-5 dimensions correspond to the compatible values of the vital parameters (temperatures between T1 and T2, metabolisms between R1 and R2, etc.): all intervals of extended criticality in which the limits are those of viability. The metrics of the volume's space would correspond roughly to the correlation lengths; the metrics of time would maximally correspond to the life-spans (for $\mathbf{R}$) and to pure maximal numbers (maximum endogenous frequencies) for $\mathbf{S_1}$. One will notice that the endogenous rhythmicities and cyclicities are not so much rhythms or cycles as much as they are *iterations* of which the total number is set (please refer to the quoted articles and book for the technical details).

Let's finally return to the play between retention and protention. We propose to situate the primordial or even minimal protentional gesture/experience in the *expectation of the return of a vital rhythm*, as we hinted above. Protention, therefore, presents itself as a consequence of the act intrinsic to life phenomena: as soon as there is life, from its very evolutive or embryonic origin, a rhythm is established, a metabolic rhythm at the least, the other ones afterward. We describe this process as the sudden formation, a sort of "big bang", of a new temporal dimension that characterizes life phenomena, the dimension of biological rhythms. They generate the anticipation of their own return, therefore the primary protention/anticipation, which justifies, without teleonomy as such nor retrograde physical causality, this inverted biological causality we mentioned earlier, that which modifies present action by the preparation of the protentional gesture.

6 Conclusion

Knowledge construction is based on a complex friction between the knowing subject and the "real world". An essential component of its scientific objectivity depends on the invariance of the proposed concepts and abstract thought structures, under transformations of reference system and its metrics (the scale and tools for measuring). This is the main epistemological lesson we draw for the physics of relativity and, even more so, from quantum mechanics.

We proposed above some invariant schemata for the description of the phenomenal time of life. They derive from observations and experiments (from paramecia to neurological measurement), and are based on an explicit philosophical commitment to a phenomenological analysis of life processes and their temporality. Mathematics, even the elementary notions we referred to here, provides useful symbolic tools for describing a relatively stable conceptual invariance. The question remains of the actual generality, effectiveness and independence from the knowing subject of the proposal of those abstract concepts. "But perhaps this question can be answered by pointing toward the essentially historical nature of that life of the mind of which my own existence is an integral but not autonomous part. It is light and darkness, contingency and necessity, bondage and freedom, and it cannot be expected that a symbolic construction of the world in some final form can ever be detached from it." (Weyl 1927, 1949, p. 62)

References

Bailly, F., & Longo, G. (2008). Extended critical situation: The physical singularity of life phenomena. *Journal of Biological Systems, 16*(02), 309–336. https://doi.org/10.1142/S0218339008002514.

Bailly, F., & Longo, G. (2009). Biological organization and anti-entropy. *Journal of Biological Systems, 17*(01), 63–96. https://doi.org/10.1142/S0218339009002715.

Bailly, F., & Longo, G. (2011). *Mathematics and the natural sciences: The physical singularity of life*. London: Imperial College Press.

Bailly, F., Longo, G., & Montévil, M. (2011). A 2-dimensional geometry for biological time. *Progress in Biophysics and Molecular Biology, 106*(3), 474–484.

Bitbol, M. (1996). *Schrödinger's philosophy of quantum mechanics*. Dordrecht, Boston: Kluwer Academic Publishers.

Bitbol, M. (1998). *L'Aveuglante proximité du réel anti-réalisme et quasi-réalisme en physique*. Paris: Flammarion.

Bitbol, M. (2000). Le corps matériel et l'objet de la physique quantique. In F. Monnoyeur (Ed.), *Qu'est-ce que la matière?: Regards scientifiques et philosophiques*. Paris: Librairie générale française.

Bitbol, M., Kerszberg, P., & Petitot, J. (2009). *Constituting objectivity: Transcendental perspectives on modern physics*. Dordrecht: Springer.

Botzung, A., Denkova, E., & Manning, L. (2008). Experiencing past and future personal events: Functional neuroimaging evidence on the neural bases of mental time travel. *Brain and Cognition, 66*(2), 202–212. https://doi.org/10.1016/j.bandc.2007.07.011.

Buiatti, M., & Longo, G. (2013). Randomness and multi-level interactions in biology. *Theory in Biosciences, 132*(3), 139–158.

Cassirer, E. (2004). In W. C. Swabey & M. C. Swabey (Eds.), *Substance and function & Einstein's theory of relativity* (p. 480). Dover: Courier Dover Publications.

Chaline, J. (1999). *Les horloges du vivant: Un nouveau stade de la théorie de l'évolution*. Paris: Hachette.

Chibbaro, S., Rondoni, L., & Vulpiani, A. (2014). *Reductionism, emergence and levels of reality*. Berlin: Springer.

Depraz, N. (2001). *La conscience: Approches croisées: des classiques aux sciences cognitives*. Paris: Armand Colin.

Gallagher, S., & Varela, F. J. (2003). Redrawing the map and resetting the time: Phenomenology and the cognitive sciences. *Canadian Journal of Philosophy, 29*, 93–132.

Gould, S. J. (1989). *Wonderful life: The Burgess shale and the nature of history*. New York: W. W. Norton.

Husserl, E. (1964). *The phenomenology of internal time-consciousness*. (J. S. Churchill, Trans.). The Hage: M. Nijhoff.

Kant I. (1995). *Opus postumum. The Cambridge edition of the works of Immanuel Kant*. Cambridge: Cambridge University Press.

Kant I. (2000). Critique of pure reason. In P. Guyer & A. W. Wood (Eds.), *The Cambridge edition of the works of Immanuel Kant*. Cambridge: Cambridge University Press.

Kauark-Leite, P. (2012). *Théorie quantique et philosophie transcendantale: Dialogues possibles*. Paris: Hermann.

Lassègue, J. (2015). *Les formes symboliques, du transcendantal à la culture* (collection M.A. thesis). Vrin, Paris.

Longo, G., & Montévil, M. (2011a). From physics to biology by extending criticality and symmetry breakings. *Progress in Biophysics and Molecular Biology, 106*(2), 340–347. https://doi.org/10.1016/j.pbiomolbio.2011.03.005.

Longo, G., & Montévil, M. (2011b). Protention and retention in biological systems. *Theory in biosciences = Theorie in den Biowissenschaften, 130*(2), 107–117. https://doi.org/10.1007/s12064-010-0116-6.

Longo, G., & Montévil, M. (2012). Randomness increases order in biological evolution. In M. J. Dinneen, B. Khoussainov, & A. Nies (Eds.), *Computations, physics and beyond'* (Vol. 7318, pp. 289–308). Auckland, New Zealand.

Longo, G., & Montévil, M. (2014). *Perspectives on organisms: Biological time, symmetries and singularities*. Berlin: Springer.

Misslin, R. (2003). Une vie de cellule. *Revue de Synthèse, 124*(1), 205–221. https://doi.org/10.1007/BF02963405.

Nicolas, F. (2006). Quelle unité pour l'œuvre musicale? In A. Lautman, J. Lautman, & F. Zalamea (Eds.), *Les mathématiques, les idées et le réel physique*. Paris: Vrin.

Perfetti, C. A., & Goldman, S. R. (1976). Discourse memory and reading comprehension skill. *Journal of Verbal Learning and Verbal Behavior, 15*(1), 33–42. https://doi.org/10.1016/S0022-5371(76)90004-9.

Perret, N., Sonnenschein, C., & Soto, A. M. (2017). Metaphysics: The proverbial elephant in the room. *Organisms. Journal of Biological Sciences, 1*(1), 1–5.

Petitot, J., Varela, F. J., & Pachoud, B. (1999). In J. Petitot, F. J. Varela, & B. Pachoud (Eds.), *Naturalizing phenomenology: Issues in contemporary phenomenology and cognitive*. Stanford Calif.: Stanford university press.

Soto, A., & Longo, G. (Eds.). (2016). From the century of the genome to the century of the organism: New theoretical approaches [Special issue]. *Progress in Biophysics & Molecular Biology, 122*.

Vogeley, K., & Kupke, C. (2007). Disturbances of time consciousness from a phenomenological and a neuroscientific perspective. *Schizophrenia Bulletin, 33*(1), 157–165. https://doi.org/10.1093/schbul/sbl056.

Weyl, H. (1918). *Das Kontinuum* (Translated: *The continuum, a critical examination of the foundation of analysis*). NY: Dover (1987).

Weyl, H. (1927). *Philosophy of Mathematics and of Natural Sciences* (English Trans.). Princeton, New Jersey: Princeton University Press (1949).

How to Build New Hypotheses

Ἀπαγωγή and the Optimization of the Eco-cognitive Situatedness

Lorenzo Magnani

Abstract The process of building new hypotheses can be clarified by the *eco-cognitive model* (EC-Model) of abduction I have recently introduced. I will take advantage of three examples: (1) a new interpretation of Aristotle's seminal work on abduction, which stresses the need, to build creative and selective abductive hypotheses, of a situation of eco-cognitive openness, (2) a philosophical example of building new hypotheses, in the case of phenomenology, in which we can take advantage of an abductive interpretation of the concept of adumbration and anticipation, and (3) the abductive discovery in geometry, which illustrates in both a semiotic and distributed perspective, crucial aspects of what I have called manipulative abduction. The first example will also help us to introduce the concept of *optimization of the eco-cognitive situatedness* as one of the main characters of the abductive inferences to new hypotheses. Thanks to these examples we can gain a new vivid perspective on the "constitutive" eco-cognitive character of building hypotheses through abduction.

Keywords Abduction · Hypothesis building · Eco-cognitive model of abduction
Eco-cognitive openness · Anticipation · Adumbration · Diagrammatic reasoning
Manipulative abduction

1 Eco-cognitive Model of Abduction (EC-Model), Aristotle's Ἀπαγωγή, and Its Eco-cognitive Openness

At the center of my perspective on cognition is the emphasis on the "practical agent", of the individual agent operating "on the ground", that is, in the circumstances of real life. In all its contexts, from the most abstractly logical and mathematical to the most roughly empirical, I always emphasize the cognitive nature of

L. Magnani (✉)
Department of Humanities, Philosophy Section and Computational
Philosophy Laboratory, University of Pavia, Pavia, Italy
e-mail: lmagnani@unipv.it

© Springer International Publishing AG 2018   261
D. Danks and E. Ippoliti (eds.), *Building Theories*, Studies in Applied Philosophy,
Epistemology and Rational Ethics 41, https://doi.org/10.1007/978-3-319-72787-5_13

abduction. Reasoning is something performed by cognitive systems. At a certain level of abstraction and as a first approximation, a cognitive system is a triple (A, T, R), in which A is an *agent*, T is a *cognitive target* of the agent, and R relates to the *cognitive resources* on which the agent can count in the course of trying to meet the target-information, time and computational capacity, to name the three most important. My agents are also *embodied distributed cognitive systems*: cognition is embodied and the interactions between brains, bodies, and external environment are its central aspects. Cognition is occurring taking advantage of a constant exchange of information in a complex distributed system that crosses the boundary between humans, artifacts, and the surrounding environment, where also instinctual and unconscious abilities play an important role. This interplay is especially manifest and clear in various aspects of abductive cognition, that is in reasoning to hypotheses.

My perspective adopts the wide Peircean philosophical framework, which approaches "inference" *semiotically* (and not simply "*logically*"): Peirce distinctly says that all inference is a form of sign activity, where the word sign includes "feeling, image, conception, and other representation" (Peirce 1931–1958, 5.283). It is clear that this semiotic view is considerably compatible with my perspective on cognitive systems as embodied and distributed systems. It is in this perspective that we can fully appreciate the role of abductive cognition, which not only refers to propositional aspects but it is also performed in a framework of distributed cognition, in which also models, artifacts, internal and external representations, manipulations play an important role.

The backbone of this approach can be found in the manifesto of my eco-cognitive model (EC-model) of abduction in Magnani (2009).[1] It might seem awkward to speak of "abduction of a hypothesis in literature," but one of the fascinating aspects of abduction is that not only it can warrant for scientific discovery, but for other kinds of creativity as well. We must not necessarily see abduction as a *problem solving device* that sets off in response to a cognitive irritation/doubt: conversely, it could be supposed that esthetic abductions (referring to creativity in art, literature, music, games, etc.) arise in response to some kind of esthetic irritation that the author (sometimes a *genius*) perceives in herself or in the public. Furthermore, not only esthetic abductions are free from empirical constraints in order to become the "best" choice: many forms of abductive hypotheses in traditionally-perceived-as-rational domains (such as the setting of initial conditions, or axioms, in physics or mathematics) are relatively free from the need of an empirical assessment. The same could be said of moral judgements: they are eco-cognitive abductions, inferred upon a range of internal and external cues and, as soon as the judgment hypothesis has been abduced, it immediately becomes prescriptive and "true," informing the agent's behavior as such. Assessing that there is a common ground in all of these works of what could be broadly defined as

---

[1]Further details concerning the EC-model of abduction can be found in Magnani (2015a, 2016).

"creativity" does not imply that all of these forms of selective or creative abduction[2] with their related cognitive strategies are the same, contrarily it should spark the need for firm and sensible categorization: otherwise it would be like saying that to construct a doll, a machine-gun and a nuclear reactor are all the same thing because we use our hands in order to do so!

Aristotle presents a seminal perspective on abduction, which is in tune with my EC-Model: indeed Aristotle's abduction (ἀπαγωγή) exhibits a clear eco-cognitive openness. As I have illustrate in detail in Magnani (2015a), we have to remember that in the cultural heritage of the Aristotle's passages on abduction of chapter B25 of *Prior Analytics* we can trace the method of analysis and of the middle terms in Plato's dialectic argumentation, considered as related to the diorismic/poristic process in ancient geometry. Thanks to Aristotle we can gain a new positive perspective about the "constitutive" eco-cognitive character of abduction.

First of all, we have to take note that it seems Peirce was not satisfied with the possible Apellicon's correction of Aristotle's text about abduction: "Indeed, I suppose that the three [abduction, induction, deduction] were given by Aristotle in the *Prior Analytics*, although the unfortunate illegibility of a single word in his MS, and its replacement by a wrong word by his first editor, the 'stupid' [Apellicon],[3] has completely altered the sense of the chapter on Abduction. At any rate, even if my conjecture is wrong, and the text must stand as it is, still Aristotle, in that chapter on Abduction, was even in that case evidently groping for that mode of inference which I call by the otherwise quite useless name of Abduction—a word which is only employed in logic to translate the [ἀπαγωγή] of that chapter" (Peirce 1931–1958, 5.144–145, *Harvard Lectures on Pragmatism*, 1903).

At this point I invite the reader to carefully follow Aristotle's chapter from the *Prior Analytics* quoted by Peirce. In this case the discussion turns arguments that transmit the uncertainty of the minor premise to the conclusion, rather than the certainty of the major premise. If we regard uncertainty as an epistemic property, then it is reasonably sound also to say that this transmission can be effected by truth-preserving arguments: by the way, it has to be said that this is not at all shared by the overall Peirce's view on abduction, which is not considered as truth preserving.

I want first of all to alert the reader that in the case of the Aristotelian chapter, abduction does not have to be discussed keeping in mind the schema of the fallacy of affirming the consequent. What is at stake is abduction considered either (1) the classification of a certain "unclear" dynamic argument in a *context-free* sequence of three propositions; or (2) the introduction in a similar "unclear" dynamic

---

[2]For example, selective abduction is active in diagnostic reasoning, where it is merely seen as an activity of "selecting" from an encyclopedia of pre-stored hypotheses; creative abduction instead refers to the building of new hypotheses. I have proposed the dichotomic distinction between selective and creative abduction in Magnani (2001). A recent and clear analysis of this dichotomy and of other classifications emphasizing different aspects of abduction is given in Park (2015).

[3]Apellicon was the ancient editor of Aristotle's works. Amazingly, Peirce considers him, in other passages from his writings, "stupid" but also "blundering" and "scamp" (Kraus 2003, p. 248).

three-propositions argument (in this case no longer *context-free*) of few new middle terms. Hence, ἀπαγωγή—(that the translator of the Prior Analytics I am adopting usefully renders with "leading away" (abduction)—is, exactly (in the Aristotelian words we will soon entirely report below)

1. the feature of an argument in which "it is clear δῆλον that the first term belongs to the middle and unclear (ἄδηλον) that the middle belongs to the third, though nevertheless equally convincing (πιστόν) as the conclusion, or more so" (Aristotle 1989, B25, 69a, 20–22, p. 100);
2. the introduction of suitable middle terms able to make the argument capable of guiding reasoning to substantiate an already available conclusion in a more plausible way: Aristotle says in this way we "are closer to scientific understanding": "if the middles between the last term and the middle are few (ὀλίγα) (for in all these ways it happens that we are closer to scientific understanding [πάντως γάρ ἐγγύτερον εἶναι συμβαίνει τῆς ἐπιστήμης)]" (Aristotle 1989, B25, 69a, 22–24, p. 100).

It is clear that the first case merely indicates a certain status of the uncertainty of the minor premiss and of the conclusion and of the related argument; the second case, from the perspective of the eco-cognitive model of abduction, is much more interesting, because directly refers to the need, so to speak, of "additional/external" interventions in reasoning. It has to be said that Aristotle does not consider the case of the creative reaching of a *new* conclusion (that is of a creative abductive reasoning, instantly knowledge-enhancing or simply presumptive): however, I have illustrate in (Magnani 2015a) that this case appears evident if we consider the method of analysis in ancient geometry, as a mathematical argument which mirrors the propositional argument given by Aristotle, provided we consider it in the following way: *we do not know the conclusion/hypothesis, but we aim at finding one thanks to the introduction of further "few" suitable middle terms.*

The following is the celebrated Chapter B25 of the *Prior Analytics* concerning abduction. The translator usefully avoids the use of the common English word *reduction* (for ἀπαγωγή): some confusion in the literature, also remarked by Otte (2006, p. 131), derives from the fact reduction is often rigidly referred to the hypothetical deductive reasoning called *reductio ad absurdum*, unrelated to abduction, at least if intended in Peircean sense. Indeed, the translator chooses, as I have anticipated, the bewitching expression "leading away".

XXV. It is leading away (ἀπαγωγή) when it is clear (δῆλον) that the first term belongs to the middle and unclear (ἄδηλον) that the middle belongs to the third, though nevertheless equally convincing (πιστόν) as the conclusion, or more so; or, next, if the middles between the last term and the middle are few (ὀλίγα) (for in all these ways it happens that we are closer to scientific understanding (πάντως γάρ ἐγγύτερον εἶναι συμβαίνει τῆς ἐπιστήμης)). For example, let A be teachable, B stand for science [otherwise translated as "knowledge"], and C justice [otherwise translated as "virtue"]. That science is teachable, then, is obvious, but it is unclear whether virtue is a science. If, therefore, BC is equally convincing (πιστόν) as AC, or more so, it is a leading away (ἀπαγωγή) (for it is closer to scientific understanding (ἐγγύτερον γάρ τον ἐπίστασθαι) because of taking something in addition, as we previously did not have scientific understanding (ἐπιστήμη) of AC).

Or next, it is leading away (ἀπαγωγή) if the middle terms between B and C are few (ὀλίγα) (for in this way also it is closer to scientific understanding (εἰδέναι)). For instance, if D should be "to be squared," E stands for rectilinear figure, F stands for circle. If there should only be one middle term of E and F, to wit, for a rectilinear figure together with lunes to become equal to a circle, then it would be close to knowing (ἐγγύς ἂν εἴη τοῦ εἰδέναι). But when BC is not more convincing (πιστότερον) than AC and the middles are not few (ὀλίγα) either, then I do not call it leading away (ἀπαγωγή). And neither when BC is unmiddled: for this sort of case is scientific understanding (ἐπιστήμη) (Aristotle 1989, B25, 69a, 20–36, pp. 100–101).

This passage is very complicated and difficult. I have indicated words and expressions in ancient Greek because they stress, better than in English, some of the received distinctive characters of abductive cognition:

1. ἄδηλον [unclear] refers to the lack of clarity we are dealing with in this kind of reasoning; furthermore, it is manifest that we face with a situation of ignorance —something is not known—to be solved;
2. πιστόν [convincing, credible] indicates that degrees of uncertainty pervade a great part of the argumentation;
3. the expression "then it would be close to knowing (ἐγγύς ἂν εἴη τοῦ εἰδέναι)", which indicates the end of the conclusion of the syllogism,[4] clearly relates to the fact we can only reach credible/plausible results and not ἐπιστήμη; Peirce will say, similarly, that abduction reaches plausible results and/or that is "akin to the truth";
4. the adjective ὀλίγα [few] dominates the passage: for example, Aristotle says, by referring to the hypotheses/terms that have to be added—thanks to the process of leading away—to the syllogism: "Or next, it is leading away (ἀπαγωγή) if the middle terms between B and C are few (ὀλίγα) [for in this way also it is closer to scientific understanding (εἰδέναι)]". The term ὀλίγα certainly resonates with the insistence on "minimality" that dominates the first received models of abduction of the last decades of XX century.

I favor the following interpretation (Phillips 1992, p. 173): abduction denotes "the method of argument whereby in order to explain an obscure or ungrounded proposition one can lead the argument away from the subject to one more readily acceptable".

In the passage above Aristotle gives the example of the three terms "science" [knowledge], "is teachable", and "justice" [virtue], to exhibit that justice [virtue] is teachable: Aristotle is able to conclude that justice [virtue], is teachable, on the basis of an abductive reasoning, that is ἀπαγωγή. A second example of *leading away* is also presented, which illustrates that in order to make a rectilinear figure equal to a circle only one additional middle term is required; that is the addition of half circles to the rectilinear figure.

---

[4]Aristotle insists that all syllogisms are valid; there is no such thing as an invalid syllogism. The syllogistic tradition began to relax this requirement: here I will use the term syllogism in this modern not strictly Aristotelian sense.

I do not think appropriate to consider, following Kraus (2003, p. 247), the adumbrated syllogism (first Aristotelian example in the passage above)

AB Whatever is knowledge, can be taught
BC Virtue (e.g., justice) is knowledge
AC Therefore virtue can be taught

Just an example of a valid deduction, so insinuating Peirce's interpretation failure. Indeed, it seems vacuous to elaborate on the syntactic structure of the involved syllogism, as Kraus does: the problem of abduction in Chapter B25 is embedded in the activity of the inferential mechanism of "leading away" performed thanks to the introduction of new terms, as I explained above. He also says that the second Aristotelian example

Whatever is rectilinear, can be squared
A circle can be transformed into a rectilinear figure by the intermediate of lunes
Therefore, a circle can be squared

Still a simple deduction, was questionably supposed by Peirce to be fruit of the correction of Aristotle's original text due to the "stupid" Apellicon, considered responsible of blurring Aristotle's reference to abduction. Indeed, Kraus suggests that, following Peirce, the original text would have to be the following:

Whatever is equal to a constructible rectilinear figure, is equal to a sum of lunes
The circle is equal to a sum of lunes
Therefore, the circle is equal to a constructible rectilinear figure

Which indeed fits the Peircean abductive schema. At this point Kraus (2003, p. 248) ungenerously—and, in my opinion, erroneously, as I have already said—concludes "Peirce's argument surely is bad. It begs the question". I disagree with this skeptical conclusion.

We need a deeper and better interpretation of Aristotle's passage. To this aim we would need analyze some aspects of Plato's dialectic,[5] ancient geometrical cognition, and the role of middle terms: by illustrating these aspects in Magnani (2015a) I tried to convince the reader that we can gain a new positive perspective about the constitutive eco-cognitive character of abduction, just thanks to Aristotle himself. In the present section it was sufficient to stress the eco-cognitive openness indicated by Aristotle with his emphasis on the need in abduction of cognitive externalities—leading away—able to permit reasoners to go beyond that eco-cognitive *immunization* he himself considered crucial for founding syllogism.

---

[5]I agree with the following claim by Woods: "Whatever else it is, a dialectical logic is a logic of consequence-drawing" (Woods 2013a, p. 31), that is not merely a logic of "consequence-having".

1.1 Selective and Creative Abductions and the Optimization of Eco-cognitive Situatedness

When I say that abduction can be knowledge-enhancing[6] I am referring to various types of new produced knowledge of various novelty level, from that new piece of knowledge about an individual patient we have abductively reached (a case of selective abduction, no new biomedical knowledge is produced) to the new knowledge produced in scientific discovery, which Paul Feyerabend emphasized in *Against Method* (Feyerabend 1975), as I have illustrated in Magnani (2016). However, also knowledge produced in an artificial game thanks to a smart application of strategies or to the invention of new strategies and/or heuristics has to be seen as the fruit of knowledge enhancing abduction.

I contend that to reach selective or creative good abductive results efficient strategies have to be exploited, but it is also necessary to count on an environment characterized by what I have called *optimization of eco-cognitive situatedness*, in which eco-cognitive openness is fundamental (Magnani 2016). To favor good creative and selective abduction reasoning strategies must not be "locked" in an external restricted eco-cognitive environment (an artificial game, Go or Chess, for example) such as in a scenario characterized by fixed definitory rules and finite material aspects, which would function as cognitive mediators able to constrain agents' reasoning.

It is useful to provide a short introduction to the concept of eco-cognitive openness. The new perspective inaugurated by the so-called naturalization of logic (Magnani 2015b) contends that the normative authority claimed by formal models of ideal reasoners to regulate human practice on the ground is, to date, unfounded. It is necessary to propose a "naturalization" of the logic of human inference. Woods holds a naturalized logic to an adequacy condition of "empirical sensitivity" (Woods 2013b). A naturalized logic is open to study many ways of reasoning that are typical of actual human knowers, such as for example fallacies, which, even if not truth preserving inferences, nonetheless can provide truths and productive results. Of course one of the best examples is the logic of abduction, where the naturalization of the well-known fallacy "affirming the consequent" is at play. Gabbay and Woods (2005, p. 81) clearly maintain that Peirce's abduction, depicted as both (a) a surrender to an idea, and (b) a method for testing its consequences, perfectly resembles central aspects of practical reasoning but also of creative scientific reasoning.

It is useful to refer to my recent research on abduction (Magnani 2016), which stresses the importance in good abductive cognition of what has been called

---

[6]This means that abduction is not necessarily ignorance-preserving (reached hypotheses would always be "presumptive" and to be accepted they always need empirical confirmation). Abduction can creatively build new knowledge by itself, as various examples coming from the area of history of science and other fields of human cognition clearly show. I better supported my claim about the knowledge enhancing character of abduction in the recent (Magnani 2015a, 2016).

optimization of situatedness: abductive cognition is for example very important in scientific reasoning because it refers to that activity of creative hypothesis generation which characterizes one of the more valued aspects of rational knowledge. The study above teaches us that situatedness is related to the so-called eco-cognitive aspects, referred to various contexts in which knowledge is "traveling": to favor the solution of an inferential problem—not only in science but also in other abductive problems, such as diagnosis—the richness of the flux of information has to be maximized.

It is interesting to further illustrate this problem of optimization of eco-cognitive situatedness taking advantage of simple logical considerations. Let $\Theta = \{\Gamma_1, \ldots, \Gamma_m\}$ be a theory, $P = \{\Delta_1, \ldots, \Delta_n\}$ a set of true sentences corresponding—for example—to phenomena to be explained and $\Vdash$ a consequence relation, usually—but not necessarily—the classical one. In this perspective an abductive problem concerns the finding of a suitable improvement of $A_1, \ldots, A_k$ such that $\Gamma_1, \ldots \Gamma_m, A_1, \ldots, A_k \Vdash_L \Delta_1, \ldots, \Delta_n$ is *L-valid*. It is obvious that an improvement of the inputs can be reached both by additions of new inputs but also by the modification of inputs already available in the given inferential problem in Magnani (2016). I contend that to get good abductions, such as for examples the creative ones that are typical of scientific innovation, the input and output of the formula $A_1, \ldots, A_i, ?_I \Vdash_L^X \Upsilon_1, \ldots, .\Upsilon_j$, (in which $\Vdash_L^X$ indicates that inputs and outputs do not stand each other in an expected relation and that the modification of the inputs $?_I$ can provide the solution) have to be thought as *optimally positioned*. Not only, this optimality is made possible by a *maximization of changeability* of both input and output; again, not only inputs have to be enriched with the possible solution but, to do that, other inputs have usually to be changed and/or modified.[7]

Indeed, in our eco-cognitive perspective, an "inferential problem" can be enriched by the appearance of new outputs to be accounted for and the inferential process has to restart. This is exactly the case of abduction and the cycle of reasoning reflects the well-known nonmonotonic character of abductive reasoning. Abductive consequence is ruptured by new and newly disclosed information, and so defeasible. In this perspective abductive inference is *not only* the result of the modification of the inputs, but, in general, actually involves the intertwined modification of both input and outputs. Consequently, abductive inferential processes are highly *information-sensitive*, that is the flux of information which interferes with them is continuous and systematically human (or machine)-promoted and enhanced when needed. This is not true of traditional inferential settings, for example proofs in classical logic, in which the modifications of the inputs are *minimized*, proofs are usually taken with "given" inputs, and the burden of proofs is dominant and charged on rules of inferences, and on the smart choice of them together with the choice of their appropriate sequentiality. This changeability first of all refers to a wide psychological/epistemological openness in which knowledge transfer has to be maximized.

---

[7]More details are illustrated in Magnani, (2016), section three.

In sum, considering an abductive "inferential problem" as symbolized in the above formula, a suitably anthropomorphized logic of abduction has to take into account a continuous flux of information from the eco-cognitive environment and so the constant modification of both inputs and outputs on the basis of both

1. the *new information available*,
2. the *new information inferentially generated*, for example new inferentially generated inputs aiming at solving the inferential problem.

To conclude, optimization of situatedness is the main general property of logical abductive inference, which—from a general perspective—defeats the other properties such as minimality, consistency, relevance, plausibility, etc. These are special subcases of optimization, which characterize the kind of situatedness required, at least at the level of the appropriate abductive inference to generate the new inputs of the above formula.

2   Anticipations as Abductions

2.1   Adumbrations and the Generation of the Three-Dimensional Space: Abduction in Embodiment and in Distributed Cognition Environments

As I promised in the abstract of this article this second section is devoted to study—in the light of abductive cognition—the so-called "anticipations": they will help us to delineate both the role of abduction in distributed hypothetical reasoning.[8] Indeed, in 2001 (Magnani 2001) I have introduced the concept of manipulative abduction,[9] which is particularly appropriate to stress the eco-cognitive aspects of hypothetical reasoning, In manipulative abduction cognitive processes (for example strategic reasoning) not only refer to propositional aspects but they are also performed in a distributed cognition framework, in which models, artifacts, internal and external representations, sensations, and manipulations play an important role:

---

[8]I have to add that the concept of anticipation is also useful characterize the role of what I have called "unlocked" strategies. I have introduced and illustrated the concepts of locked and unlocked strategies in Magnani (2018).

[9]The concept of *manipulative abduction*—which also takes into account the external dimension of abductive reasoning in an eco-cognitive perspective—captures a large part of scientific thinking where the role of action and of external models (for example diagrams) and devices is central, and where the features of this action are implicit and hard to be elicited. Action can provide otherwise unavailable information that enables the agent to solve problems by starting and by performing a suitable abductive process of generation and/or selection of hypotheses. Manipulative abduction happens when we are thinking through doing and not only, in a pragmatic sense, about doing [cf. (Magnani 2009, Chap. 1)].

indeed the phenomenological example illustrated in this section also shows that abductive cognition can involve, when clearly seen in embodied and distributed systems, visual, kinesthetic, and motor sensations.

Looking at the philosophical explanations of the ways humans perform to build "idealities", geometrical idealities, and the objective space, Husserl contends that "facticities of every type [...] have a root in the essential structure of what is generally human", and that "human surrounding world is the same today and always" (Husserl 1978, p. 180). However, the horizon of the rough surrounding pre-predicative world of appearances and primordial and immediately given experiences—which is at the basis of the constructive cognitive activity—is a source of potentially infinite data,[10] which cannot "lock" cognitive strategies related to the multiple strategic abductive generation of idealities, geometrical ideal forms, and spatiality. Indeed, step by step, ideal objects in Husserlian sense are constructed and become *traditional* objects, and so they possess historicity as one of their multiple eidetic components. They become, Husserl says, "sedimentations of a truth meaning", which describe the cumulative character of human experience (not every "abiding possession" of mine is traceable to a self-evidence of my own). The research which takes advantage of the already available sedimented idealities (sedimentations of someone else's already accomplished experience) is at the basis of further abductive work to the aim, for example, of discovering new mathematical knowledge in the field of geometry.

Let us follow some Husserlian speculations that lead us to consider the important strategic role of anticipations as abductions. In Magnani (2009, Chap. 1, Sect 1.5.2). I have already illustrated the constitutive abductive character of perception in the light of Peircean philosophy. Now we will see the strategic abductive role of both perception and kinesthetic data in the Husserlian philosophical framework, integrating it with a reference to some of the current results of neuroscience. Indeed, the philosophical tradition of phenomenology fully recognizes the protogeometrical role of kinesthetic data in the generation of the so-called "idealities" (and of geometrical idealities). The objective space we usually subjectively experience has to be put in brackets by means of the transcendental reduction, so that pure lived experiences can be examined without the compromising intervention of any psychological perspective, any "doxa". By means of this transcendental reduction, we will be able to recognize perception as a structured "intentional constitution" of the external objects, established by the rule-governed activity of consciousness (similarly, space and geometrical idealities, like the Euclidean ones, are "constituted" objective properties of these transcendental objects).

The modality of appearing in perception is already markedly structured: it is not that of concrete material things immediately given, but it is mediated by sensible schemata constituted in the temporal continual mutation of adumbrations. So at the

---

[10]The pre-predicative world is not yet characterized by predications, values, empirical manipulations and techniques of measurement as instead the Husserl's prescientific world is.

level of "presentational perception" of pure lived experiences, only partial aspects [*adumbrations* (*Abschattungen*)] of the objects are provided. Therefore, an activity of unification of the different adumbrations to establish they belong to a particular and single object (noema) is further needed.[11]

The analysis of the generation of idealities (and geometrical idealities) is constructed in a very thoughtful philosophical scenario. The noematic appearances are the objects as they are intuitively and immediately given (by direct acquaintance) in the constituting multiplicity of the so-called adumbrations, endowed with a morphological character. The noematic meaning consists of a syntactically structured categorical content associated with judgment. Its ideality is "logical". The noema consists of the object as deriving from a constitutive rule or synthetic unity of the appearances, in the transcendental sense (Petitot 1999). To further use the complex Husserlian philosophical terminology—which surely motivates an interpretation in terms of abduction—we can say: hyletic data (that is immediate given data) are vivified by an intentional synthesis (a noetic apprehension) that transforms them into noematic appearances that adumbrate objects, etc.

As illustrated by Husserl in *Ding und Raum* [1907] (Husserl 1973) the geometrical concepts of point, line, surface, plane, figure, size, etc., used in eidetic descriptions are not spatial "in the thing-like sense": rather, in this case, we deal with the problem of the generation of the objective space itself. Husserl observes: it is "senseless" to believe that "the visual field is [...] in any way a surface on objective space" (Sect. 48, p. 166), that is, to act "as if the oculomotor field were located, as a surface, in the space of things" (Sect. 67, p. 236).[12] What about the phenomenological genesis of geometrical global three-dimensional space?

The process of making adumbrations represents a strategy which is distributed in visual, kinesthetic, and motor activities usually involving the manipulations of some parts of the external world. The adumbrative aspects of things are part of the visual field. To manage them a first requirement is related to the need of gluing different fillings-in of the visual field to construct the temporal continuum of perceptive adumbrations in a global space: the visual field is considered not translation-invariant, because the images situated at its periphery are less differentiated than those situated at its center (and so resolution is weaker at the periphery than at the center), as subsequently proved by the pyramidal algorithms in neurophysiology of vision research.

---

[11]On the role of adumbrations in the genesis of ideal space and on their abductive and nonmonotonic character cf. below Sect. 2.2. An interesting article (Overgaard and Grünbaum 2007) deals with the relationship between perceptual intentionality, agency, and bodily movement and acknowledges the abductive role of adumbrations. In the remaining part of this section I will try to clarify their meaning.

[12]Moreover, Husserl thinks that space is endowed with a double function: it is able to constitute a phenomenal extension at the level of sensible data and also furnishes an intentional moment. Petitot says: "Space possesses, therefore, a noetic face (format of passive synthesis) and a noematic one (pure intuition in Kant's sense)" (Petitot 1999, p. 336).

Perceptual intentionality basically depends on the ability to realize kinesthetic situations and sequences. In order for the subject to have visual sensations of the world, he/she must be able not only to possess kinesthetic sensations but also to freely initiate kinesthetic strategic "sequences": this involves a bodily sense of agency and awareness on the part of the doer (Overgaard and Grünbaum 2007, p. 20). The kinesthetic control of perception is related to the problem of generating the objective notion of three-dimensional space, that is, to the phenomenological constitution of a "thing",[13] as a single body unified through the multiplicity of its appearances. The "meaning identity" of a thing is of course related to the continuous flow of adumbrations: given the fact that the incompleteness of adumbrations implies their synthetic consideration in a temporal way, the synthesis in this case, *kinetic*, involves eyes, body, and objects.

Visual sensations are not sufficient to constitute objective spatiality. Kinesthetic sensations[14] (relative to the movements of the perceiver's own body)[15] are required. Petitot observes, de facto illustrating the abductive role of kinesthetic sensations:

> Besides their "objectivizing" function, kinesthetic sensations share a "subjectivizing" function that lets the lived body appear as a proprioceptive embodiment of pure experiences, and the adumbrations as subjective events. [...] There exists an obvious equivalence between a situation where the eyes move and the objects in the visual field remain at rest, and the reciprocal situation where the eyes remain at rest and the objects move. But this trivial aspect of the relativity principle is by no means phenomenologically trivial, at least if one does not confuse what is constituting and what is constituted. Relativity presupposes an *already* constituted space. At the preempirical constituting level, one must be able to discriminate the two equivalent situations. The kinesthetic control paths are essential for achieving such a task (Petitot 1999, pp. 354–355).

Multidimensional and hierarchically organized, the space of kinesthetic controls includes several degrees of freedom for movements of eyes, head, and body. Kinesthetic controls are kinds of *spatial* gluing operators. They are able to compose, in the case of visual field, different partial aspects—identifying them as belonging to the same object, that is constituting an ideal and transcendent "object". They are realized in the pure consciousness and are characterized by an intentionality that demands a temporal lapse of time.

With the help of very complex eidetic descriptions, that further develop the strategic operations we sketched, Husserl is able to explain the constitution of the objective parametrized time and of space, dealing with stereopsis, three-dimensional space and three-dimensional things inside it. Of course, when the three-dimensional space (still inexact) is generated (by means of two-dimensional gluing and

---

[13]Cf. also (Husserl 1931, Sect. 40, p. 129) [originally published in 1913].

[14]Husserl uses the terms "kinestetic sensations" and "kinesthetic sequences" to denote the subjective awareness of position and movement in order to distinguish it from the position and movement of perceived objects in space. On some results of neuroscience that corroborate and improve several phenomenological intuitions cf. (Pachoud 1999, pp. 211–216; Barbaras 1999; Petit 1999).

[15]The ego itself is only constituted thanks to the capabilities of movement and action.

stereopsis) it is possible to invert the phenomenological order: the visual field is so viewed as a portion of surface in $\mathbf{R}^3$, and the objective constituted space comes first, instead of the objects as they are intuitively and immediately given by direct acquaintance. So the space is in this case an objective datum informing the cognitive agent about the external world where she can find objects from the point of view of their referentiality and denotation. The kinesthetic system "makes the oculomotor field (eventually enlarged to infinity) the mere projection of a three spatial thingness" (Husserl 1973, Sect. 63, p. 227). Adumbrations now also appear to be consequences of the objective three-dimensional space, as continuous transformations of two-dimensional images as if the body were embedded in the space $\mathbf{R}^3$.[16]

2.2 Anticipations as Abductions

Of course adumbrations, the substrate of gluing operations that give rise to the two-dimensional space, are multiple and infinite, and there is a potential co-givenness of some of them (those potentially related to single objects). They are incomplete and partial so for the complete givenness of an object a temporal process is necessary. Adumbrations, not only intuitively presented, can be also represented at the level of *imagination*. Just because incomplete, *anticipations* instead correspond to a kind of non-intuitive intentional expectation: when we see a spherical form from one perspective (as an adumbration), we will assume that it is effectively a sphere, but it could be also a hemisphere (an example already employed by Locke).

Anticipations share with visual and manipulative abduction various features: they are highly conjectural and nonmonotonic, so wrong anticipations have to be replaced by other plausible ones. Moreover, they constitute an activity of "generate and test" as a kind of action-based cognition: the finding of adumbrations involves kinesthetic controls, sometimes in turn involving manipulations of objects; but the activity of testing anticipations also implies kinesthetic controls and manipulations. Finally, not all the anticipations are informationally equivalent and work like attractors for privileged individuations of objects. In this sense the whole activity is toward "the best anticipation", the one that can display the object in an optimal way. Prototypical adumbrations work like structural-stable systems, in the sense that they can "vary inside some limits" without altering the apprehension of the object.

---

[16]The role of adumbrations in objectifying entities can be hypothesized in many cases of non-linguistic animal cognition dealing with the problem of reification and the formation of a kind of "concept", cf. chapter five of Magnani (2009). In human adults objects are further individuated and reidentified by using both spatial aspects, such as place and trajectory information and static-property information (in this last case exploiting what was gained through previous adumbration activity); adults use this property information to explain and predict appearances and disappearances: "If the same large, distinctive white rabbit appears in the box and later on in the hat, I assume it's the same rabbit" (Gopnik and Meltzoff 1997).

As in the case of selective abduction, anticipations are able to select possible paths for constituting objects, actualizing them among the many that remain completely tacit. As in the case of creative abduction, they can construct new ways of aggregating adumbrations, by delineating the constitution of new objects/things. In this case they originate interesting "attractors" that give rise to *new* "conceptual" generalizations.

Some of the wonderful, philosophical Husserlian speculations are being further developed scientifically from the neurological and cognitive perspective in current cognitive science research. Grush (2004a, 2007) has built an emulation theory based on control theory where forward models as emulators (shared by humans and many other animals) are used to illustrate, in the case of humans, various cognitive processes like perception, imagery, reasoning, and language. He contends that simulation circuits are able to hypothesize forward mapping from control signals to the anticipated—and so abduced—consequences of executing the control command. In other words, they mimic the body and its interaction with the environment, enhancing motor control through sensorimotor abductive hypotheticals: "For example, in goal-directed hand movements the brain has to plan parts of the movement before it starts. To achieve a smooth and accurate movement proprioceptive/kinesthetic (and sometimes visual) feedback is necessary, but sensory feedback per se is too slow to affect control appropriately" (Desmurget and Grafton 2002). The "solution" is an emulator/forward model that can predict the sensory feedback resulting from executing a particular motor command" (Svensson and Ziemke 2004, p. 1310). The control theory framework is also useful to describe the emergence of implicit and explicit agency (Grush 2007). The humans' understanding of themselves as explicit agents is accomplished through an interplay between the standard egocentric point of view and the so-called "simulated alter-egocentric" point of view, which represents the agent itself as an entity in the environment.

Given the fact that motor imagery can be seen as the off-line driving force of the emulator via efference copies, it is noteworthy that the emulation theory can be usefully extended to account for visual imagery as the off-line operator behind an emulator of the motor-visual loop. In these systems a kind of *amodal* spatial imagery can be hypothesized: "Modal imagery [...] is imagery based on the operation of an emulator of the sensory system itself, whereas amodal imagery is based on the operation of an emulator of the organism and its environment: something like arrangements of solid objects and surfaces in egocentric space. I show how the two forms of emulation can work in tandem" (Grush 2004a, p. 386). [17]

---

[17]It is important to note that amodal imagery is neither sentential nor pictorial because the amodal environment space/objects emulators are closely tied to the organism's sensorimotor engagement with the environment. An interesting example of amodal abduction, in our terms, "where an object cannot currently be sensed by any sensory modality (because it is behind an occluder, is silent and odorless, etc.) yet it is represented as being at a location. I think it is safe to say that our representation of our own behavioral (egocentric) space allows for this, and it is not clear how a multisensory system, in which tags for specific modalities were always present, could accomplish this" (Grush 2004b, p. 434). On Grush's approach cf. the detailed discussion illustrated in Clark

The Husserlian phenomenological explanation of the generation of "idealities" leads to a moment in which, once the space as an objective datum is settled, it informs the cognitive agent about the external world where she can find objects from the point of view of their referentiality and denotation, like it is happening—currently—to beings like us.

Let us abandon the phenomenological speculative story regarding the building of precious philosophical hypotheses about that external world in which we simply find objects from the point of view of their referentiality and denotation. We can now turn our attention to those kinds of geometrical reasoning which represent cases of manipulative abduction and still concern visual, kinesthetic, and motor sensations and actions, but also involve a strong role of visual, iconic, and propositional representations (both internal and external).

3   Diagram Construction in Geometry Is a Kind of Manipulative Abduction

In this last section I will address the example of geometrical reasoning, which clearly illustrates the role of some aspects of visualizations and diagrammatization in abductive reasoning taking advantage of both a semiotic and a distributed cognition perspective.

Let's quote Peirce's passage about mathematical constructions. Peirce says that mathematical and geometrical reasoning "[...] consists in constructing a diagram according to a general precept, in observing certain relations between parts of that diagram not explicitly required by the precept, showing that these relations will hold for all such diagrams, and in formulating this conclusion in general terms. All valid necessary reasoning is in fact thus diagrammatic" (Peirce 1931–1958, 1.54). This kind of reasoning is also called by Peirce "theorematic" and it is a kind of "deduction" necessary to derive significant theorems (Necessary Deduction]: "[...] is one which, having represented the conditions of the conclusion in a diagram, performs an ingenious experiment upon the diagram, and by observation of the diagram, so modified, ascertains the truth of the conclusion" (Peirce 1931–1958, 2.267). The experiment is performed with the help of "[...] imagination upon the image of the premiss in order from the result of such experiment to make corollarial deductions to the truth of the conclusion" (Peirce 1976, IV, p. 38). The "corollarial" reasoning is mechanical (Peirce thinks it can be performed by a "logical machine")

(2008, chapter seven) in the framework of the theory of the extended mind; a treatment of current cognitive theories, such as the sensorimotor theory of perception, which implicitly furnish a scientific account of the phenomenological concept of anticipation, is given in chapter eight of the same book. A detailed treatment of recent neuroscience achievements which confirm the abductive character of perception is given in the article "Vision, thinking, and model-based inferences" (Raftopoulos 2017), recently published in the *Handbook of Model-Based Science* (Magnani and Bertolotti 2017).

and not creative, "A Corollarial Deduction is one which represents the condition of the conclusion in a diagram and finds from the observation of this diagram, as it is, the truth of the conclusion" (Peirce 1931–1958, 2.267) (cf. also Hoffmann 1999).

In summary, the point of theorematic reasoning is the transformation of the problem by establishing an *unnoticed* point of view to get interesting—and possibly new—insights. The demonstrations of "new" theorems in mathematics are examples of theorematic deduction.

Not dissimilarly Kant says that in geometrical construction of external diagrams "[…] I must not restrict my attention to what I am actually thinking in my concept of a triangle (this is nothing more than the mere definition); I must pass beyond it to properties which are not contained in this concept, but yet belong to it" (Kant 1929, A718–B746, p. 580).

Theorematic deduction can be easily interpreted in terms of manipulative abduction. I have said that manipulative abduction is a kind of abduction, mainly model-based (that is not fundamentally based on propositions but on "models" of various kinds, from visualizations to thoughts experiments, to complicated artifacts, etc.), that exploits external models endowed with delegated (and often implicit) cognitive and semiotic roles and attributes:

1. the model (diagram) is *external* and the strategy that organizes the manipulations is unknown a priori;
2. the result achieved is *new* (if we, for instance, refer to the constructions of the first creators of geometry), and adds properties not contained before in the concept [the Kantian to "pass beyond" or "advance beyond" the given concept (Kant 1929, A154–B193/194, p. 192)]. Of course in the case we are using diagrams to demonstrate already known theorems (for instance in didactic settings), the strategy of manipulations is not necessary unknown and the result is not new, like in the Peircean case of corollarial deduction.

Iconicity in theorematic reasoning is central. Peirce, analogously to Kant, maintains that "[…] philosophical reasoning is reasoning with words; while theorematic reasoning, or mathematical reasoning is reasoning with specially constructed schemata" (Peirce 1931–1958, 4.233); moreover, he uses diagrammatic and schematic as synonyms, thus relating his considerations to the Kantian tradition where schemata mediate between intellect and phenomena. The following is the famous related passage in the *Critique of Pure Reason* ("Transcendental Doctrine of Method"):

Suppose a philosopher be given the concept of a triangle and he be left to find out, in his own way, what relation the sum of its angles bears to a right angle. He has nothing but the concept of a figure enclosed by three straight lines, and possessing three angles. However long he meditates on this concept, he will never produce anything new. He can analyse and clarify the concept of a straight line or of an angle or of the number three, but he can never arrive at any properties not already contained in these concepts. Now let the geometrician take up these questions. He at once begins by constructing a triangle. Since he knows that the sum of two right angles is exactly equal to the sum of all the adjacent angles which can be constructed from a single point on a straight line, he prolongs one side of his triangle and obtains two adjacent angles, which together are equal to two right angles. He then divides the external angle by drawing a line parallel to the opposite side of the triangle, and observes that he has thus obtained an external adjacent angle which is equal to an internal angle—and so on. In this fashion, through a chain of inferences guided throughout by intuition, he arrives at a fully evident and universally valid solution of the problem (Kant 1929, A716–B744, pp. 578–579).

We can depict the situation of the philosopher described by Kant at the beginning of the previous passage taking advantage of some ideas coming from the catastrophe theory. As a human being who is not able to produce anything new relating to the angles of the triangle, the philosopher experiences a feeling of frustration (just like the Kölher's monkey which cannot keep the banana out of reach). The bad affective experience "deforms" the organism's regulatory structure by complicating it and the cognitive process stops altogether. The geometer instead "at once constructs the triangle", that is, he makes an external representation of a triangle and acts on it with suitable manipulations. Thom thinks that this action is triggered by a "sleeping phase" generated by possible previous frustrations which then change the cognitive status of the geometer's available and correct the internal idea of triangle (like the philosopher, he "has nothing but the concept of a figure enclosed by three straight lines, and possessing three angles", but his action is triggered by a sleeping phase). Here the idea of the triangle is no longer the occasion for "meditation", "analysis" and "clarification" of the "concepts" at play, like in the case of the "philosopher". Here the inner concept of triangle—symbolized as insufficient—is amplified and transformed thanks to the sleeping phase (a kind of Kantian imagination active through schematization) in a prosthetic triangle to be put outside, in some external support. The instrument (here an external diagram) becomes the extension of an organ:

> What is strictly speaking the end […] [in our case, to find the sum of the internal angles of a triangle] must be set aside in order to concentrate on the means of getting there. Thus the problem arises, a sort of vague notion altogether suggested by the state of privation. […] As a science, heuristics does not exist. There is only one possible explanation: the affective trauma of privation leads to a folding of the regulation figure. But if it is to be stabilized, there must be some exterior form to hold on to. So this anchorage problem remains whole and the above considerations provide no answer as to why the folding is stabilized in certain animals or certain human beings whilst in others (the majority of cases, needless to say!) it fails (Thom 1990, pp. 63–64).

As I have already said, for Peirce the whole mathematics consists in building diagrams that are "[…] (continuous in geometry and arrays of repeated signs/letters in algebra) according to general precepts and then [in] observing in the parts of these diagrams relations not explicitly required in the precepts" (Peirce 1931–1958, 1.54). Peirce contends that this diagrammatic nature is not clear if we only consider syllogistic reasoning "which may be produced by a machine" but becomes extremely clear in the case of the "logic of relatives, where any premise whatever will yield an endless series of conclusions, and attention has to be directed to the particular kind of conclusion desired" (Peirce 1987, pp. 11–23).

In ordinary geometrical proofs auxiliary constructions are present in terms of "conveniently chosen" figures and diagrams where strategic moves are important aspects of deduction. The system of reasoning exhibits a dual character: deductive and "hypothetical". Also in other—for example logical—deductive frameworks there is room for strategic moves which play a fundamental role in the generations of proofs. These strategic moves correspond to particular forms of abductive reasoning.

We know that the kind of reasoned inference that is involved in creative abduction goes beyond the mere relationship that there is between premises and conclusions in valid deductions, where the truth of the premises guarantees the truth of the conclusions, but also beyond the relationship that there is in probabilistic reasoning, which renders the conclusion just more or less probable. On the contrary, we have to see creative abduction as formed by the application of *heuristic procedures* that involve all kinds of good and bad inferential actions, and not only the mechanical application of rules. It is only by means of these heuristic procedures that the acquisition of *new* truths is guaranteed. Also Peirce's mature view illustrated above on creative abduction as a kind of inference seems to stress the strategic component of reasoning.[18]

Many researchers in the field of philosophy, logic, and cognitive science have maintained that deductive reasoning also consists in the employment of logical rules in a heuristic manner, even maintaining the truth preserving character: the application of the rules is organized in a way that is able to recommend a particular course of actions instead of another one. Moreover, very often the heuristic procedures of deductive reasoning are performed by means of model-based abductive steps where iconicity is central.

We have seen that the most common example of manipulative creative abduction is the usual experience people have of solving problems in elementary geometry in a model-based way trying to devise proofs using diagrams and illustrations: of course the attribute of creativity we give to abduction in this case does not mean that it has never been performed before by anyone or that it is original in the history of some knowledge (they actually are cases of Peircean corollarial deduction).

3.1   Iconic Brain and External Diagrammatization Coevuolution

Following our previous considerations it would seem that diagrams can be fruitfully seen from a semiotic perspective as external representations expressed through

---

[18]On the interesting interplay involved in the cooperation between heuristic procedures see the recent (Ulazia 2016): the multiple roles played by analogies in the genesis of fluid mechanics is illustrated together with the fact they can cooperate with other heuristic strategies.

icons and symbols, aimed at simply "mimicking" various humans' internal images. However, we have seen that they can also play the role of creative representations human beings externalize and manipulate not just to mirror the internal ways of thinking of human agents but to find room for concepts and new ways of inferring which cannot—at a certain time—be found internally "in the mind".

In summary, we can say that

– diagrams as external iconic (often enriched by symbols) representations are formed by external materials that either mimic (through reification) concepts and problems already internally present in the brain or creatively express concepts and problems that do not have a semiotic "natural home" in the brain;
– subsequent internalized diagrammatic representations are internal re-projections, or recapitulations (learning), in terms of neural patterns of activation in the brain ("thoughts", in Peircean sense), of external diagrammatic representations. In some simple cases complex diagrammatic transformations—can be "internally" manipulated *like* external objects and can further originate new internal reconstructed representations through the neural activity of transformation and integration.

This process explains—from a cognitive point of view—why human agents seem to perform both computations of a connectionist type such as the ones involving representations as

– (I Level) patterns of neural activation that arise as the result of the interaction (also presemiotic) between body and environment (and suitably shaped by the evolution and the individual history): pattern completion or image recognition,

and computations that use representations as

– (II Level) derived combinatorial syntax and semantics dynamically shaped by the various artificial external representations and reasoning devices found or constructed in the semiotic environment (for example iconic representations); they are—more or less completely—neurologically represented contingently as patterns of neural activations that "sometimes" tend to become stabilized meaning structures and to fix and so to permanently belong to the I Level above.

It is in this sense we can say the "System of Diagrammatization", in Peircean words, allows for a self-controlled process of thought in the fixation of originally vague beliefs: as a system of learning, it is a process that leads from "absolutely undefined and unlimited possibility" (Peirce 1931–1958, 6.217) to a fixation of belief and "by means of which any course of thought can be represented with exactitude" (Peirce 1931–1958, 4.530). Moreover, it is a system which could also improve other areas of science, beyond mathematics, like logic, it "[...] greatly facilitates the solution of problems of Logic. [...] If logicians would only embrace this method, we should no longer see attempts to base their science on the fragile foundations of metaphysics or a psychology not based on logical theory" (Peirce 1931–1958, 4.571).

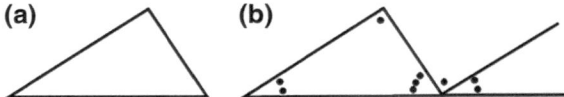

Fig. 1 Diagrammatic demonstration that the sum of the internal angles of any triangle is 180°.
a Triangle. **b** Diagrammatic manipulation/construction

An example of fixation of belief and cognitive manipulating through icons is the diagrammatic demonstration illustrated in Fig. 1, taken from the field of elementary geometry. In this case a simple manipulation of the triangle in Fig. 1a gives rise to an external configuration—Fig. 1b—that carries relevant semiotic information about the internal angles of a triangle "anchoring" new meanings.[19] It is worth noting that Kant exactly refers to this geometrical cognitive process in the passage of the *Critique of Pure Reason* ("Transcendental Doctrine of Method") I have quoted in the previous section.

As already stressed the I Level originates those sensations (they constitute a kind of "face" we think the world has), that provide room for the II Level to reflect the structure of the environment, and, most important, that can follow the computations suggested by the iconic external structures available. It is clear that in this case we can conclude that the growth of the brain and especially the synaptic and dendritic growth are profoundly determined by the environment. Consequently we can hypothesize a form of coevolution between what we can call the *iconic brain* and the development of the external diagrammatic systems. Brains build iconic signs as diagrams in the external environment learning from them new meanings through interpretation (both at the spatial and sentential level) after having manipulated them.

When the fixation is reached—imagine for instance the example above, that fixes the sum of the internal angles of the triangle—the pattern of neural activation no longer needs a direct stimulus from the external spatial representation in the environment for its construction and can activate a "final logical interpretant", in Peircean terms. It can be neurologically viewed as a fixed internal record of an external structure (a fixed belief in Peircean terms) that can exist also in the absence of such external structure. The pattern of neural activation that constitutes the I Level Representation has kept record of the experience that generated it and, thus, carries the II Level Representation associated to it, even if in a different form, the form of *semiotic memory* and not the form of the vivid *sensorial experience* for example of the triangular construction drawn externally, over there, for instance in a blackboard. Now, the human agent, via neural mechanisms, can retrieve that II Level Representation and use it as an internal representation (and can use it to

---

[19]The reader interested in further analysis of visual thinking in mathematics can refer to the classical (Giaquinto 2007). The book adopts an epistemological rather than cognitive perspective, also related to the discussion of the status of the Kantian so-called synthetic a priori judgments.

construct new internal representations less complicated than the ones previously available and stored in memory).

At this point we can easily understand the particular *mimetic* and *creative* role played by external diagrammatic representations in mathematics:

1. some concepts, meanings, and "ways of [geometrical] inferring" performed by the biological human agents appear hidden and more or less tacit and can be rendered explicit by building external diagrammatic mimetic models and structures; later on the agent will be able to pick up and use what was suggested by the constraints and features intrinsic and immanent to their external semiotic materiality and the relative established conventionality: artificial languages, proofs, new figures, examples, etc.;
2. some concepts, meanings, and "new ways of inferring" can be discovered only through a problem solving process occurring in a distributed interplay between brains and external representations. I have called this process externalization (or disembodiment) of the mind: the representations are mediators of results obtained and allow human beings

   (a) to re-represent in their brains new concepts, meanings, and reasoning devices picked up outside, externally, previously absent at the internal level and thus impossible: first, a kind of alienation is performed, second, a recapitulation is accomplished at the neuronal level by re-representing internally that which has been "discovered" outside. We perform cognitive geometric operations on the structure of data that synaptic patterns have "picked up" in an analogical way from the explicit diagrammatic representations in the environment;

   (b) to re-represent in their brains portions of concepts, meanings, and reasoning devices which, insofar as explicit, can facilitate inferences that previously involved a very great effort because of human brain's limited capacity. In this case the thinking performance is not completely processed internally but in a hybrid interplay between internal (both tacit and explicit) and external iconic representations. In some cases this interaction is between the internal level and a computational tool which in turn can exploit iconic/geometrical representations to perform inferences.

An evolved mind is unlikely to have a natural home for complicated concepts like the ones geometry introduced, as such concepts do not exist in a definite way in the natural (not artificially manipulated) world: so whereas evolved minds could construct spatial frameworks and perform some simple spatial inferences in a more or less tacit way by exploiting modules shaped by natural selection, how could one think exploiting explicit complicated geometrical concepts without having picked them up outside, after having produced them?

Let me repeat that a mind consisting of different separated implicit templates of thinking and modes of inferences exemplified in various exemplars expressed through natural language cannot come up with certain mathematical and geometrical entities without the help of the external representations. The only way is to extend the mind into the material world, exploiting paper, blackboards, symbols, artificial languages, and other various semiotic tools, to provide *semiotic anchors* for finding ways of inferring that have no natural home within the mind, that is for finding ways of inferring and concepts that take us beyond those that natural selection and previous cultural training could enable us to possess at a certain moment.

Hence, we can hypothesize—for example—that many valid spatial reasoning habits which in human agents are performed internally have a deep origin in the past experience lived in the interplay with iconic systems at first represented in the environment. As I have just illustrated other recorded thinking habits only partially occur internally because they are hybridized with the exploitation of already available or suitably constructed external diagrammatic artifacts.

4 Conclusion

I this article I have centered the attention on three heterogeneous examples which are appropriate to stress, in an interdisciplinary perspective and taking advantage of my EC-model of abduction, the philosophical, logical, cognitive, and semiotic aspects of building hypotheses thanks to abductive cognition. First of all I illustrated a new interpretation of Aristotle's seminal work on abduction and stressed his emphasis on the need of a related situation of eco-cognitive openness, beyond that kind of eco-cognitive immunization Aristotle himself considered necessary in syllogism. The second example concerns a thoughtful case of building philosophical hypotheses in the field of phenomenology: I interpreted the Husserlian concepts of "adumbrations" and "anticipations" in terms of abduction, showing that the illustrated cognitive process of building objects/things involves embodied and distributed perspectives. The final example is devoted to illustrating the problem of the extra-theoretical dimension of cognition from the perspective of geometrical reasoning. This case study is particularly appropriate because it shows relevant aspects of diagrammatic abduction, which involve intertwined processes of internal and external representations, crucial to understand the eco-cognitive character of abductive distributed cognitive processes. Moreover, especially thanks to the first example. I introduced and explained the concept of *optimization of the eco-cognitive situatedness*, describing it as one of the main characters of the abductive inferences to new hypotheses. I contended that thanks to the three examples we can gain a new positive perspective about the "constitutive" eco-cognitive character of abduction.

Acknowledgements Parts of this article were originally published in chapters five, six and seven of the book *The Abductive Structure of Scientific Creativity. An Essay on the Ecology of Cognition*, Springer, Switzerland, 2017 and in chapters one and two of the book *Abductive Cognition. The Epistemological and Eco-Cognitive Dimensions of Hypothetical Reasoning*, Springer, Heidelberg, 2009. For the instructive criticisms and precedent discussions and correspondence that helped me to develop my analysis of abductive cognition I am indebted and grateful to John Woods, Atocha Aliseda, Woosuk Park, Luís Moniz Pereira, Paul Thagard, Ping Li, Athanassios Raftopoulos, Michael Hoffmann, Gerhard Schurz, Walter Carnielli, Akinori Abe, Yukio Ohsawa, Cameron Shelley, Oliver Ray, John Josephson, Ferdinand D. Rivera and to my collaborators Tommaso Bertolotti and Selene Arfini.

References

Aristotle. (1989). *Prior analytics*. (R. Smith, Trans.). Indianapolis, Cambridge: Hackett Publishing Company.

Barbaras, R. (1999). The movement of the living as the originary foundation of perceptual intentionality. In J. Petitot, F. J. Varela, B. Pachoud, & J.-M. Roy (Eds.), *Naturalizing phenomenology* (pp. 525–538). Stanford, CA: Stanford University Press.

Clark, A. (2008). *Supersizing the mind: Embodiment, action, and cognitive extension*. Oxford, New York: Oxford University Press.

Desmurget, M., & Grafton, S. (2002). Forward modeling allows feedback control for fast reaching movements. *Trends in Cognitive Sciences, 4*, 423–431.

Feyerabend, P. (1975). *Against method*. London-New York: Verso.

Gabbay, D. M., & Woods, J. (2005). *The reach of abduction*. Amsterdam: North-Holland.

Giaquinto, M. (2007). *Visual thinking in mathematics: An epistemological study*. Oxford, New York: Oxford University Press.

Gopnik, A., & Meltzoff, A. (1997). *Words, thoughts and theories (learning, development, and conceptual change)*. Cambridge, MA: The MIT Press.

Grush, R. (2004a). The emulation theory of representation: Motor control, imagery, and perception. *Behavioral and Brain Sciences, 27*, 377–442.

Grush, R. (2004). Further explorations of the empirical and theoretical aspects of the emulation theory. *Behavioral and Brain Sciences, 27*, 425–435. Author's Response to Open Peer Commentary to R. Grush, The emulation theory of representation: Motor control, imagery, and perception.

Grush, R. (2007). Agency, emulation and other minds. *Cognitive Semiotics, 0*, 49–67.

Hoffmann, M. H. G. (1999). Problems with Peirce's concept of abduction. *Foundations of Science, 4*(3), 271–305.

Husserl. E. (1931). *Ideas: General introduction to pure phenomenology* (W. R. Boyce Gibson, Trans.). London, New York: Northwestern University Press. (First book 1913).

Husserl, E. (1970). *The crisis of European sciences and transcendental phenomenology* [1954] (D. Carr, Trans.). London and New York: George Allen and Unwin and Humanities Press.

Husserl, E. (1973). *Ding und Raum: Vorlesungen* (1907). In U. Claesges (Ed.), *Husserliana* (Vol. 16). The Hague: Nijhoff.

Husserl, E. (1978). The origin of geometry. In J. Derrida (Ed.), *Edmund Husserl's "The Origin of Geometry"* (pp. 157–180) (D. Carr, Trans. pp. 353–378). Stony Brooks, NY: Nicolas Hays (Also published in Husserl, 1970, pp. 353–378).

Kant, I. (1929). *Critique of pure reason* (N. Kemp Smith, Trans.). London: MacMillan. (originally published 1787, reprint 1998).

Kraus, M. (2003). Charles S. Peirce theory of abduction and the Aristotelian enthymeme from signs. In F. H. Eemeren, J. A. Blair, C. A. Willard, & A. F. Snoeck Henkemans (Eds.), *Anyone* who has a view: Theoretical contributions to the study of argumentation (pp. 237–254). Dordrecht, Boston, London: Kluwer Academic Publishers.

Magrani, L. (2001). Abduction, reason, and science: Processes of discovery and explanation. New York: Kluwer Academic/Plenum Publishers.

Magrani, L. (2009). Abductive cognition: The epistemological and eco-cognitive dimensions of hypothetical reasoning. Heidelberg, Berlin: Springer.

Magrani, L. (2015a). The eco-cognitive model of abduction. Ἀπαγωγή now: Naturalizing the logic of abduction. Journal of Applied Logic, 13, 285–315.

Magnani, L. (2015b). Naturalizing logic. Errors of reasoning vindicated: Logic reapproaches cognitive science. Journal of Applied Logic, 13, 13–36.

Magnani, L. (2016). The eco-cognitive model of abduction: Irrelevance and implausibility exculpated. Journal of Applied Logic, 15, 94–129.

Magnani, L. (2018). Playing with anticipations as abductions. Strategic reasoning in an eco-cognitive perspective, November 3–4, 2016. Forthcoming in If Colog Journal of Logics and their Applications, Special Issue on "Logical Foundations of Strategic Reasoning" (guest editor W. Park).

Magnani, L., & Bertolotti, T. (Eds.). (2017). Handbook of model-based science. Switzerland: Springer.

Otte, M. (2006). Proof-analysis and continuity. Foundations of Science, 11, 121–155.

Overgaard, S., & Grünbaum, T. (2007). What do weather watchers see? Perceptual intentionality and agency. Cognitive Semiotics, 0, 8–31.

Pachoud, B. (1999). The teleological dimension of perceptual and motor intentionality. In J. Petitot, F. J. Varela, B. Pachoud, & J.-M. Roy (Eds.), Naturalizing phenomenology (pp. 196–219). Stanford, CA: Stanford University Press.

Park W. (2015). On classifying abduction. Journal of Applied Logic, 13, 215–238.

Peirce, C. S. (1931–1958). In C. Hartshorne, P. Weiss, & A. W. Burks (Eds.), Collected papers of Charles Sanders Peirce (Vol. 1–6). Cambridge, MA: Harvard University Press.

Peirce, C. S. (1976). In C. Eisele (Ed.) The new elements of mathematics by Charles Sanders Peirce (Vols. I–IV). The Hague-Paris/Atlantic Highlands, NJ: Mouton/Humanities Press.

Peirce, C. S. (1987). In C. Eisele (Ed.) Historical perspectives on Peirce's logic of science: A history of science (Vols. I–II). Mouton, Berlin.

Petit, J.-L. (1999). Constitution by movement: Husserl in the light of recent neurobiological findings. In J. Petitot, F. J. Varela, B. Pachoud, & J.-M. Roy (Eds.), Naturalizing phenomenology (pp. 220–244). Stanford, CA: Stanford University Press.

Petitot, J. (1999). Morphological eidetics for a phenomenology of perception. In J. Petitot, F. J. Varela, B. Pachoud, & J.-M. Roy (Eds.), Naturalizing phenomenology (pp. 330–371). Stanford, CA: Stanford University Press.

Phillips, J. (1992). Aristotle's abduction: The institution of frontiers. The Oxford Literary Review, 14 (1–2):171–196. Special Issue on "Frontiers", G. Bennington and B. Stocker (Eds.).

Raftopoulos, A. (2017). Vision, thinking, and model-based inferences. In L. Magnani & T. Bertolotti (Eds.), Handbook of model-based science. Switzerland: Springer.

Svensson, H., & Ziemke, T. (2004). Making sense of embodiment: Simulation theories and the sharing of neural circuitry between sensorimotor and cognitive processes. In K. D. Forbus, D. Gentner & T. Regier (Eds.) CogSci 2004, XXVI Annual Conference of the Cognitive Science Society, Chicago, IL, CD-Rom.

Thom, R. (1990). Esquisse d'une sémiophysique, InterEditions, Paris, 1988 (English trans. by V. Meyer, Semio physics: A sketch, Redwood City, CA: Addison Wesley).

Ulazia, A. (2016). Multiple roles for analogies in the genesis of fluid mechanics: How analogies can cooperate with other heuristic strategies. Foundations of Science, 21(4), 543–565.

Woods, J. (2013a). Against fictionalism. In L. Magnani (Ed.), Model-based reasoning in science and technology: Theoretical and cognitive issues (pp. 9–42). Heidelberg, Berlin: Springer.

Woods, J. (2013b). Errors of reasoning: Naturalizing the logic of inference. London: College Publications.